Sustainable Drying Technologies of Food

This book is an updated reference about the fundamentals, applications, and modeling of sustainable energy-based drying systems in the food industry. It includes the economic assessment, quality control, and safety parameters of the drying techniques. Drying is one of the most important food preservation techniques. Sustainable drying uses little or no fossil fuel input and low-cost technologies. This book reviews the alternative sources of energy for drying fresh produce. It includes relevant case studies from some developing countries. The book is meant for researchers and industry experts in food processing, food engineering, and agriculture.

Key Features:

- Covers the drying technology on a vast array of food types like fruits, vegetables, fish, and meat
- Highlights the alternative energy sources for drying fresh produce
- Discusses modeling and simulation approaches
- Reviews energy and exergy assessment of sustainable drying

Sustainable Drying Technologies of Food

Edited by
Pankaj B. Pathare and Hemanatha Jayasuriya

CRC Press
Taylor & Francis Group
Boca Raton London New York

CRC Press is an imprint of the
Taylor & Francis Group, an **informa** business

Designed cover image: Shutterstock

First edition published 2026
by CRC Press
2385 NW Executive Center Drive, Suite 320, Boca Raton FL 33431

and by CRC Press
4 Park Square, Milton Park, Abingdon, Oxon, OX14 4RN

CRC Press is an imprint of Taylor & Francis Group, LLC

ISBN: 9781032378343 (hbk)
ISBN: 9781032382302 (pbk)
ISBN: 9781003344087 (ebk)

DOI: 10.1201/9781003344087

Typeset in Times
by KnowledgeWorks Global Ltd.

Contents

Contributors

M. Adnouni
Key Laboratory of Refrigeration and Cryogenic Technology of Zhejiang Province
Institute of Refrigeration and Cryogenics
Zhejiang University
Hangzhou, China

Rohit Biswas
Agricultural and Food Engineering Department
Indian Institute of Technology Kharagpur
Kharagpur, West Bengal, India

Darshana Deka
Institute of Chemical Technology Mumbai, ICT-IOC Odisha Campus
Bhubaneswar, India

Mulugeta A. Delele
Faculty of Chemical and Food Engineering
Bahir Dar Institute of Technology
Bahir Dar University
Bahir Dar, Ethiopia

Piverjeet Kaur Dhillon
Krishi Vigyan Kendra
Guru Angad Dev Veterinary and Animal Sciences University
Ludhiana, Punjab, India

Senanur Durgut
Department of Food Engineering
Faculty of Agriculture
Bursa Uludağ University
Bursa, Turkey

Eshetu Getahun
Faculty of Chemical and Food Engineering
Bahir Dar Institute of Technology
Bahir Dar University
Bahir Dar, Ethiopia
Bahir Dar Energy Center
Bahir Dar Technology Institute
Bahir Dar University
Bahir Dar, Ethiopia

Nagamaniammai Govindarajan
Department of Food Process Engineering
School of Bioengineering
College of Engineering and Technology
SRM Institute of Science and Technology
Kattankulathur, India

Hemanatha Jayasuriya
Department of Soils, Water and Agricultural Engineering
College of Agricultural & Marine Sciences
Sultan Qaboos University
Muscat, Oman

L. Jiang
Key Laboratory of Refrigeration and Cryogenic Technology of Zhejiang Province
Institute of Refrigeration and Cryogenics
Zhejiang University
Hangzhou, China

Azime Özkan Karabacak
Department of Food Processing
Gemlik Asım Kocabıyık Vocational School
Bursa Uludağ University
Bursa, Turkey
Science and Technology Application and Research Center (BITUAM)
Bursa Uludağ University
Bursa, Turkey

Navneet Kaur
Department of Chemical Engineering
Chandigarh University
Mohali, Punjab, India

Preetinder Kaur
Department of Processing and Food Engineering
Punjab Agricultural University
Ludhiana, Punjab, India

Vishnuvardini Narayanan
Department of Food Process Engineering
School of Bioengineering
College of Engineering and Technology
SRM Institute of Science and Technology
Kattankulathur, India

Pankaj B. Pathare
Department of Soils, Water and Agricultural Engineering
College of Agricultural and Marine Sciences
Sultan Qaboos University
Muscat, Oman

Abhipriya Patra
Department of Food Process Engineering
National Institute of Technology
Rourkela, India

Ashok Prabhakar
Department of Chemical Engineering
Chandigarh University
Mohali, Punjab, India

V. Arun Prasath
Department of Food Process Engineering
National Institute of Technology
Rourkela, India

Deep Raj
Department of Environmental Science and Engineering
SRM University
Guntur, Andhra Pradesh, India

Poonam Rani
Food Chemistry & Technology Department
Teagasc Food Research Centre
Fermoy, Cork, Ireland

Shivanand S. Shirkole
Institute of Chemical Technology Mumbai, ICT-IOC Odisha Campus
Bhubaneswar, India

Manpreet Singh
Department of Renewable Energy Engineering
Punjab Agricultural University
Ludhiana, Punjab, India

Prashant M. Singh
Institute of Chemical Technology Mumbai, ICT-IOC Odisha Campus
Bhubaneswar, India

Priya Singla
Desh Bhagat University
District Fategarh Sahib, Punjab, India

Özge Süfer
Food Engineering Department
Faculty of Engineering
Osmaniye Korkut Ata University
Osmaniye, Turkey

Canan Ece Tamer
Department of Food Engineering
Faculty of Agriculture
Bursa Uludağ University
Bursa, Turkey

P. P. Tripathy
Agricultural and Food Engineering Department
Indian Institute of Technology Kharagpur
Kharagpur, West Bengal, India

Ruchika Zalpouri
ICAR-Central Institute of Post-Harvest Engineering and Technology
Ludhiana, Punjab, India

X. J. Zhang
Key Laboratory of Refrigeration and Cryogenic Technology of Zhejiang Province
Institute of Refrigeration and Cryogenics
Zhejiang University
Hangzhou, China

About the Editors

Pankaj B. Pathare is Associate Professor of Postharvest Technology at Sultan Qaboos University. He graduated with a B.Tech. degree (Agricultural Engineering) from India and Ph.D. (Process & Chemical Engineering) from the University College Cork (Ireland). Before joining SQU, he worked as a researcher at Newcastle University, UK, and Stellenbosch University, South Africa. His work focuses on postharvest loss quantification, computer vision applications, and renewable energy in food processing. Dr. Pathare addresses critical food industry challenges, including minimizing mechanical damage, enhancing product quality, and improving energy efficiency. An internationally recognized expert, he has published over 100 peer-reviewed articles and numerous keynote conference presentations. He has also edited three books. He received the Distinguished Researcher Award from SQU in 2025, and his team earned the National Research Award (Oman) in 2024. Dr. Pathare serves on the editorial boards of several leading SCI journals, demonstrating his commitment to advancing knowledge in food engineering and postharvest technology. His expertise has been acknowledged by Stanford University, which recognized him as a World Top 2% Scientist from 2022 to 2024.

Hemanatha Jayasuriya is a Faculty Member at the Department of Soils, Water and Agricultural Engineering, College of Agricultural and Marine Sciences, Sultan Qaboos University, Oman. He obtained his B.Sc. Eng. (Hons.) in Mechanical Engineering from the University of Peradeniya, Sri Lanka, in 1983. His M.Eng. and Ph.D. in Agricultural Engineering were from the Asian Institute of Technology (AIT) Thailand. He is a chartered mechanical engineer by profession with over seven years of industrial experience and over 30 years of academic experience in teaching, research, and administrative activities during his career. He has supervised and completed nearly 40 postgraduate students, including several Ph.D. students, and published over 100 scientific publications in peer-reviewed journals and international conference proceedings. He has also published edited books, technical reports, and book chapters, and he possesses patents. He has membership credentials with international professional bodies such as ASABE, ISPA, ISTVS, and AAAE and is currently Vice President of AAAE for Energy, Environment, and Emerging Technologies.

1 Recent Advances in Sustainable Food Drying Technologies

M. Adnouni, L. Jiang, and X. J. Zhang
Key Laboratory of Refrigeration and Cryogenic Technology of Zhejiang Province, Institute of Refrigeration and Cryogenics, Zhejiang University, Hangzhou, China

Pankaj B. Pathare
Department of Soils, Water and Agricultural Engineering, College of Agricultural and Marine Sciences, Sultan Qaboos University, Oman

1.1 INTRODUCTION

Food preservation is essential for ensuring global food security, reducing postharvest losses, and enhancing the shelf life of agricultural products [1]. Drying is one of the oldest and most widely used methods among various preservation techniques. Drying inhibits microbial growth and enzymatic reactions by significantly lowering moisture content, thereby preventing spoilage and improving product stability [2]. Additionally, drying facilitates storage, transportation, and trade, making it a crucial process in the global food supply chain [3].

Traditionally, drying has been carried out using sun drying and open-air drying techniques, which rely on natural solar radiation. While cost-effective and simple, these methods are highly dependent on weather conditions, slow, and susceptible to contamination. Mechanical drying methods such as hot-air drying, freeze-drying (FD), and vacuum drying (VD) have been developed to overcome these drawbacks. These techniques offer faster drying rates and better process control but come with high operational costs, energy demands, and environmental concerns, primarily due to reliance on fossil fuels [4].

Given the increasing emphasis on climate change mitigation, energy efficiency, and sustainable food production, developing efficient and environment-friendly drying technologies has become essential [5]. Recent advancements in solar drying, bioenergy-based drying, hybrid drying systems, and novel techniques offer promising alternatives that reduce energy consumption while preserving food quality. Additionally, advanced modelling and simulation techniques, including computational fluid dynamics (CFD), AI-based models, and digital twins (DTs), are emerging as powerful tools to optimise drying processes, enabling real-time monitoring, predictive analysis, and precise control over heat and mass transfer [6].

This chapter explores traditional and sustainable food drying technologies, examining their advantages, limitations, and environmental impact. It also discusses modelling approaches for optimising drying processes and assessing their effects on food quality, energy efficiency, and sustainability. The insights presented in this chapter contribute to the broader objective of developing sustainable food processing methods that align with global food security and climate resilience goals. The energy–quality–environment-modelling nexus related to the sustainable drying process of food products is illustrated in Figure 1.1.

DOI: 10.1201/9781003344087-1

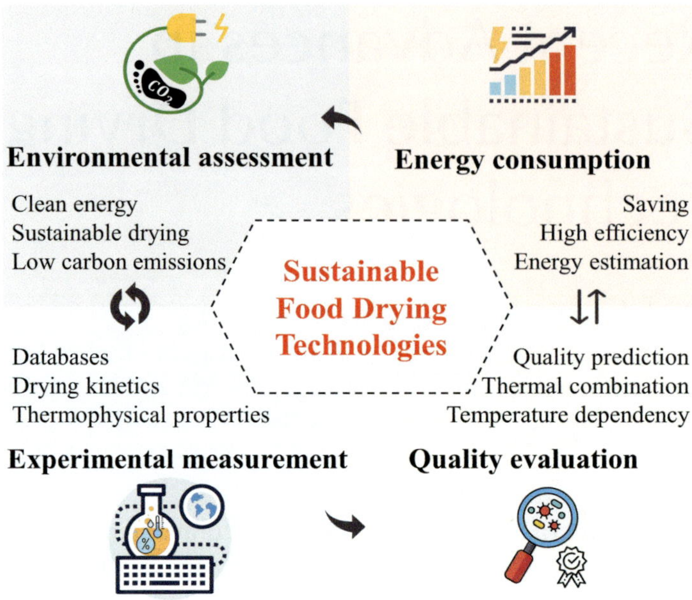

FIGURE 1.1 A conceptual schema of energy–quality–environment-modelling nexus related to the sustainable drying process of food products.

1.2 TRADITIONAL DRYING METHODS

Drying has been practised for centuries as a fundamental method to preserve food products, ensuring their availability across cultures and beyond the harvest season during periods of scarcity. Long before refrigeration and modern preservation techniques, communities worldwide relied on drying to extend the shelf life of their local products. Beyond its role in food security, drying has significantly influenced culinary traditions and trade. Dried commodities such as fruits, fish, meats, spices, and grains have long been transported across long distances without spoilage, enabling economic exchange and cultural diffusion. Despite the emergence of industrial-scale food processing, traditional drying practices remain prevalent in many regions, particularly in rural and resource-limited areas, due to their low cost and accessibility [7].

Traditional drying methods primarily depend on natural heat sources, such as sunlight, ambient air, and wind. These methods incorporate natural, passive techniques as they do not require external energy input other than what nature provides. However, long-established conventional methods, often known as mechanical drying, are also considered traditional food preservation techniques. These methods involve external heat sources to accelerate the drying process and provide better control over temperature, humidity, and airflow, reducing drying time and improving hygiene and food safety. Although conventional drying represents an advancement over passive techniques, it comes with trade-offs such as high energy consumption, increased operational costs, and environmental concerns. This section explores passive and conventional drying methods, highlighting their historical significance, working principles, advantages, limitations, and applications.

1.2.1 Sun Drying

Sun drying is one of the oldest and most widely used drying techniques. It exposes food directly to sunlight, allowing solar energy to heat the product and evaporate moisture. Sun drying is especially common in warm, dry climates where abundant radiation ensures efficient moisture removal. This method is commonly used for drying fruits, vegetables, fish, and meats. Despite its simplicity and low cost, sun drying suffers from weather dependency, significantly affected by temperature, humidity, and cloud cover, risks of contamination from dust, insects, and birds, and uneven drying due to lack of control over conditions, leading to inconsistent moisture content and increased spoilage risk [8].

1.2.2 Shade Drying

Shade drying is a variation of sun drying used primarily for heat-sensitive products such as herbs, spices, medicinal plants, and highly perishable fruits. In this method, food is dried in a well-ventilated but shaded area to prevent direct exposure to sunlight, which can degrade colour, flavour, and essential nutrients. Shade drying helps retain the natural properties of delicate food products, making it preferable for high-value products such as green tea leaves, medicinal herbs, and certain vegetables and fruits. However, it is slower than sun drying and still susceptible to microbial growth if humidity levels are not controlled properly.

A well-documented example of shade drying is the traditional Qunje drying chambers used for grape drying in Xinjiang Province in China, as presented in Figure 1.2. These unique mud-brick structures facilitate natural air circulation while protecting grapes from direct sunlight, ensuring a gradual drying process that preserves the colour, flavour, and nutriments of the raisins [9]. This traditional drying method is particularly effective in maintaining the quality of the final product, making it an essential part of the local raisin industry.

1.2.3 Smoke Drying

Smoke drying is a traditional method that combines drying and smoking to preserve food, mainly fish and meats. In this process, food is exposed to low heat and smoke from burning wood or plant materials. The dual effect of dehydration and antimicrobial compounds in smoke enhances shelf life while imparting a distinct flavour. Common examples of smoke-dried foods include smoked fish (e.g. herring, salmon) and meat products like jerky. However, this drying

FIGURE 1.2 Sun-dried grapes and shade drying of grapes in traditional Qunje drying houses, Xinjiang, China.

method requires careful management to avoid excessive drying, charring, or the formation of harmful compounds such as polycyclic aromatic hydrocarbons (PAHs), which are associated with serious health risks.

1.2.4 Open-Air Drying

Open-air drying involves spreading food in a well-ventilated environment, allowing ambient air circulation to remove moisture. This method is widely used for grains, cereals, and seeds, often carried out on mats, rooftops, or raised platforms to minimise contact with contaminants.

While open-air drying is inexpensive and requires minimal equipment, it is highly susceptible to environmental factors such as rainfall, temperature fluctuations, and airborne contaminants. As a result, spoilage and quality degradation are common concerns.

1.2.5 Hot-Air Drying

Hot-air drying involves circulating heated air over food products to accelerate moisture removal. The process allows for greater control over drying conditions, reducing drying time and improving hygiene and food safety. Typical applications include drying grains, fruits, vegetables, and dairy products. Hot-air drying enhances efficiency, requiring significant energy input and increasing costs. Additionally, excessive heating can degrade nutritional and sensory properties, such as vitamin content, colour, and texture [10].

1.2.6 Drum Drying

Drum drying is a continuous technique in which food slurry or puree is spread over a heated drum, where moisture rapidly evaporates, leaving behind a dry product layer. This method produces dried powders such as instant mashed potatoes, baby food, and cereal products. This drying method is highly efficient but can lead to texture changes and the loss of heat-sensitive nutrients. The high energy demand also makes it a less sustainable option compared to emerging drying technologies [11].

1.2.7 Spray Drying

Spray drying is widely used in the food industry to convert liquid food products into powdered form. It involves spraying a liquid food suspension into a hot air chamber, where moisture evaporates almost instantly, leaving behind dry powder particles. This method is essential for producing milk powder, coffee powder, and instant soup bases. Spray drying provides excellent control over product quality and is ideal for large-scale production. However, the process is highly energy-intensive and expensive, limiting its accessibility in small-scale or rural settings [12].

Whether passive or mechanical, traditional drying methods have unique advantages and inherent limitations that affect efficiency, food quality, and environmental sustainability, influencing their suitability for different food products. Table 1.1 provides a comparative overview.

1.3 SUSTAINABLE DRYING TECHNOLOGIES

While traditional drying continues to be frequently adopted and marketed, concerns regarding its efficiency, quality, and environmental sustainability are increasing. Many of these methods are energy-intensive, weather-dependent, or prone to contamination, prompting the development of innovative drying technologies that reduce energy use, improve food quality, and minimise environmental impact. Sustainable food drying technologies, representing the future of food preservation,

TABLE 1.1
Comparison of Standard Traditional Drying Methods

Drying Method	Dried Products	Advantages	Limitations	Reference
Sun drying	Fruits, vegetables	Low cost, no energy input	Weather-dependent, contamination risk, slow	[13]
Shade drying	Grapes, herbs, tea leaves	Preserves colour and nutrients	Slow drying, mould risk in high humidity	[14]
Smoke drying	Fish, meat, tomatoes	Enhances flavour, antimicrobial properties	Potential contamination. Requires careful temperature control	[15]
Open-air drying	Grains, cereals, seeds	Sunshine independent, simple, low-cost	Susceptible to rain, contamination	[16]
Hot-air drying	Fruits, vegetables, dairy	Faster drying, better hygiene	High energy consumption, nutrient loss	[17]
Drum drying	Baby food, cereal powders	Efficient for pastes and purees	Texture changes, nutrient degradation	[11]
Spray drying	Milk powder, coffee, soups	High-quality powdered products	Expensive, energy-intensive	[18]

primarily focus on two strategies [19]: enhancing the efficiency of existing dryers and enhancing the energy supply of drying systems (Figure 1.3). This section explores sustainable drying technology advancements, particularly in solar drying, bioenergy-based drying, and hybrid drying systems, which offer higher efficiency, environmental benefits, and cost-effectiveness than traditional methods.

1.3.1 Renewable Drying Systems

Providing consistent and low-cost energy resources for industrial applications is a global challenge, especially in the drying sector, one of the most energy-intensive processes in food production. As fossil fuel reserves deplete and concerns about climate change intensify, renewable energy–powered drying systems have gained considerable attention due to their sustainability and economic benefits [20]. Several renewable energy sources have been explored for drying

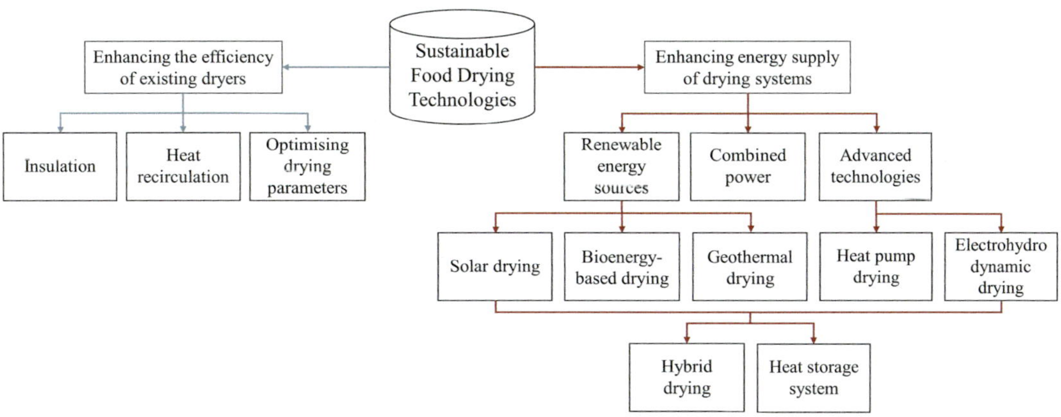

FIGURE 1.3 An overview of sustainable drying technologies.

applications, with solar energy being the most widely used due to its affordability, availability, and environment-friendliness. However, other clean energy options, such as bioenergy-based drying and wind-powered drying systems, are also being investigated to ensure reliable energy supplies for food dehydration. These sustainable solutions can reduce fossil fuels dependence while ensuring efficient drying operations.

1.3.1.1 Solar Drying Systems

Sun drying has been used since time immemorial as a natural and low-cost method for food dehydration. However, traditional open-sun drying is often inefficient due to long drying times, contamination risks, and dependence on weather conditions. Modern solar drying systems aim to overcome these limitations by incorporating solar collectors, photovoltaic panels, and thermal energy storage, thereby improving drying efficiency and ensuring greater control over the process.

Solar dryers can generally be classified based on their airflow mechanism and heat transfer method. Passive solar dryers rely on natural air circulation through buoyancy forces or wind pressure, whereas active solar dryers need fans or blowers to enhance airflow and improve heat distribution. Regarding heat transfer, direct solar dryers expose food directly to sunlight within an enclosed chamber, reducing contamination risks while accelerating the drying process [21]. Conversely, indirect solar dryers utilise solar collectors to heat air before directing it into the drying chamber, providing a more controlled environment suitable for heat-sensitive foods [22].

Hybrid solar dryers integrate both approaches, combining direct solar exposure with preheated air. In contrast, some hybrid systems incorporate backup energy sources such as biomass combustion, geothermal energy, or electrical heating to ensure uninterrupted drying, even during cloudy conditions or nighttime. Despite their advantages, solar drying systems remain influenced by climatic variations, limiting their effectiveness in regions with inconsistent sunlight. However, ongoing research and technological advancements continue to enhance their performance, making them an increasingly viable solution for sustainable food preservation. The classification of solar dryers based on their operational principles is summarised in Table 1.2.

Most previous studies focused on solar-based drying systems, while wind energy deployment in drying is scanty and requires much attention given its enormous potential, particularly among smallholder farmers. Wind-powered systems can operate all day long as the wind blows, suitable for developing countries with unreliable and expensive electricity supplies. Weather change, drying unit parameters, and dried material properties contribute to the complexity and non-linearity of renewable energy systems used in the drying industry. As a result, solving complex differential equations becomes onerous or time-consuming. Thus, accurate models are vital for forecasting and regulating the performances of these technologies.

TABLE 1.2
Classification of Solar Drying Systems

Classification	Type	Description
Airflow mechanism	Passive	Uses natural air circulation (buoyancy/wind pressure)
	Active	Uses fans/blowers for forced airflow
Heat transfer	Direct	Food exposed directly to sunlight inside a chamber
	Indirect	Sunlight heats air separately before reaching the food
Hybrid Variations	Mixed-mode	Combines direct exposure and preheated air
	Hybrid	Mixed-mode system with a backup energy source

1.3.1.2 Bioenergy-Based Drying Systems

Bioenergy, derived from organic materials such as wood, agricultural plants, and biogas, provides a sustainable and continuous energy source for food drying applications. Unlike solar drying, which depends on sunlight availability, bioenergy-based drying systems offer a reliable and weather-independent alternative, making them particularly beneficial in regions with inconsistent solar radiation. These systems not only ensure a steady energy supply but also contribute to environmental sustainability by recycling agricultural waste, thereby reducing waste accumulation and mitigating carbon emissions.

One of the most common implementation of bioenergy-based drying involves biomass combustion to generate hot air, which is then circulated within the drying chamber. Biomass dryers generally operate on two fundamental principles: direct biomass drying and indirect biomass drying. In direct biomass dryers, combustion gases contact food materials, facilitating heat transfer primarily through convection. As the hot air interacts with the food surface, moisture evaporates from the outer layer, creating a moisture gradient that drives internal water migration towards the surface through osmosis. This process continues until the food reaches the desired moisture level. While direct biomass drying is an efficient method, it poses potential contamination risks due to the presence of combustion products such as carbon monoxide (CO), nitrogen oxides (NO_X), sulphur dioxide (SO_2), and particulate matter, which can degrade food quality and safety.

Indirect biomass dryers have been developed to address these concerns, where the heat generated from biomass combustion is transferred to a secondary working fluid, typically fresh air, through a heat exchanger. This heated air is then directed into the drying chamber, ensuring food products are not exposed to combustion gases. The indirect drying approach is preferred for its superior hygiene and product quality, as it prevents direct contact between food and potentially harmful emissions. Moreover, integrating bioenergy with other renewable sources, such as solar-assisted biomass drying, can enhance efficiency and sustainability by optimising energy utilisation and reducing reliance on a single fuel source.

Beyond solid biomass combustion, biogas-based drying systems offer an additional renewable energy alternative for food dehydration. Biogas digesters utilise organic waste materials, such as livestock manure, food scraps, and crop residues, to produce methane gas (CH_4) through anaerobic digestion (Figure 1.4). This methane gas can then be burned to generate heat, which is used in

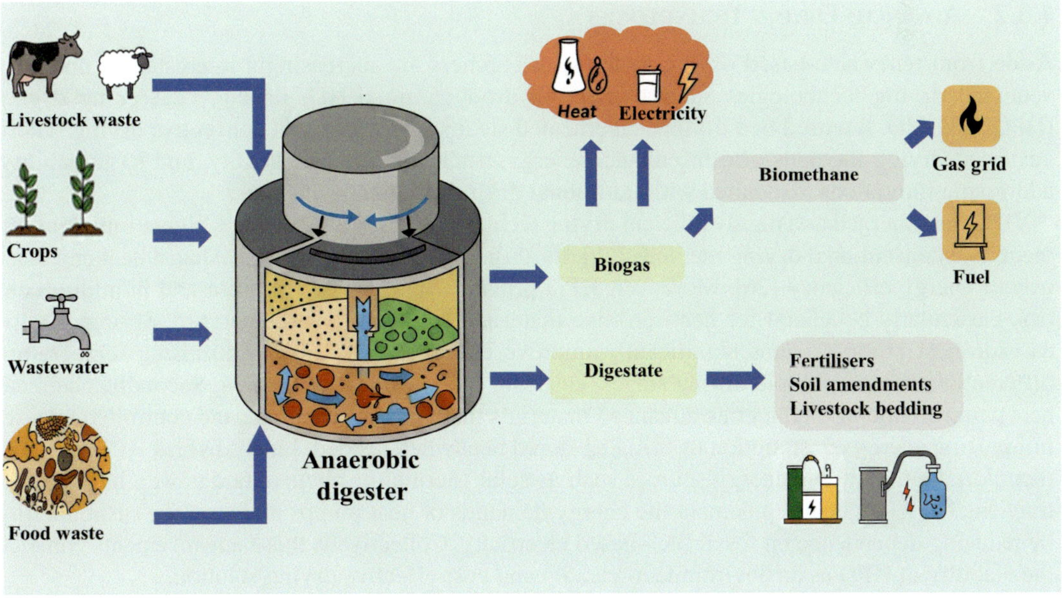

FIGURE 1.4 Biogas production and utilisation system.

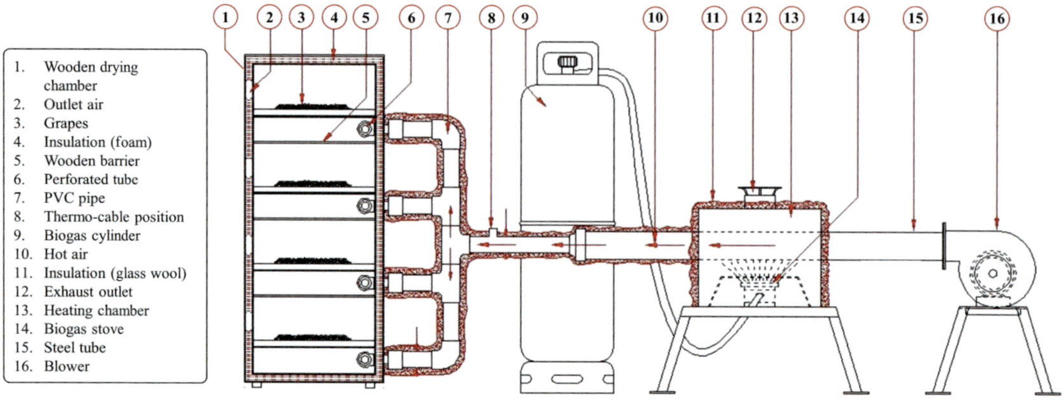

FIGURE 1.5 A biogas-powered grape dryer diagram [24].

drying applications through direct combustion or a heat exchanger. Compared to solid biomass, biogas combustion produces fewer air pollutants and can be more efficiently controlled, making it an attractive low-emission alternative for food drying in rural and off-grid areas [23].

A practical illustration of a biogas-powered drying system is the biogas-powered grape dryer (Figure 1.5), which utilises methane gas to generate heat for controlled food drying. This system ensures a continuous and clean energy supply, reducing dependence on fossil fuels while maintaining dried grapes' quality and nutritional value. Advancements in biomass gasification, biogas digesters, and high-efficient burner designs continue to improve bioenergy-driven food drying technologies' feasibility and environmental performance, making them a viable alternative to fossil-fuel-dependent drying systems. Future innovations in intelligent control systems, artificial intelligence (AI)-based optimisation, and advanced heat recovery mechanisms will further enhance the performance of bioenergy-based dryers, ensuring long-term sustainability in food drying applications.

1.3.2 Advanced Drying Technologies

Aside from renewable-based drying systems, researchers are increasingly interested in non-conventional drying technologies such as electrohydrodynamic (EHD) drying, heat pump drying (HPD), FD, VD, fluidised bed drying, superheated steam, and waste heat convective drying. These advanced drying methods aim to enhance energy efficiency, product quality, and sustainability, addressing limitations associated with traditional drying systems.

HPD has emerged as a highly efficient drying technology despite requiring a higher initial investment than conventional drying methods [25]. Its ability to recover waste and exhaust heat enhances overall energy efficiency [26]. Moreover, HPD provides superior temperature and humidity control, particularly beneficial for heat-sensitive materials requiring low-temperature drying. Firstly, cascade heat pump systems significantly improve energy efficiency by optimising temperature differentials across multiple drying stages, ensuring effective heat utilisation. Secondly, chemical heat pump technologies integrate advanced materials that allow heat storage and controlled release, minimising energy consumption by utilising stored heat when needed. Lastly, hybrid HPD systems, incorporating renewable energy sources such as solar thermal or photovoltaic power, have gained traction. These systems supplement the energy demands of heat pumps and improve sustainability by reducing dependence on fossil fuel–based electricity. Collectively, these advancements enhance the viability of HPD as an environment-friendly and cost-effective drying solution.

VD is another efficient method operating at reduced pressure and low drying temperatures. This results in a higher drying rate, better preservation of volatile compounds, and minimal oxidation,

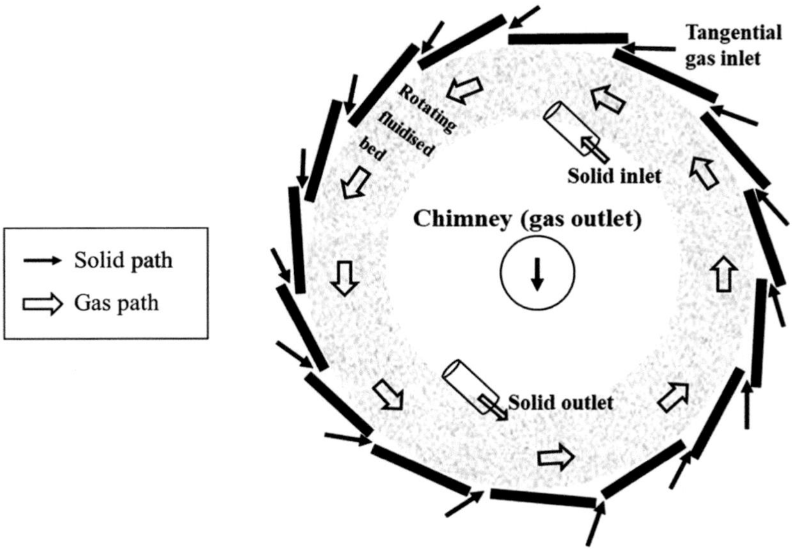

FIGURE 1.6 Schematic of a rotating fluidised bed drying system for biomass drying [29].

which is crucial for maintaining the taste, aroma, and nutritional value of dried products [27]. Similarly, superheated steam drying (SSD) has gained popularity due to its superior product quality retention compared to other drying methods, such as convective drying and VD [28]. This technology eliminates oxidative reactions, prevents hazardous gas formation, and allows efficient steam recovery and utilisation, making it an energy-efficient and environmentally sustainable drying solution.

These emerging technologies include fluidised bed drying, particularly vortex chamber-based rotating fluidised bed systems, as illustrated in Figure 1.6. This system provides enhanced heat and mass transfer efficiency, overcoming the limitations of traditional spray and rotary dryers. It enables rapid and uniform drying by generating high turbulence and ensuring close contact between solid particles and drying air. Unlike spray drying, which may cause particle shrinkage and loss of volatile compounds, fluidised bed drying minimises material degradation, preserving product quality. Moreover, its ability to optimise air distribution and minimise heat losses results in lower energy consumption. These advantages make fluidised bed dryers particularly suitable for biomass drying applications, offering superior drying kinetics, enhanced moisture removal rates, and reduced emissions compared to conventional biomass drying methods such as rotary drum dryers [30].

EHD drying has also emerged as a promising low-temperature and energy-efficient drying technology. Its system consists of a high-voltage power supply and an array of electrodes (emitter and collector), where a strong electric field generates an ionic wind that accelerates moisture removal from drying materials. Unlike traditional drying methods, EHD operates at significantly lower temperatures, preserving the dried product's nutritional content, bioactive compounds, and sensory properties. A recent study by Iranshahi et al. [31] on apple slice drying indicated that EHD drying outperforms traditional drying techniques regarding energy efficiency, exergy efficiency, and environmental impact. Their findings demonstrate that EHD drying achieves 60% higher energy efficiency, 220% higher exergy efficiency, and a 66% reduction in carbon footprint.

Additionally, it retains 50% more antioxidant capacity, 37% higher total phenolic content, and 21% less colour degradation compared to hot-air drying and FD techniques. These advantages highlight EHD drying as a promising approach for preserving nutritional and sensory properties in food products. However, its scalability remains challenging, as drying kinetics must be improved

TABLE 1.3
Some Selected Works about Advanced Drying Technologies

No.	Drying Method	Material	Main Findings	Reference
1	HPD	Banana slices	The higher temperature reduced drying time and increased effective moisture diffusivity. The best drying performance was achieved at the highest measured temperature (43°C), consuming 3.586 kWh	[32]
2	HPD	Ginger	The air's specific moisture extraction (SMER) was 0.06 $kg_{water}{\cdot}MJ^{-1}$ relative to 0.07 $kg_{water}{\cdot}MJ^{-1}$ in nitrogen. The external condenser gained 37.92% and 28.79% heat output for air and nitrogen drying, respectively	[33]
3	HPD	Kiwi	The difference in humidity between the dry air and kiwifruit slices was increased, leading to more efficient moisture transfer. An auxiliary heat pump system improved drying kinetics and reduced energy usage from 21.7 kWh kg^{-1} to 7.5 kWh kg^{-1}	[34]
4	Vacuum drying	Yacon	The drying temperature influenced the fractal dimension of the slices, whereas the slice thickness influenced the rehydration ratio. Fractal dimension was highly correlated with moisture content, hue angle, and rehydration ratio	[35]
5	Ultrasound (US)	Raspberry	Energy-saving and rehydration were improved. Porosity, total pore volume, and drying rate increased	[36]
6	EHD field	Quince slices	Minimal shrinkage. Increased voltage caused increased water absorption capacity, high-quality colour, and decreased shear strength	[37]
7	EHD field	Blueberry leather	Degradation of bioactive compounds and colour. Considerable reduction in enzyme activity	[38]

for large-scale food processing. While solar dryers are the most affordable option, their prolonged drying time limits their industrial feasibility. In contrast, EHD drying presents an optimal balance between cost and net present value (NPV). Nonetheless, hot-air drying remains the most efficient choice in large-scale applications where drying speed is the primary concern. When preserving bioactive compounds is the priority, EHD drying offers a more sustainable and economical alternative to FD.

Despite promising laboratory-scale results, commercial adoption remains limited, as large-scale studies comparing EHD drying's overall performance, affordability, and sustainability with existing drying technologies are still lacking. Further research is crucial to enhance its feasibility and industrial implementation. A summary of EHD drying, alongside other advanced drying techniques, is provided in Table 1.3.

1.3.3 Hybrid Drying Technologies

Hybrid drying technologies represent the next-generation drying method that eliminates various shortcomings. These systems combine the advantages of relevant drying technologies while minimising the drawbacks of using a single technology. Numerous drying technologies can be used together to achieve sustainability without compromising food quality. Given their economic feasibility and applicability in large-scale food industries, hybrid drying has garnered special research attention, particularly for agricultural and food processing applications.

Recent studies indicate that microwave and infrared drying (IRD) are among the most widely adopted hybrid technologies, often paired with convective drying to optimise drying kinetics and

TABLE 1.4
Selected Research on Hybrid Drying Technologies

No.	Drying Method	Material	Main Findings	Reference
1	Vacuum and heat pump	*Moringa* leaves	Reducing the pressure and increasing the velocity of circulating air reduced the drying time	[42]
2	Infrared and hot air	Tomato slices	The increase in radiation intensity decreased specific energy consumption (SEC), unlike the increase of air velocity, which increased it	[43]
3	Solar and heat pump	Copra	Better quality compared to other drying methods with a maximum coefficient of performance (COP) of 2.77	[44]
4	Solar and vacuum	Banana slices	The solar heater can fully supply the drying process at 70°C, but increases the drying time by 1 hour	[45]
5	Solar and biomass	Banana slices	A payback period of less than one year is less than the expected life (four years)	[46]

energy efficiency [39]. IRD primarily controls surface moisture transfer, while microwave drying enhances internal moisture diffusion. When combined, these technologies increase moisture diffusivity, leading to faster drying times and lower energy consumption [40]. Similarly, ultrasound-assisted drying has emerged as a promising non-thermal technique for agricultural products. Its integration with other drying methods enhances mass transfer, reduces drying time, and preserves product quality [41].

Beyond technological combinations, alternative energy sources can be incorporated successfully into hybrid drying systems to improve overall energy efficiency. Due to their potential for continuous replenishment, integration of such sources can be suitable for smallholder farmers, particularly in remote areas with difficulty accessing the electricity grid [19]. Among these, solar-assisted hybrid dryers are widely recognised for their cost-effectiveness and sustainability.

Table 1.4 provides an overview of selected research on hybrid drying technologies, highlighting their benefits across different food products.

Solar drying integrated with phase change material (PCM) energy storage systems has been widely explored to mitigate the effects of solar energy intermittency, ensuring heat supply. However, integrating sorption-based heat recovery systems into hybrid drying technologies has not yet been explored, remains largely theoretical, and requires further investigation. The concept revolves around capturing and reusing waste heat from the drying process, aiming to enhance energy efficiency and increase the coefficient of performance (COP). If successfully implemented, this approach could yield substantial energy savings while maintaining or improving drying performance. Sorption-assisted drying theoretically allows for more effective heat recovery, reducing fuel consumption and operational costs. By leveraging sorption materials capable of adsorbing and releasing moisture efficiently, these systems may enable better thermal management, leading to improved drying efficiency and reduced energy losses. Additionally, optimised moisture control within these hybrid systems could contribute to higher product stability and quality retention, particularly for heat-sensitive materials.

Future research and development (R&D) efforts should improve hybrid drying systems incorporating renewable energy, infrared, ultrasound, microwave, heat pumps, and sorption-based drying technologies. While traditional drying technologies provide advantages such as higher energy efficiency, reduced oxidative degradation, and lower environmental impact, further studies are required to optimise drying kinetics for heat-sensitive food materials. Moving forward, the focus must shift towards sustainable drying innovations, particularly in the field of numerical modelling and process optimisation. As advanced drying systems grow in complexity, modelling techniques are becoming essential in predicting, optimising, and scaling up sustainable systems, ensuring their practical viability for large-scale food processing. The following section delves into sustainable drying approaches, emphasising the role of numerical modelling in achieving energy-efficient and environmentally responsible dehydration technologies.

1.4 MODELLING APPROACHES FOR SUSTAINABLE DRYING

Modelling plays a crucial role in predicting the performance of drying systems. A wide variety of modelling techniques have been utilised to analyse and control different drying processes. This section discusses key modelling approaches, including CFD and mathematical modelling, AI modelling, and DTs, which are increasingly used to optimise and enhance drying processes.

1.4.1 Computational Fluid Dynamics and Mathematical Modelling

Theoretical models are mathematically complex and involve many parameters since they are established from fundamental drying physics [47], which couples transport equations with thermodynamically interactive fluxes and phase equilibrium expressions. This time-consuming modelling has led to the development of robust numerical methods. The food drying sector has increasingly adopted CFD approaches to analyse processes design and optimise drying systems [6]. CFD is a discipline that has progressed from a mathematical curiosity to become an indispensable tool in nearly every branch of fluid dynamics [48]. The use of numerical solution methods is trendy, and the popularity of such an advanced modelling approach is due to its plentiful advantages. CFD modelling plays a pivotal role in understanding and solving complex heat and transport equations and other associated phenomena such as chemical reactions, providing an attractive visualisation of temperature fields, velocity vectors, shrinkage, and moisture distribution inside the food or within the drying chamber to improve final product quality.

The user-friendly interface of the computer simulation tool allows for assessing the performance of the drying system with a qualitative flow visualisation throughout the process. Data from CFD simulation can be analysed to improve efficiency and product quality and thus design a new optimised drying system. CFD results give more assured scale-up, better reliability, higher productivity, and the ability to investigate systems under hazardous conditions at and above their normal performance limits with a reduced workload, cost, and drying space required in laboratory experiments. However, numerical simulation should be combined with physical experiments for validation and analytical solutions to accurately depict the kinetics of the heat and mass transfer mechanisms. CFD modelling relies on involving a precise physical definition of variables and proper initial and boundary conditions, as well as a selection of mathematical equations such as Navier–Stokes equations in line with suitable turbulence models using numerical methods and empirical correlations [49].

1.4.1.1 Mathematical Modelling Approaches

Mathematical models play a crucial role in understanding the underlying heat and mass transfer mechanisms governing food drying. These models facilitate the development of numerical simulations, enabling precise analysis and optimisation of drying parameters. Drying behaviour prediction is typically based on three categories of mathematical models: theoretical, semi-theoretical, and empirical models.

Semi-theoretical and empirical models offer practical and straightforward methods for describing drying kinetics. Many of these models, particularly thin-layer models, are derived from Fick's law of diffusion and Newton's law of cooling. While they effectively simulate drying curves and predict drying outcomes, they lack the physical significance of theoretical models. These simplified models often disregard important factors such as food shrinkage, complex geometries, and mass diffusivity/conductivity.

To accurately represent the dynamic drying behaviour of materials, selecting a model capable of capturing the water diffusion mechanism is crucial. Typically, moisture content data is converted

TABLE 1.5
Mathematical Models Applied for Regression Modelling of Drying Kinetics

No.	Model Name	Model Equation	Reference
1	Page	$MR = \exp\left(-k \times t^{n}\right)$	[50]
2	Lewis	$MR = \exp(-k \times t)$	[50]
3	Wang and Singh	$MR = 1 + a \times t + b \times t^{2}$	[50]
4	Logarithmic	$MR = a \times \exp(-k \times t) + c$	[51]
5	Diffusion approach	$MR = a \times \exp(-k \times t) + (1-a) \times \exp(-k \times b \times t)$	[51]
6	Midilli and Kucuk	$MR = a \times \exp\left(-k \times t^{n}\right) \times b \times t$	[51]
7	Two terms	$MR = a \times \exp(-k1 \times t) + b \times \exp(-k2 \times t)$	[51]
8	Weibull	$MR = \exp\left(-(t/\beta)^{n}\right)$	[52]
9	Henderson and Pabis	$MR = a \times \exp(-k \times t)$	[52]
10	Verma et al.	$MR = a \times \exp(-k \times t) + (1-\mathrm{a}) \times \exp(-g \times t)$	[52]

into moisture content ratio expressions before subjecting the models to curve-fitting calculations based on drying time. The models listed in Table 1.5 represent some of the most frequently cited approaches. Various statistical techniques can be applied to assess the accuracy of experimental data fitting within these models.

Theoretical models are based on fundamental physics involved in the drying process, primarily expressed by coupling fundamental transport equations with thermodynamically interactive fluxes and phase equilibrium expressions. However, due to their reliance on complex mathematical formulations and numerous parameters, theoretical models require extensive computational resources and expertise. The modelling process generally follows several key steps, as illustrated in Figure 1.7:

Conceptualisation: Establishing mathematical representations of transport phenomena to reflect real drying systems.

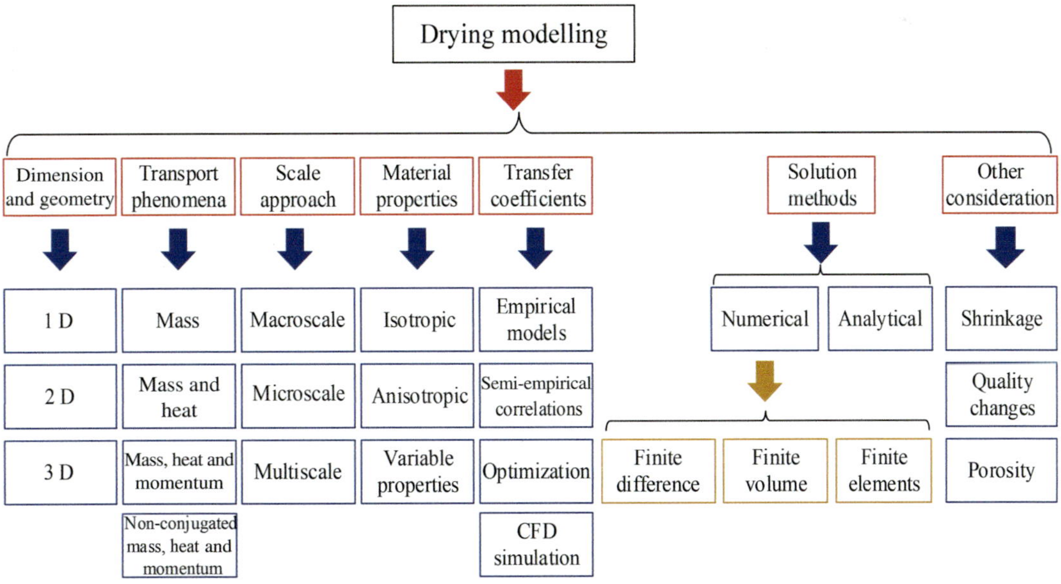

FIGURE 1.7 Drying modelling parameters of agricultural products drying [2].

TABLE 1.6
Classification of Mathematical Models for Drying Kinetics

Model Type	Description	Strengths	Limitations
Empirical models	Regression-based equations derived from experimental data	Simple, easy to implement	Limited physical significance
Semi-theoretical models	Based on Fick's law of diffusion and Newton's law of cooling	Good balance between simplicity and accuracy	Often neglect shrinkage and geometry changes
Theoretical models	Derived from fundamental transport equations	High accuracy, physically meaningful	Computational demanding requires precise input data

Formulation of Governing Equations: Using partial differential equations (PDEs) to describe heat, mass, and momentum transfer across two domains (air and food).
Determination of Material Properties: Identifying thermophysical properties and transfer coefficients.
Solution Method Selection: Depending on the complexity of the equations and system geometry, PDEs are solved analytically or numerically using methods such as the finite difference method (FDM), finite volume method (FVM), or finite element method (FEM).
Model Validation: Comparing simulation results with experimental data ensures that the model accurately represents real-world drying processes.

A comparative overview of different mathematical models used for drying kinetics is presented in Table 1.6, highlighting their descriptions, strengths, and limitations. As drying process modelling continues to evolve, recent advancements in numerical methods have improved the practicality of theoretical models, making them invaluable for understanding food drying. These models provide a comprehensive framework for analysing heat and mass transfer phenomena, capturing multiple coupled mechanisms and source terms.

1.4.1.2 Computational Fluid Dynamics for Drying

CFD modelling is a numerical approach in fluid mechanics that involves solving fundamental governing equations to describe fluid motion within a given geometry. This process can also integrate heat and mass transfer equations when necessary. The solved equations are mainly the continuity or conservation of mass (Eq. 1.1), Navier–Stokes or conservation of momentum (Eq. 1.2), and conservation of energy equations (Eq. 1.3) [53]. They are described as follows:

$$\frac{\partial \rho}{\partial t} + \nabla \cdot (\rho \vec{u}) = 0 \tag{1.1}$$

$$\frac{\partial (\rho u)}{\partial t} + \nabla \cdot (\rho u \vec{u}) = -\nabla P + +\mu \nabla^2 \vec{u} + \rho g \tag{1.2}$$

$$\rho C_p \frac{\partial T}{\partial t} + \rho C_p \vec{u} \cdot \nabla T = \nabla \cdot (k \nabla T) + S \tag{1.3}$$

where ρ is the density (kg·m^{-3}), $\vec{u}$ is the velocity (m·s^{-1}), P is the pressure (Pa), g is the acceleration of gravity (m·s^{-2}), C_p is the specific heat of the material (J·kg^{-1}·K^{-1}), T is the temperature (°C), k is the thermal conductivity (W·m^{-1}·K^{-1}), and S is the heat source (W·m^{-3}).

Numerical methods approximate solutions by discretising the computational domain into finite elements or grids, converting complex governing equations into simpler linear or polynomial forms. The choice of solution method depends on the problem's nature and the specific information required. Factors influencing this selection include domain geometry, initial and boundary conditions, and the classification of differential equations. The most commonly applied numerical techniques in convective drying studies are the FDM, accounting for 46% of previous research, the FVM at 39%, and the FEM at 15% [54].

CFD modelling of convective heat and mass transfer in porous materials during drying is categorised into two main approaches – non-conjugated and conjugated models. In the non-conjugated method, heat and mass transfer within the material and the surrounding air are treated separately. This approach assumes that airflow properties remain unaffected by internal heat and moisture transfer, neglecting the influence of internal water vaporisation on external fluid flow. Typically, momentum transfer is first analysed by determining temperature and velocity profiles to estimate convective transfer coefficients (CTCs) at the air–product interface, which are then used as input parameters for simulating heat and mass transport inside the product. Esfahani et al. demonstrated that in non-conjugated models, CTCs can vary across the surface during drying simulations, revealing differences between the front and rear faces of the product [55].

While non-conjugated models are simpler, computationally efficient, and flexible, conjugated models are more complex, requiring a deep understanding of mathematical equations and advanced computational tools. In the conjugated approach, heat and mass transfer in both the porous material and the surrounding airflow are solved simultaneously in a transient approach. Unlike non-conjugated models, which rely on predefined CTCs, conjugated models ensure the continuity of heat and mass fluxes, temperature, and moisture fraction at the air–food interface, eliminating the need for predefined CTC values. Instead, these parameters emerge as part of the simulation process. This method provides a more detailed representation of the morphological characteristics of food products, offering more profound insights into the drying process while reducing energy consumption and experimental costs.

Joardder and Karim [56] proposed a multiphase porous media model for intermittent microwave convective drying (IMCD) of apple slices, demonstrating that advanced CFD simulations can improve energy efficiency while maintaining food quality. Their research emphasised the importance of incorporating shrinkage impact in improving model accuracy. Khan and Straatman [57] demonstrated that conjugated models accurately depict real drying kinetics in porous materials by solving mass, momentum, energy, and moisture transport equations across two domains, considering water and vapour exchanges between components. This approach, known as the non-equilibrium method, assigns separate energy and moisture transfer equations to the fluid and solid components of the porous material. Although simplifications are advantageous and only a few authors have employed the conjugated approach, the authors showed that using the complete formulation is more valuable to simulate non-equilibrium energy and mass transport in computational domains of all scales. It is also necessary to take the next step towards the dynamic phase-coupled approach.

The fluid flow can be considered as either laminar or turbulent. Given the complex geometry of dried products and high airflow rates, drying processes often involve turbulence. Choosing an appropriate turbulence model is critical, significantly affecting CFD accuracy and performance. Commonly used turbulence models include k-ε models, k-w models, and Reynolds stress models. For a more detailed discussion on turbulence models, Malekjani and Jafari [58] extensively reviewed their applications in CFD simulations, including multiphase modelling approaches such as Eulerian–Eulerian, Eulerian–Lagrangian, and the volume of fluid (VOF) method [59]. A basic flow chart of the CFD simulation process is outlined in Figure 1.8.

1.4.1.3 Limitations of Numerical Modelling Approaches

Experimental approaches to drying analysis have several challenges and limits, including difficulties in obtaining data from inaccessible points, constraints on the number of tests and variations in

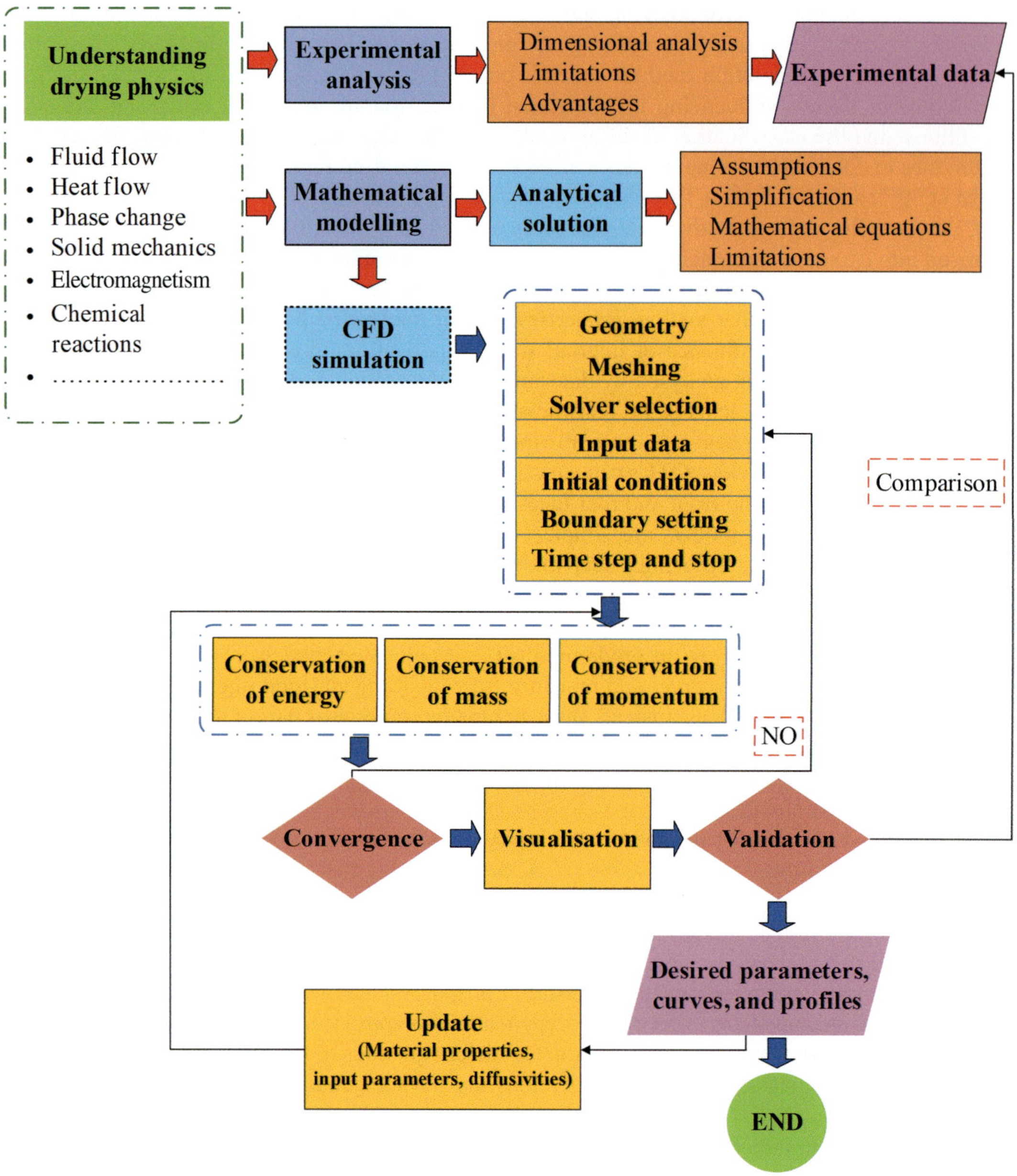

FIGURE 1.8 CFD simulation flowchart [2].

process conditions, and the requirement for extensive laboratory facilities with skilled operators for system maintenance. Researchers have increasingly turned to CFD simulations for drying process modelling to address these challenges. However, numerical methods also present their own difficulties, particularly when estimating parameters such as heat and mass transfer coefficients, temperature distribution, airflow velocity, streamlines, and moisture content.

Due to their heterogeneous structure, composition, and biological variability, the complexity of food products further complicates the modelling process. Generating accurate geometric representations of food items is particularly challenging for small-scale, irregularly shaped, or porous materials. Additionally, drying involves multiple interdependent parameters that are often non-linear and difficult to decouple. Incorporating many variables into a CFD model increases computational complexity, often necessitating custom user-defined functions (UDFs) or coding. As a result, simulation time can become excessive, requiring powerful computing resources. Given the expertise and expensive computational infrastructure required for low-value products, the cost-effectiveness of such high-level modelling techniques remains a concern.

Another significant obstacle in CFD-based drying analysis is the lack of comprehensive material property data, especially for food products whose physicochemical characteristics are highly temperature- and moisture-content dependent and change dynamically during drying. The challenge lies in efficiently obtaining and applying these properties while ensuring sufficient precision for reliable model predictions. Additionally, large-scale three-dimensional CFD simulations demand advanced and user-friendly software capable of handling complex multiphysics interactions while maintaining ease of setup and execution. A disparity exists between academic research and industrial needs, where industries require fast, practical 3D modelling tools, while researchers often focus on developing open-source codes for more fundamental studies beyond the scope of commercial software.

Another challenge in drying simulations arises from the complex interactions of dehydration, airflow, and material shrinkage. Many models simplify the process by considering only mass transfer, while fewer simultaneously integrate heat, mass, and momentum transport. Furthermore, integrating energy sources and real-world climatic conditions into existing drying models is rarely addressed, limiting their applicability in sustainability-focused research. Achieving a balance between simplifying assumptions and maintaining sufficient accuracy is crucial for modelling food drying with CFD. Table 1.7 summarises the strengths and limitations of CFD modelling in this context.

Despite its challenges, CFD remains a powerful tool for drying system optimisation, offering detailed insights into heat and mass transfer phenomena. However, further advancements are needed to enhance model accuracy, integrate real-world variables, and streamline computational processes for industrial-scale applications.

TABLE 1.7
Strengths and Weaknesses of CFD Modelling

Strengths	Weaknesses
• Detailed and scaled representation of product geometry and morphology • Capability to analyse complex systems where experiments may be hazardous or impractical • Ability to study drying conditions beyond experimental limits • Provides data at any point within the system, unlike experiments, which are limited to specific measurement points • Generally lower costs compared to physical experiments • Enables visual tracking of flow characteristics and drying behaviour • Digital data storage requires minimal laboratory space	• Potential errors due to simplified boundary conditions or flow models • Possible uncertainties arising from inadequate computing resolution per cell, leading to interpolation errors • Computationally expensive and time-consuming, particularly for large-scale simulations • Incorrect model configurations can lead to significant errors, sometimes more expensive than physical experiments • CFD users with limited expertise tend to believe that the simulation results are always correct without proper validation

1.4.2 Artificial Intelligence Modelling for Drying

The drying process's complexity, non-linearity, and multivariable nature have driven significant research efforts towards applying advanced soft-computing techniques, such as AI-based models. AI has gained rapid traction due to its ability to handle complex, ill-defined processes, making it a powerful and promising approach for improving food drying efficiency and control. AI integrates science, technology, and computing power, utilising techniques like cloud computing and big data to tackle non-linearities and uncertainties inherent in drying processes. Several AI-based techniques, including artificial neural networks (ANNs), adaptive-network-based fuzzy inference systems (ANFIS), genetic algorithms (GAs), and fuzzy modelling, have been employed to optimise drying parameters and enhance process control:

ANNs, inspired by biological neural networks, are widely used for establishing complex input–output relationships in drying systems. Their remarkable information-processing capabilities make them suitable for handling non-linear function approximation, noise reduction, clustering, pattern recognition, and process control.

Fuzzy modelling transforms linguistic descriptions into computational algorithms, allowing the division of a system into fuzzy subsets based on expert knowledge. This technique is highly effective for handling uncertainty and imprecision in complex non-linear systems.

ANFIS combines ANN and fuzzy logic principles, leveraging the learning capabilities of neural networks with the interpretability of fuzzy inference systems. This hybrid approach enhances prediction accuracy and adaptability.

GAs are inspired by Darwinian evolutionary principles and are designed to find optimal solutions through evolutionary processes. Instead of evaluating individual solutions in isolation, GAs work with a population of potential solutions, referred to as individuals, making it a promising tool for drying process optimisation.

Numerous studies have demonstrated the effectiveness of AI-based models in predicting and optimising drying parameters across different food products. Chokphoemphun and Chokphoemphun [60] found that ANN performed best in predicting moisture content during paddy drying in a fluidised bed dryer with a vortex flow generator. Jafari et al. [61] conducted a comparative study on the drying kinetics of onions using non-linear regression techniques, fuzzy logic, and ANNs. The fuzzy logic and ANN deserved the highest accuracy in determining the drying kinetics of onions. Interpolation results of the fuzzy model theoretically improved moisture ratio data and trained neural networks.

Aktas et al. [62] developed an ANN model to control drying air temperature, air velocity, and relative humidity for bay leaves' closed-loop heat pump dryer. Changes in energy consumption and moisture content were successfully observed, compared, and predicted. Table 1.8 presents an overview of selected AI-based drying studies, highlighting their key applications and findings.

Combining various AI approaches is essential for effectively modelling and controlling the food drying process. However, limited research has been conducted on AI modelling techniques due to particular inherent challenges that may hinder their application in real-world industrial drying systems. Issues such as overtraining, overfitting, and validation difficulties remain prevalent. Additionally, selecting appropriate input variables poses a significant challenge, as redundant or irrelevant data can lead to overly complex networks, large weight matrices, and increased computational demands.

AI has been successfully integrated into multi-objective optimisation in drying processes, significantly improving computational efficiency and decision-making. A widely adopted approach involves coupling ANNs with optimisation algorithms, such as the Non-Dominated Sorting Genetic Algorithm II (NSGA-II). This hybrid methodology enhances optimisation by leveraging ANNs as surrogate models, reducing the need for time-intensive simulations or experimental runs. Instead of conducting numerous physical trials, ANNs provide fast, data-driven predictions, effectively minimising computational costs.

TABLE 1.8
Selected Studies on AI Modelling in Food Drying

Drying Method	Material	Main Findings	Modelling Method	Reference
Solar dryer	Brewer's Spent Grains	A natural convection solar dryer (NCSD) was compared to a conventional electric convective drying system (CECD) The drying time for the CECD and NCSD were 30–85 minutes (363.15 and 333.15 K) and 345–430 minutes (depending on the process starting), respectively NCSD is 40–45% cheaper than the CECD NCSD offers a superior product quality and is eco-friendlier	Empirical ANN CFD Combined models	[63]
Heat pump	Thyme	The logarithmic model showed the best fit among empirical models, while diffusion modelling estimated effective diffusivity and activation energy ANNs provided the most accurate prediction of thyme drying kinetics (Variance ≥99.3%, Error ≤8.7%) ANN-based models are ideal for real-time optimisation and predictive control of drying processes	Empirical Theoretical ANN GA	[64]
Vacuum drying	Culinary banana (Musa ABB)	The influence of drying parameters on quality attributes during VD was evaluated Determination of rehydration ratio, scavenging activity, colour (non-enzymatic browning), and texture (hardness)	ANN GA	[65]
Convective drying	Three-leaved yam starch	ANN and SVM gave the most accurate prediction of exergy efficiency ANN is the only method to predict the sustainability index	ANFIS ANN and support vector machine (SVM)	[66]
Fluidised bed drying	Green tea leaves	Defined the drying time, total colour difference, and total phenolic content ANN model was trained using the feed-forward back-propagation method. Optimal conditions for the best quality were obtained	GA–ANN	[67]
Greenhouse solar dryer	Jaggery	The ANFIS model outperformed the thermal model, demonstrating better moisture content and air temperature accuracy with a correlation coefficient of 0.999–1.00 and a percentage error within 0.7–1.0%	MATLAB-based ANFIS modelling	[68]

The optimisation workflow follows a systematic training and validation process, as illustrated in Figure 1.9. The first step involves data collection from experiments, where input parameters such as drying temperature, air velocity, and relative humidity are recorded. The data undergoes normalisation to ensure consistency and improve training performance, which scales input variables to a uniform range. Once pre-processed, the dataset is divided into training and testing subsets – the training data is used to develop the ANN model, while the test set assesses the model's ability to generalise unseen conditions.

During the training phase, the ANN learns patterns in the dataset using a supervised learning algorithm, where network weights and biases are adjusted iteratively to minimise prediction errors. Training continues until a convergence criterion, such as a predefined error threshold or a maximum number of iterations, is met. After training, model performance is validated using the testing dataset, ensuring its reliability as a surrogate model for optimisation. The predictions are then

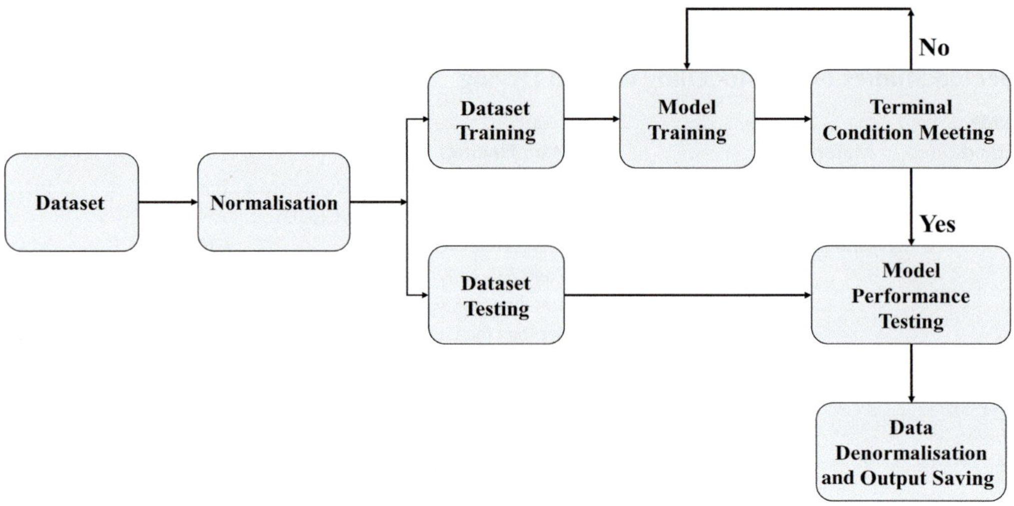

FIGURE 1.9 Flowchart of ANN training and validation process.

denormalised to their original scale for practical interpretation. The accuracy of the ANN model is typically evaluated using statistical indicators like the coefficient of determination (R^2) and the root mean square error (RMSE). These metrics help assess the model's effectiveness in capturing complex relationships between input parameters and key drying performance indicators.

The ANN architecture often follows a multi-layer perceptron (MLP) structure, as depicted in Figure 1.10. This model consists of interconnected layers of neurons that process and learn patterns

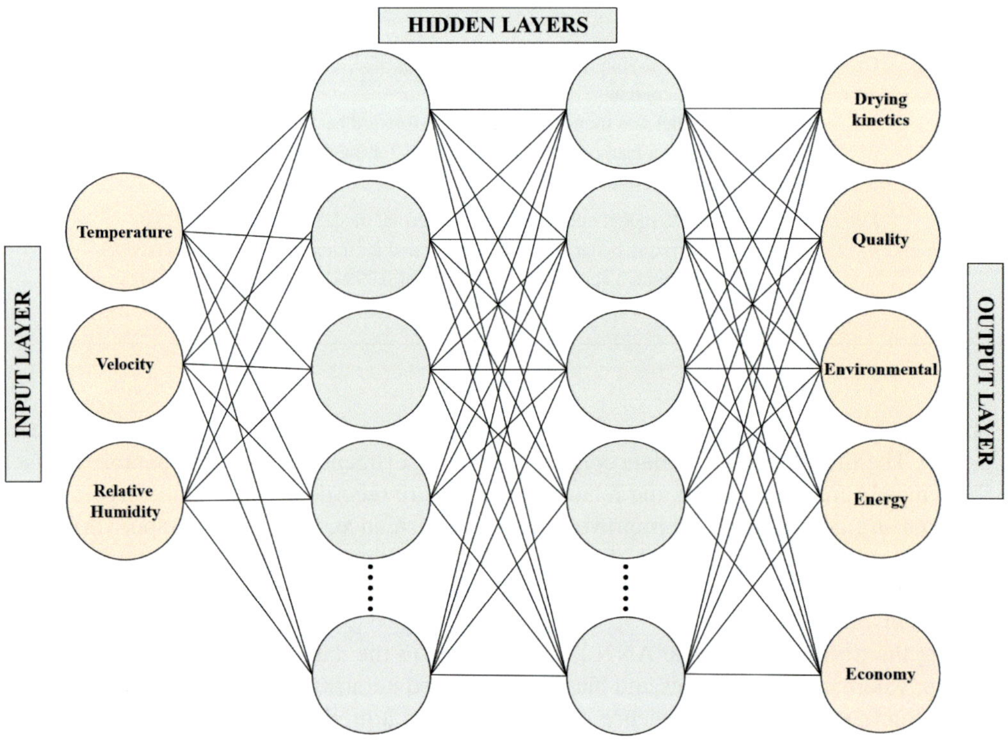

FIGURE 1.10 A representative structure of the multi-layer perceptron (MLP) used in food drying optimization.

in data, enabling it to capture non-linear relationships in drying behaviour and improve optimisation outcomes. A commonly used approach in drying optimisation, MLP allows for accurate prediction of key performance indicators such as drying time, product quality, energy consumption, and environmental impact. By leveraging AI-driven optimisation, researchers and industry professionals can develop more efficient and sustainable drying strategies, reducing costs while maintaining high product standards. These advancements are particularly valuable in agriculture, energy, and environmental science, where optimising drying processes plays a crucial role in sustainability and operational efficiency.

1.4.3 Digital Twins in Drying Systems

DTs are transforming the efficiency and precision of drying systems across various industries, including food processing, pharmaceuticals, and manufacturing. A DT is a virtual representation of a physical drying system that continuously receives real-time sensor data to simulate, predict, and optimise drying performance [69]. The implementation of DTs in drying systems typically follows a multilayered structure, integrating sensor data acquisition, real-time simulation, predictive analytics, and optimisation. A typical DT architecture consists of the following:

1. *Physical Layer:* Includes sensors and IoT (Internet of Things) devices embedded in the drying equipment to collect real-time data on key parameters such as temperature, humidity, airflow, and moisture content
2. *Data Processing Layer:* Aggregates and processes raw data using cloud computing, edge computing, or local processing units
3. *Simulation Layer:* Uses physics-based models or AI-driven machine learning algorithms to replicate the drying process virtually, predicting system behaviour under varying conditions.
4. *Optimisation Layer:* Adjusts drying parameters dynamically to enhance energy efficiency, minimise waste, and ensure uniform drying.
5. *Decision-Making Layer:* Provides real-time feedback to operators or autonomously adjusts system parameters to maintain optimal drying conditions.

DTs can optimise sustainable drying cycles through predictive modelling, preventing over-drying, under-drying, or uneven drying [70]. Integrating these layers, the DT forms a closed-loop control system where real-world data continuously informs simulations and optimisations, leading to an intelligent, self-improving drying process (Figure 1.11). However, several challenges remain in the widespread adoption of DTs in drying systems:

High Initial Investment: Integrating IoT, AI, and computational modelling requires significant investment.

Limited Availability of Advanced Sensors: While temperature and humidity sensors are widely available, moisture content and gas sensors remain costly and less commercially accessible.

Complexity in Model Development: Creating an accurate virtual twin of a drying system demands detailed physics-based models or extensive machine learning training.

Data Management and Integration: Handling large-scale sensor data while ensuring real-time responsiveness is a technical hurdle.

To overcome these challenges, future research and development must focus on different key areas. Firstly, integrating machine learning and deep learning into AI-enhanced DTs is crucial for improving predictive accuracy and automation. Secondly, developing affordable, compact sensors that can be embedded within drying products or machinery will enable more widespread data collection. Finally, shifting from cloud-based analytics to edge computing, where data is processed

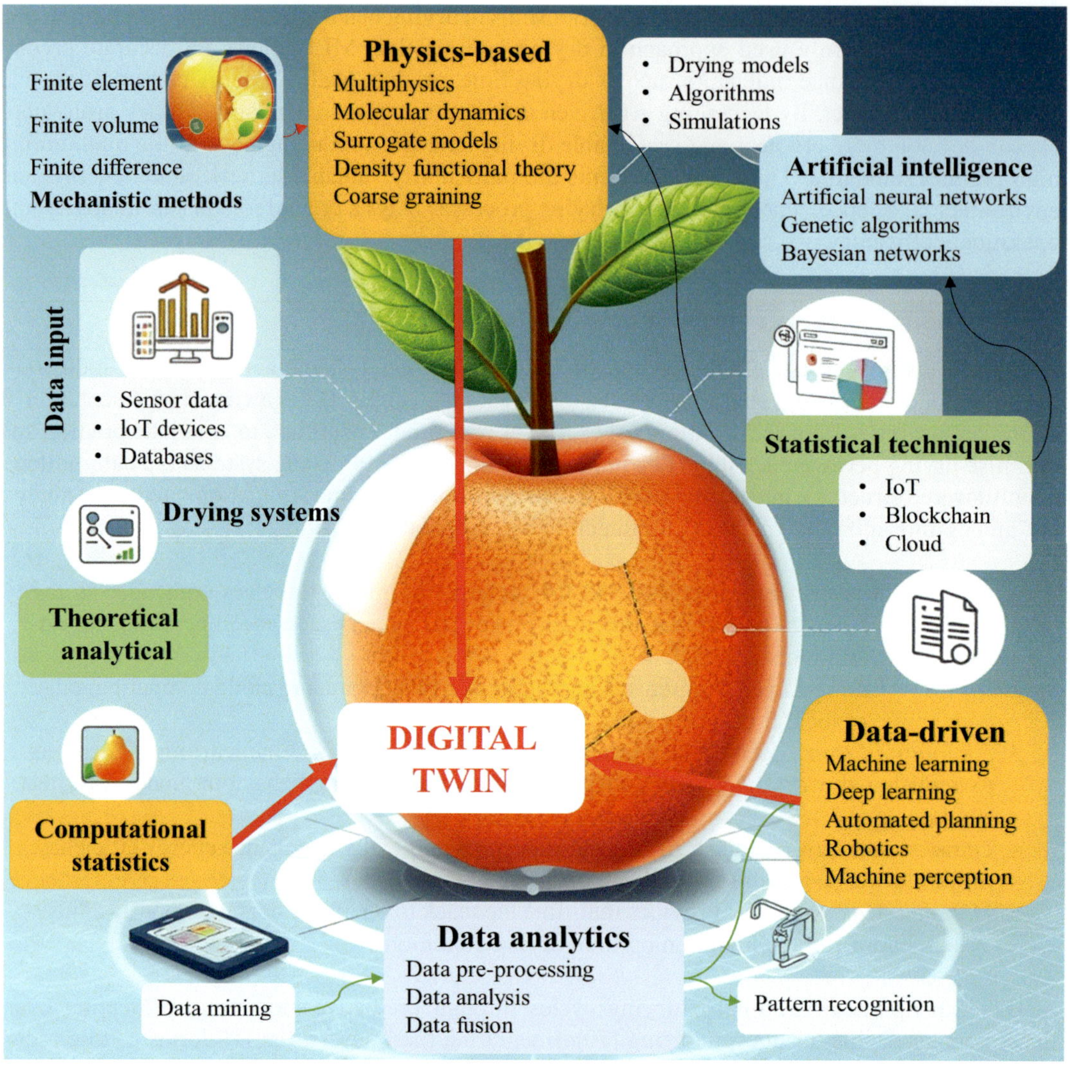

FIGURE 1.11 Conceptual framework of DTs in drying systems.

directly on-site, will facilitate real-time processing and faster decision-making. Addressing these challenges will make DTs more accessible, cost-effective, and widely adopted in next-generation drying systems.

1.5 EFFECT ON FOOD QUALITY

The assessment of food quality is critical in determining suitable drying methods and process parameters. With growing consumer concerns regarding food safety and hygiene, maintaining the quality of agricultural products has become increasingly vital [71]. The quality attributes of dried products should closely resemble those of fresh agricultural commodities. Consequently, the food industry must support research on alternative drying technologies and modern quality evaluation techniques. However, conventional quality testing methods are often labour-intensive, time-consuming, and destructive.

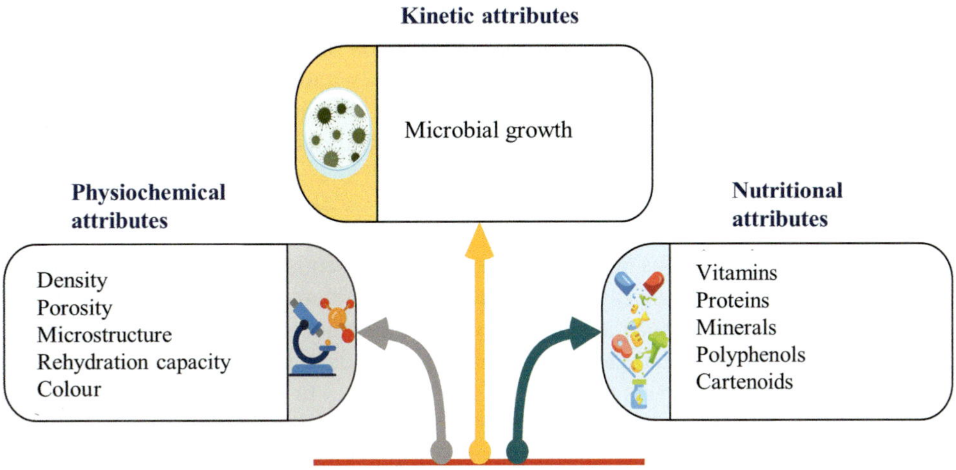

FIGURE 1.12 Key attributes affecting the quality of dried products.

In contrast, emerging technologies offer non-destructive, rapid, and highly accurate quality assessment methods. For instance, Hussain et al. recently explored the potential of novel imaging and spectroscopy techniques such as MRI, soft X-ray imaging, hyperspectral imaging, laser light backscattering, and fluorescence imaging for cereal quality evaluation, which could be extended to other crops [72]. The developed technologies, such as MRI, soft X-ray imaging, hyperspectral imaging, laser light backscattering, and fluorescence imaging, have proven to be faster, non-destructive, and more accurate than classical techniques. The quality attributes can be classified into three categories: physiochemical, microbiological, and nutritional (Figure 1.12).

1.5.1 Physiochemical Quality

The physiochemical characteristics of dried food, including rehydration capacity, porosity, structure, bulk density, case hardening, shrinkage, and collapse, significantly impact the quality of the final product. Rehydration is a crucial indicator among these factors, as it reflects the extent of physical and chemical changes that occur during the drying process. The efficiency of rehydration is influenced by drying technology, pre-treatments, initial moisture content, and drying conditions [73]. Visual appearance is a primary factor in consumer acceptance of dried foods, making colour retention a significant challenge in food dehydration. Advanced drying methods aim to reduce heat exposure and shorten drying time to achieve optimal moisture content while preserving colour and structural integrity. To enhance food quality assessment, Al-Dairi et al. introduced a machine vision-based approach utilising fractal dimension analysis to detect and quantify mechanical damage in bananas [74]. Their study demonstrated that fractal dimension analysis provides an accurate, non-destructive method for evaluating bruise severity, which can be applied to monitor food quality deterioration during storage and transportation. By integrating such techniques into the food industry, the quality assessment of dried products can become more precise, enabling better control over drying conditions and minimising physical degradation.

1.5.2 Microbial Quality

Dried food products are prone to microbial contamination during storage due to dust, insects, or rodents exposure. To ensure food safety, dried foods must be stored under appropriate conditions to inhibit microbial growth. Below a specific water activity level, microbial proliferation is suppressed.

Fruits, mainly, are dominated by moulds and yeasts, whereas vegetables primarily harbour Gram-negative bacteria [75]. Several factors influence dried foods' microbial quality, including product composition, pH, pre-treatments, and drying techniques. Alp and Bulantekin conducted a comprehensive review of the impact of different drying methods on microbial stability, highlighting the importance of selecting appropriate drying conditions to minimise microbial risks [76].

1.5.3 Nutritional Quality

Dried produce provides numerous health benefits, offering essential vitamins, minerals, and phytochemicals such as polyphenols and carotenoids. However, the retention of these bioactive compounds is highly dependent on factors such as temperature, pre-treatments, drying methods, and exposure to light and oxygen. The degradation of nutritional compounds during drying remains a significant challenge. Hence, precise drying control measures are necessary to preserve valuable nutrients while ensuring food safety [77]. Emerging kinetic models have been developed to predict the impact of drying on nutrient stability. These models assess the degradation kinetics of vitamins and other bioactive compounds, aiding in optimising drying conditions to balance efficiency and quality preservation.

1.5.4 Quality Modelling in Drying Processes

Technological advancements in CFD have significantly contributed to modelling food drying kinetics. However, only a few studies have attempted to predict the evolution of the quality of dried products. Quality modelling is complex, as it involves multiple interdependent factors, including heat and mass transfer, physical deformations, and microstructural changes. Devahastin and Niamnuy [78] reviewed advances in modelling quality changes in fruits and vegetables during drying, including methods to predict colour, nutritional content, microstructural integrity, and physical changes. They highlighted that quality attributes could be effectively modelled as a function of temperature and moisture content using reliable drying simulation techniques such as CFD and ANN. Quality modelling is a complex endeavour involving physical deformations, microstructural changes, and heat and mass transfer.

Boekel et al. [79] suggested that kinetic modelling could quantify various quality changes during drying. This approach involves mathematical models containing characteristic kinetic parameters like rate constants and activation energies. The general rate law for quality degradation follows the below equation:

$$\text{Rate} = \frac{-\mathrm{d}C}{\mathrm{d}t} = k_r C^p \tag{1.4}$$

where p is the order of reaction and k_r is the rate constant. For a single component, the differential equation represents the dependence of rate r on concentration. The equation can be integrated over time to obtain an equation linking concentration with time. The essence of kinetic modelling of quality changes during the drying process is to evaluate a quality attribute as a function of drying time at a specific temperature using temperature-dependent reaction rate constants after defining the order of reactions. The authors reported that quality changes are usually modelled using zero-, first-, or second-order reactions as follows:

$$C = C_0 - k_r t \quad \text{for zero-order reactions } (p = 0) \tag{1.5}$$

$$C = C_0 \mathrm{e}^{-k_r t} \tag{1.6}$$

$$\frac{1}{C} = k_r t + \frac{1}{C_0} \quad \text{for second-order reactions } (p = 2) \tag{1.7}$$

where C_0 is the initial value of concentration at $t = 0$. Zero-order reactions are the most frequently reported, primarily when representing browning changes resulting from Maillard reactions. First-order reactions are also frequent but are for degradation reactions such as vitamin loss. Second-order reactions are less commonly used but can model amino acid degradation due to Maillard reactions. The selection of the reaction order is often empirical, based on fitting experimental data rather than theoretical considerations. Arrhenius's law is often used to describe the temperature dependence on quality decay [80]:

$$\mathrm{k}_r(T) = Ae^{\frac{-E_a}{RT}} \tag{1.8}$$

where A is a pre-exponential factor, E_a is the activation energy, R is the ideal gas constant (8.314 $\mathrm{J{\cdot}mol^{-1}{\cdot}K^{-1}}$), and T is the absolute temperature. The activation energy is the energy barrier required to initiate a chemical reaction. The number of molecules with energy higher than activation energy increases with temperature, which qualitatively explains the influence of temperature on reaction rate. Using a high order for the Arrhenius equation by including more terms that affect food quality during drying is recommended.

Despite its potential, only a few studies have reported the application of kinetics modelling to quality prediction during drying. Prawiranto et al. [81] investigated the trade-offs between drying time, product quality, and energy use in solar drying of apples. They used CFD simulations under different weather conditions to model quality deterioration, such as browning and vitamin C loss. Similarly, Defraeye [82] optimised intermittent drying conditions using numerical hygrothermal simulations and a temperature-dependent model for quality loss assessment. Defraeye and Radu [83] further compared zero- and first-order reaction kinetics to study quality evolution under different numerical assumptions. Pham et al. [84] developed a CFD model to predict temperature and moisture distributions alongside nutrient depletion, including total phenolic content and ascorbic acid degradation. Another study by Sabarez [85] successfully simulated and validated the kinetics of colour changes in trellis-dried sultanas. These numerical models enhance the ability to predict heat and mass transfer processes while providing insights into the spatiotemporal evolution of product quality. Moreover, solving heat and mass transfer equations separately from the Arrhenius equation allows operators to monitor and adjust real-time drying conditions, improving energy efficiency and product quality.

Although kinetic modelling offers a structured approach to predicting quality degradation, food deterioration processes are inherently complex. Changes such as lipid oxidation, enzymatic degradation, and Maillard reactions involve multiple simultaneous reactions that a single equation cannot fully describe. Most existing kinetic models are deterministic, assuming the same input will always produce the same output. However, real-world drying processes are subject to variability in biological materials, making deterministic models somewhat unrealistic. More advanced models that incorporate probabilistic and stochastic approaches could improve the accuracy of quality predictions by accounting for uncertainties in material properties and process conditions.

Recent efforts have focused on developing more advanced numerical models to address these limitations. Wijerathne et al. [86] introduced a coarse-grained multiscale numerical model to simulate morphological changes in apple tissues during drying. This approach integrated mesh-free numerical methods, which offer advantages over traditional grid-based modelling techniques such as FEM and FDM. Mesh-free models are particularly useful for capturing excessive deformations in cellular structures, simulating intense moisture gradients, and modelling multiphase interactions during drying. However, despite their higher accuracy in predicting microstructural transformations, 3D numerical models remain computationally expensive, especially for large tissue simulations.

Understanding the relationships between drying parameters and food quality attributes provides opportunities to enhance product quality while optimising energy consumption. Advanced numerical models offer improved predictive capabilities, including CFD, ANN, and mesh-free simulations. However, future research should focus on developing more realistic models that integrate biological variability, multiscale drying dynamics, and uncertainty quantification. By refining these models, food drying processes can be optimised to preserve product quality more effectively while minimising energy use.

1.6 ENERGY EFFICIENCY AND ENVIRONMENTAL IMPACT

The agri-food system, much like other production systems, is not merely a collection of linear processes but a complex ecosystem where every decision shapes environmental, economic, and social impacts. As part of this system, drying processes are critical in preserving food quality and extending shelf life. However, these processes are energy-intensive and contribute significantly to greenhouse gas (GHG) emissions. This section explores the energy consumption analysis of different drying methods, their environmental impact assessment, and the life cycle assessment (LCA) application to evaluate sustainability.

1.6.1 Energy Consumption Analysis of Different Drying Methods

Energy consumption analysis is a primary step in improving drying system performance. An effective assessment provides information on energy demand and costs by examining specific data at both the input and output stages of the system. Various studies have examined the energy performance of drying systems using key performance indicators.

For instance, Taşeri et al. [87] conducted an experimental study to evaluate the impact of varying drying air temperatures (45°C and 50°C) on the efficiency of a heat pump dryer used for grape pomace. Although higher temperatures required more significant energy input, performance indicators such as SEC, COP, drying efficiency, and SMER demonstrated improved efficiency at lower temperatures. Similarly, Tohidi et al. [88] analysed the energy attributes for fixed deep-bed drying of paddy, showing that thermal and mechanical contributions were 94.48% and 5.52%, respectively. Their findings indicated that increasing air temperature while reducing air velocity and relative humidity improved overall energy efficiency.

However, energy analysis alone may not adequately capture all thermodynamic inefficiencies in drying processes. Lingayat and Chandramohan [89] examined thermal efficiency and heat loss in indirect natural convection solar drying of banana chips, emphasising the need for exergy analysis to gain deeper insights into energy degradation and irreversibility. As a valuable tool, exergy analysis allows for a more comprehensive assessment of drying efficiency by identifying key points of energy dissipation, ultimately contributing to a better understanding of exergy losses and improving overall energy efficiency.

Moreover, experimental energy analysis can be time-consuming and costly, prompting researchers to explore virtual representations. Majdi and Esfahani [90] utilised the lattice Boltzmann method (LBM) to solve coupled conjugate heat and mass transfer equations during convective drying of apple slices. They determined optimal drying parameters using response surface methodology (RSM) and Box–Behnken design, demonstrating that product thickness significantly impacts energy consumption more than velocity or temperature. Hyuk [91] developed a computer simulation for concurrent-flow drying of rapeseeds, achieving fuel energy consumption predictions of 4.62–8.57% less than measured values. Minimising energy consumption during drying can be achieved by operating at low moisture content and velocity. Modelling these conditions allows for significant energy optimisation and reduces convective drying time, thereby mitigating the environmental and economic costs of food drying.

1.6.2 Environmental Impact Assessment in Food Drying

The growing awareness of global warming and the United Nations' 2030 Sustainable Development Goals (SDGs) have spurred interest in decarbonising food drying processes. Efforts are directed

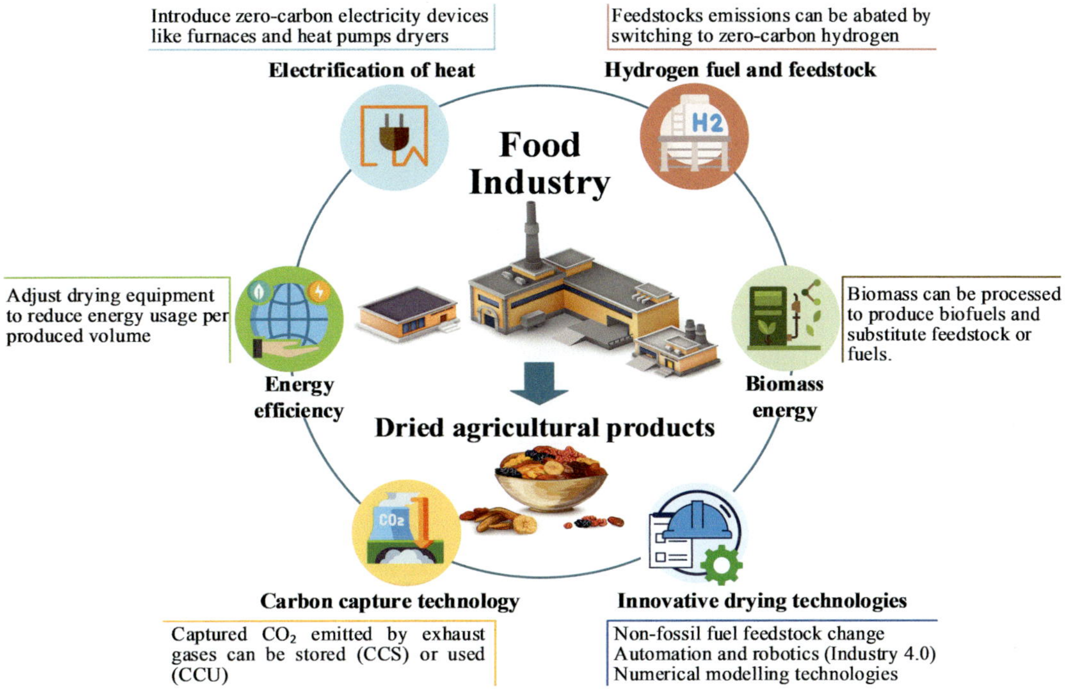

FIGURE 1.13 Options for decarbonising drying of agri-food products [2].

towards reducing GHG emissions to mitigate climate change while enhancing industrial productivity. Carbon footprint analysis, rooted in LCA principles, evaluates the environmental impacts of products and processes across their life cycle, from cultivation and processing to transport and delivery.

A range of decarbonisation strategies has been proposed to reduce emissions in food drying, including energy-efficiency improvements, automation and robotics (aligned with Industry 4.0 concepts), biomass or hydrogen fuel usage, heat electrification, and carbon capture utilisation (CCU) or storage (CCS). Figure 1.13 summarises the most promising decarbonisation options for agricultural product drying, emphasising the need for site-specific evaluations to determine feasible and economical solutions. A study by Gonzalez-Diaz et al. demonstrated that integrating carbon capture utilisation and storage (CCUS) in drying operations can reduce GHG emissions by 6–7% in the rice supply chain and 14–16% in wheat drying [92]. Despite the promising potential of these strategies, further research is needed to assess their feasibility and long-term viability in industrial settings.

Grid decarbonisation has been identified as a critical enabler in achieving long-term GHG reductions in the food and beverage industry. The electrification of heat, in particular, presents a pivotal opportunity to transition away from fossil fuel–based drying processes. A recent review by Atuonwu and Tassou [93] highlighted the critical role of grid decarbonisation in achieving long-term GHG reductions in the food and beverage industry. Electrification of heat emerges as a pivotal option, with significant reductions in GHG emissions observed when comparing electrified technologies to conventional thermal treatments. Efficient knowledge and control of the drying process mechanisms are crucial for energy savings. Although research on reducing GHG emissions is burgeoning, further exploration is needed to understand the potential of numerical modelling in decarbonising food manufacturing.

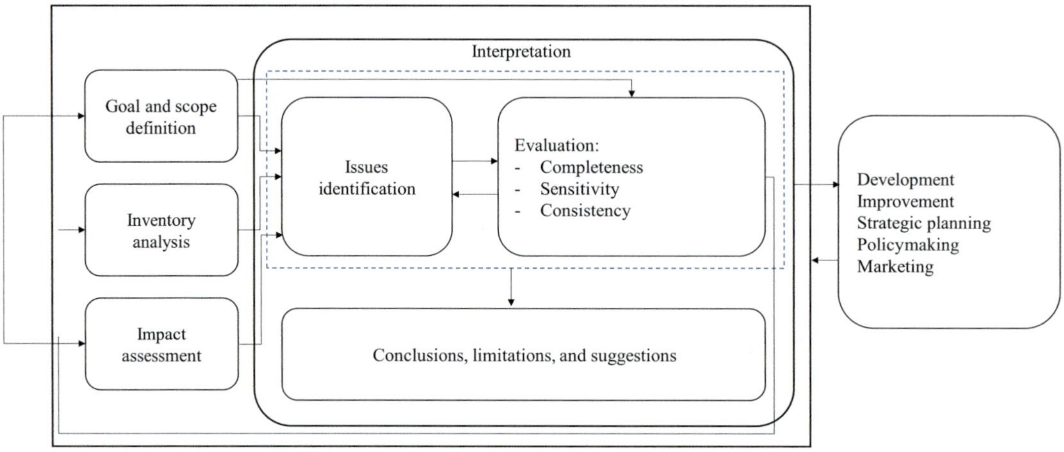

FIGURE 1.14 A LCA framework.

Given the complexity of food drying processes, environmental impact assessment methodologies such as LCA play a pivotal role in evaluating sustainability. LCA provides a comprehensive framework for analysing the environmental footprint across the entire life cycle of a product or process. LCA helps identify areas for improvement and assess the effectiveness of decarbonisation strategies. As illustrated in Figure 1.14, LCA follows a structured, iterative process that includes several key stages:

Goal and Scope Definition: Defining the boundaries and objectives of the assessment
Inventory Analysis: Quantifying resource inputs and emissions
Impact Assessment: Evaluating environmental impacts such as GHG, water use, and land use
Interpretation: Analysing results and identifying opportunities for improvement
Conclusions, Limitations, and Suggestions: Drawing conclusions and recommending actions.

Figure 1.14 depicts LCA's structured approach, highlighting the process's iterative nature and application in strategic planning and policymaking. Defining system boundaries is essential in food drying to ensure accurate environmental impact assessments. Figure 1.15 illustrates a typical system boundary for LCA applied to drying operations. This system encompasses all major components, including energy inputs, raw materials, emissions, and waste management, ensuring a holistic approach to sustainability evaluation. The iterative nature of LCA enables continuous refinement, helping industries make informed decisions for optimising drying processes.

Several studies have applied LCA to assess the sustainability of different drying techniques, identifying energy hotspots and environmental trade-offs. Zhang et al. [94] investigated LCA in milk powder spray drying and found that electricity and natural gas contribute 40–50% and 20–30% of the total environmental impact, respectively. Their study suggested that circular economy approaches, such as waste heat recovery, could enhance sustainability in the milk powder industry. Kumar et al. [95] assessed the environmental impacts of tomato drying and concluded that HPD is more sustainable than microwave vacuum drying (MVD) due to its lower energy demand and reduced emissions. Elhami et al. [96] examined the environmental footprint of sun drying versus convective drying of grapes, revealing that grape drying accounted for 90% of total vineyard energy consumption, with electricity (28%) and chemical fertilisers (19%) being the most significant contributors to environmental impact. Similarly, Merone et al. [97] analysed ultrasound-assisted freeze-drying for apples, carrots, and eggplants, demonstrating that applying atmospheric ultrasound reduced total energy consumption by up to 70% and environmental impact by 58–82%. Their

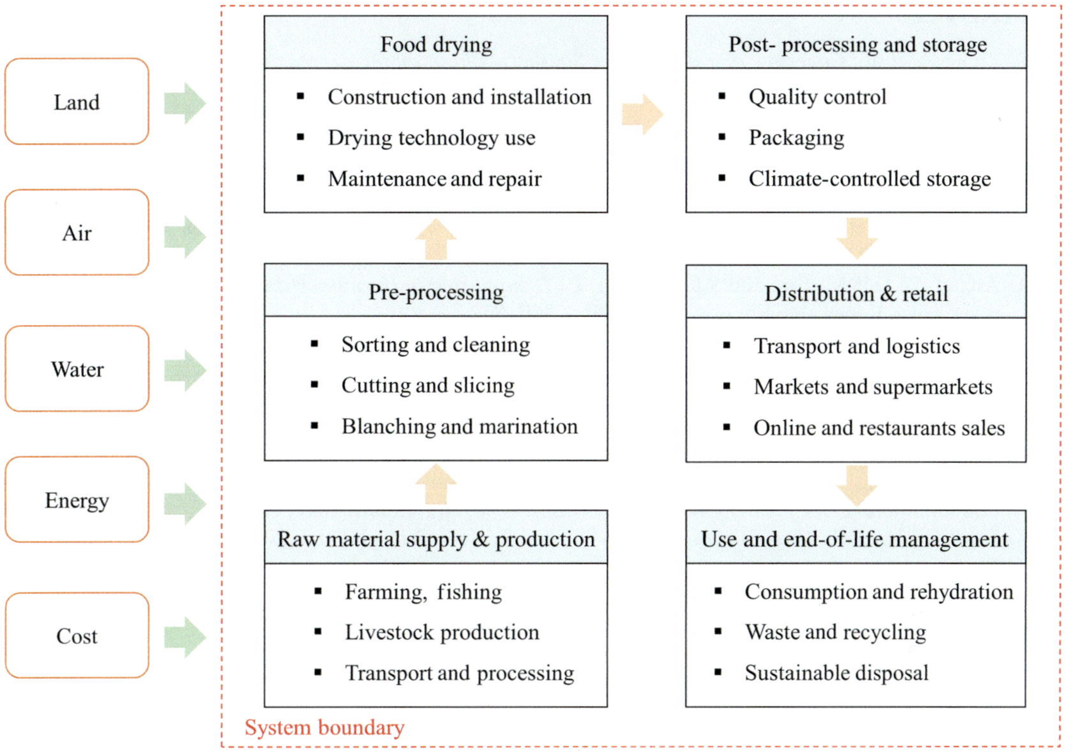

FIGURE 1.15 System boundary for food drying LCA.

study further emphasised that the internal structure of food products significantly influences drying efficiency and overall sustainability.

1.7 CONCLUSION

The transition from conventional to sustainable food drying systems demands a paradigm shift from isolated metrics to holistic systemic change. As drying technologies evolve, integrating energy-efficient methods such as hybrid solar–biomass systems, heat pump dryers, and advanced VD offers significant potential to reduce greenhouse gas emissions while preserving product quality. Computational tools like CFD and AI-driven models enable optimisation of predictive heat and mass transfer, minimising energy waste and operational costs. However, true transformation lies in embedding these innovations within a broader sustainability framework.

LCA is a cornerstone for evaluating environmental trade-offs, guiding industries towards decarbonisation through grid electrification, renewable energy integration, and waste heat recovery. The fractal-like interconnectedness of sustainability, where each technological choice reverberates across economic, social, and ecological systems, necessitates a systems-thinking approach. Future advancements must prioritise technical efficiency and contextual adaptability, ensuring scalable and equitable solutions for diverse food processing systems. As this chapter concludes, it is evident that the evolution towards energy-efficient, environmentally responsible drying systems is not merely a linear progression but a dynamic interplay of technological advancement, systemic change, and interdisciplinary collaboration. Future research and industrial adoption must prioritise integrating advanced modelling tools, renewable energy systems, and circular economy principles to transform food drying processes into sustainability benchmarks.

REFERENCES

1. Zhou, Z. et al., *Covering rice demand in Southern China under decreasing cropping intensities and considering multiple climate and population scenarios.* Sustainable Production and Consumption, 2023. **40**: p. 13–29.
2. Adnouni, M. et al., *Computational modelling for decarbonised drying of agricultural products: Sustainable processes, energy efficiency, and quality improvement.* Journal of Food Engineering, 2023. **338**: 111247.
3. Catorze, C. et al., *Study of a solar energy drying system: Energy savings and effect in dried food quality*, 2022. **8**: p. 392–398.
4. Anand, S., D.P. Mishra, and S.K. Sarangi, *CFD supported performance analysis of an innovative biomass dryer.* Renewable Energy, 2020. **159**: p. 860–872.
5. Menon, A., V. Stojceska, and S.A. Tassou, *A systematic review on the recent advances of the energy efficiency improvements in non-conventional food drying technologies.* Trends in Food Science & Technology, 2020. **100**: p. 67–76.
6. Mhd Safri, N.A. et al., *Current status of solar-assisted greenhouse drying systems for drying industry (food materials and agricultural crops).* Trends in Food Science & Technology, 2021. **114**: p. 633–657.
7. Awotide, B.A., A.A. Karimov, and A. Diagne, *Agricultural technology adoption, commercialization and smallholder rice farmers' welfare in rural Nigeria.* Agricultural and Food Economics, 2016. **4**(1): 3.
8. Esmaiili, M. et al., *Grape drying: A review.* Food Reviews International, 2007. **23**(3): p. 257–280.
9. Meng, J. et al., *Phenolic content and antioxidant capacity of Chinese raisins produced in Xinjiang Province.* Food Research International, 2011. **44**(9): p. 2830–2836.
10. Son, S.M., J.-W. Rhim, and J.H. Lee, *Dehydration characteristics of maesaengi (Capsosiphon fulvescens) in hot-air drying.* Food Science and Biotechnology, 2011. **20**(2): p. 549–553.
11. Almena, A. et al., *Optimising food dehydration processes: Energy-efficient drum-dryer operation.* Energy Procedia, 2019. **161**: p. 174–181.
12. Baker, C.G.J., and K.A. McKenzie, *Energy consumption of industrial spray dryers.* Drying Technology, 2005. **23**(1–2): p. 365–386.
13. Koyuncu, T., Tosun, and Y. Pınar, *Drying characteristics and heat energy requirement of cornelian cherry fruits (Cornus mas* L.*).* Journal of Food Engineering, 2007. **78**(2): p. 735–739.
14. Fang, Y.L. et al., *Health risk assessment of trace elements in Chinese raisins produced in Xinjiang Province.* Food Control, 2010. **21**(5): p. 732–739.
15. Taherzadeh-Shalmaei, N. et al., *Evaluating the energy use, economic and environmental sustainability for smoked fish production from life cycle assessment point of view (case study: Guilan Province, Iran).* Environmental Science and Pollution Research, 2021. **28**(38): p. 53833–53846.
16. Inyang, U. et al., *Drying and the Different Techniques.* 2017. **8**(1): p. 45–72.
17. Bai, R. et al., *Hot air convective drying of ginger slices: Drying behaviour, quality characteristics, optimisation of parameters, and volatile fingerprints analysis.* Foods, 2023. **12**(6): p. 1283.
18. Samborska, K. et al., *Innovations in spray drying process for food and pharma industries.* Journal of Food Engineering, 2022. **321**: 110960.
19. Lamidi, R.O. et al., *Recent advances in sustainable drying of agricultural produce: A review.* Applied Energy, 2019. **233-234**: p. 367–385.
20. Lingayat et al., *A review on indirect type solar dryers for agricultural crops: Dryer setup, its performance, energy storage and important highlights.* Applied Energy, 2020. **258**: 114005.
21. Sharma, S., K. Dhalsamant, and P.P. Tripathy, *Application of computer vision technique for physical quality monitoring of turmeric slices during direct solar drying.* Journal of Food Measurement and Characterization, 2018. **13**(1): p. 545–558.
22. Monicka, A.A. et al., *A comprehensive review of indirect solar drying techniques integrated with thermal storage materials and exergy-environmental analysis.* Environment, Development and Sustainability, 2024. https://doi.org/10.1007/s10668-024-04755-7
23. Lamidi, R.O. et al., *Biogas tri-generation for postharvest processing of agricultural products in a rural community: Techno-economic perspectives.* Energy Procedia, 2017. **142**: p. 63–69.
24. Zayed, M. et al., *Drying of Thompson Seedless grapes in thin-layer using a biogas dryer.* Misr Journal of Agricultural Engineering, 2018. **35**(1): p. 275–298.
25. Redko, A., O. Redko, and R. DiPippo, Heat pumps in the drying industry, in Low-Temperature Energy Systems with Applications of Renewable Energy. 2020: Academic Press. p. 135–175.
26. Patel, K.K., and A. Kar, *Heat pump assisted drying of agricultural produce: An overview.* Journal of Food Science and Technology, 2012. **49**(2): p. 142–60.

27. Nadi, F., and D. Tzempelikos, *Vacuum drying of apples (cv. Golden Delicious): Drying characteristics, thermodynamic properties, and mass transfer parameters.* Heat and Mass Transfer, 2018. **54**(7): p. 1853–1866.
28. Torki-Harchegani, M. et al., *Dehydration behaviour, mathematical modelling, energy efficiency and essential oil yield of peppermint leaves undergoing microwave and hot air treatments.* Renewable and Sustainable Energy Reviews, 2016. **58**: p. 407–418.
29. Yi, J. et al., *Drying efficiency and product quality of biomass drying: A review.* Drying Technology, 2019. **38**(15): p. 2039–2054.
30. Nandhu Lal, A.M. et al., *A comparison of the refrigerated adsorption drying of Daucus carota with fluidized bed drying.* LWT, 2022. **154**.
31. Iranshahi, K. et al., *Electrohydrodynamic drying versus conventional drying methods: A comparison of key performance indicators.* Energy Conversion and Management, 2023. **279**.
32. Tunckal, C., and İ Doymaz, *Performance analysis and mathematical modelling of banana slices in a heat pump drying system.* Renewable Energy, 2020. **150**: p. 918–923.
33. Chapchaimoh, K. et al., *Thermal characteristics of heat pump dryer for ginger drying.* Applied Thermal Engineering, 2016. **95**: p. 491–498.
34. Mohammadi, I., R. Tabatabaekoloor, and A. Motevali, *Effect of air recirculation and heat pump on mass transfer and energy parameters in drying of kiwifruit slices.* Energy, 2019. **170**: p. 149–158.
35. Reis, F.R. et al., *Vacuum drying kinetics of yacon (Smallanthus sonchifolius) and the effect of process conditions on fractal dimension and rehydration capacity.* Drying Technology, 2012. **30**(1): p. 13–19.
36. Szadzińska, J. et al., *Microwave- and ultrasound-assisted convective drying of raspberries: Drying kinetics and microstructural changes.* Drying Technology, 2018. **37**(1): p. 1–12.
37. Elmizadeh, A., M. Shahedi, and N. Hamdami, *Quality assessment of electrohydrodynamic and hot-air drying of quince slice.* Industrial Crops and Products, 2018. **116**: p. 35–40.
38. Chen, Y., and A. Martynenko, *Combination of hydrothermodynamic (HTD) processing and different drying methods for natural blueberry leather.* LWT, 2018. **87**: p. 470–477.
39. Castro, A.M., E.Y. Mayorga, and F.L. Moreno, *Mathematical modelling of convective drying of fruits: A review.* Journal of Food Engineering, 2018. **223**: p. 152–167.
40. Zhang, M. et al., *Recent developments in high-quality drying with energy-saving characteristic for fresh foods.* Drying Technology, 2015. **33**(13): p. 1590–1600.
41. Szadzińska, J. et al., *The effect of high power airborne ultrasound and microwaves on convective drying effectiveness and quality of green pepper.* Ultrasonics Sonochemistry, 2017. **34**: p. 531–539.
42. Ashok Kumar, M., G. Kumaresan, and S. Rajakarunakaran, *Experimental study of moisture removal rate in Moringa leaves under vacuum pressure in closed-loop heat pump dryer.* Materials Today: Proceedings, 2021. **45**(2): p. 1205–1210.
43. El-Mesery, H.S., and H. Mao, *Investigation of infrared assisted dryer effect on energy consumption during drying of tomato.* International Journal of Advanced and Applied Sciences, 2017. **4**(12): p. 68–72.
44. Mohanraj, M., *Performance of a solar-ambient hybrid source heat pump drier for copra drying under hot-humid weather conditions.* Energy for Sustainable Development, 2014. **23**: p. 165–169.
45. Roratto, T.B. et al., *An innovative hybrid-solar-vacuum dryer to produce high-quality dried fruits and vegetables.* LWT, 2021. **140**: p. 110777.
46. Kiburi, F. et al., *Performance evaluation and economic feasibility of a solar-biomass hybrid greenhouse dryer for drying banana slices.* Renewable Energy Focus, 2020. **34**: p. 60–68.
47. Kudra, T., and A. Martynenko, *Electrohydrodynamic drying: Theory and experimental validation.* Drying Technology, 2019. **38**(1–2): p. 168–175.
48. Kuriakose, R., and C. Anandharamakrishnan, *Computational fluid dynamics (CFD) applications in spray drying of food products.* Trends in Food Science & Technology, 2010. **21**(8): p. 383–398.
49. Getahun, *Importance of integrated CFD and product quality modeling of solar dryers for fruits and vegetables: A review.* Solar Energy, 2021. **220**: p. 88–110.
50. Page, G.E., Factors influencing the maximum rates of air drying shelled corn in thin layers. 1949: Purdue University.
51. Chandra, P.K., and R.P. Singh, *Applied Numerical Methods for Food and Agricultural Engineers.* 2017: CRC Press.
52. Akpinar, E.K., Y. Bicer, and F.J.J. Cetinkaya, *Modelling of thin layer drying of Parsley leaves in a convective dryer and under open sun.* Journal of Food Engineering, 2006. **75**(3): p. 308–315.
53. Yadav, A.S., and J.L. Bhagoria, *Heat transfer and fluid flow analysis of solar air heater: A review of CFD approach.* Renewable and Sustainable Energy Reviews, 2013. **23**: p. 60–79.

54. Chandramohan, V.P., *Convective drying of food materials: An overview with fundamental aspect, recent developments, and summary*. Heat Transfer, 2020. **49**(3): p. 1281–1313.
55. Esfahani, J. A., Hosein Majdi, and E. Barati. *Analytical two-dimensional analysis of the transport phenomena occurring during convective drying: Apple slices*. Journal of Food Engineering, 2014. **123**: 87–93.
56. Joardder, M.U.H., and M.A. Karim, *Development of a porosity prediction model based on shrinkage velocity and glass transition temperature*. Drying Technology, 2019. **37**(15): p. 1988–2004.
57. Khan, F.A., and A.G. Straatman, *A conjugate fluid-porous approach to convective heat and mass transfer with application to produce drying*. Journal of Food Engineering, 2016. **179**: p. 55–67.
58. Malekjani, N., and S.M. Jafari, *Simulation of food drying processes by computational fluid dynamics (CFD): Recent advances and approaches*. Trends in Food Science & Technology, 2018. **78**: p. 206–223.
59. Jamaleddine, T.J., and M.B. Ray, *Application of computational fluid dynamics for simulation of drying processes: A review*. Drying Technology, 2010. **28**(2): p. 120–154.
60. Chokphoemphun, S., and E. Chokphoemphun, *Moisture content prediction of paddy drying in a fluidized-bed drier with a vortex flow generator using an artificial neural network*. Applied Thermal Engineering, 2018. **145**: p. 630–636.
61. Jafari, S.M. et al., *Mathematical, fuzzy logic and artificial neural network modeling techniques to predict drying kinetics of onion*. Journal of Food Processing and Preservation, 2016. **40**(2): p. 329–339.
62. Aktaş, M. et al., *Performance analysis and modeling of a closed-loop heat pump dryer for bay leaves using artificial neural network*. Applied Thermal Engineering, 2015. **87**: p. 714–723.
63. Capossio, J.P. et al., *Sustainable solar drying of Brewer's spent grains: A comparison with conventional electric convective drying*. Processes, 2022. **10**(2): p. 339.
64. Rodriguez, J. et al., *Modelling drying kinetics of thyme (Thymus vulgaris* L.*): Theoretical and empirical models, and neural networks*. Food Science and Technology International, 2014. **20**(1): p. 13–22.
65. Khawas, P. et al., *Modeling and optimization of the process parameters in vacuum drying of culinary banana (Musa ABB) slices by application of artificial neural network and genetic algorithm*. Drying Technology, 2016. **34**(4): p. 491–503.
66. Oke, E.O., K. Nwosu-Obieogu, and E. Ude, *Experimental study and exergy efficiency prediction of three-leaved yam (Dioscorea dumetorum) starch drying*. International Journal of Exergy, 2020. **33**(4): p. 427–443.
67. Kalathingal, M.S.H., S. Basak, and J. Mitra, *Artificial neural network modeling and genetic algorithm optimization of process parameters in fluidized bed drying of green tea leaves*. Journal of Food Process Engineering, 2020. **43**(1): p. e13128.
68. Prakash, O., and A. Kumar, *ANFIS modelling of a natural convection greenhouse drying system for jaggery: An experimental validation*. International Journal of Sustainable Energy, 2012. **33**(2): p. 316–335.
69. Shrivastava, C. et al., *Digital twins enable the quantification of the trade-offs in maintaining citrus quality and marketability in the refrigerated supply chain*. Nature Food, 2022. **3**(6): p. 413–427.
70. Cetina-Quiñones, A.J. et al., *Digital twin model and global sensitivity analysis of an indirect type solar dryer with sensible heat storage material: An approach from exergy sustainability indicators under tropical climate conditions*. Journal of Energy Storage, 2023. **58**: 106368.
71. Mohd Ali, M. et al., *Principles and recent advances in electronic nose for quality inspection of agricultural and food products*. Trends in Food Science & Technology, 2020. **99**: p. 1–10.
72. Hussain, N., D.-W. Sun, and H. Pu, *Classical and emerging non-destructive technologies for safety and quality evaluation of cereals: A review of recent applications*. Trends in Food Science & Technology, 2019. **91**: p. 598–608.
73. Tao, Y. et al., *Applications of water blanching, surface contacting ultrasound-assisted air drying, and their combination for dehydration of white cabbage: Drying mechanism, bioactive profile, color and rehydration property*. Ultrasonics Sonochemistry, 2019. **53**: p. 192–201.
74. Al-Dairi, M. et al., *Banana fruit bruise detection using fractal dimension based image processing*. Food Chemistry, 2024. **455**: p. 139812.
75. Qadri, O.S. et al., *Fresh-cut fruits and vegetables: Critical factors influencing microbiology and novel approaches to prevent microbial risks: A review*. Cogent Food & Agriculture, 2015. **1**(1): 1121606.
76. Alp, D., and O. Bulantekin, *The microbiological quality of various foods dried by applying different drying methods: A review*. European Food Research and Technology, 2021. **247**: p. 1333–1343.
77. Zielinska, M., and A. Michalska, *Microwave-assisted drying of blueberry (Vaccinium corymbosum L.) fruits: Drying kinetics, polyphenols, anthocyanins, antioxidant capacity, colour and texture*. Food Chemistry, 2016. **212**: p. 671–680.

78. Devahastin, S., and C. Niamnuy, *Invited review: Modelling quality changes of fruits and vegetables during drying: A review.* International Journal of Food Science & Technology, 2010. **45**(9): p. 1755–1767.
79. Van Boekel, M.A.J.S., *Kinetic modeling of food quality: A critical review.* Comprehensive Reviews in Food Science and Food Safety, 2008. **7**(1): p. 144–158.
80. Ling, B. et al., *Kinetics of food quality changes during thermal processing: A review.* Food and Bioprocess Technology, 2014. **8**(2): p. 343–358.
81. Prawiranto, K., J. Carmeliet, and T. Defraeye, *Physics-based digital twin identifies trade-offs between drying time, fruit quality, and energy use for solar drying.* Frontiers in Sustainable Food Systems, 2021. **4**. doi: 10.3389/fsufs.2020.606845
82. Defraeye, T., *Towards more efficient intermittent drying of fruit: Insights from combined hygrothermal-quality modelling.* Innovative Food Science & Emerging Technologies, 2016. **38**: p. 262–271.
83. Defraeye, T., and A. Radu, *Insights in convective drying of fruit by coupled modeling of fruit drying, deformation, quality evolution and convective exchange with the airflow.* Applied Thermal Engineering, 2018. **129**: p. 1026–1038.
84. Pham, N.D., M.I.H. Khan, and M.A. Karim, *A mathematical model for predicting the transport process and quality changes during intermittent microwave convective drying.* Food Chemistry, 2020. **325**: p. 126932.
85. Sabarez, H.T., *Mathematical modeling of the coupled transport phenomena and color development: Finish drying of trellis-dried sultanas.* Drying Technology, 2014. **32**(5): p. 578–589.
86. Wijerathne, W. et al., *A coarse-grained multiscale model to simulate morphological changes of food-plant tissues undergoing drying.* Soft Matter, 2019. **15**(5): p. 901–916.
87. Taşeri, L. et al., *Determination of drying kinetics and quality parameters of grape pomace dried with a heat pump dryer.* Food Chemistry, 2018. **260**: p. 152–159.
88. Tohidi, M. et al., *Energy and quality aspects for fixed deep bed drying of paddy.* Renewable and Sustainable Energy Reviews, 2017. **70**: p. 519–528.
89. Lingayat, A., and Chandramohan, V.P., *Numerical investigation on solar air collector and its practical application in the indirect solar dryer for banana chips drying with energy and exergy analysis.* Thermal Science and Engineering Progress, 2021. **26**: p. 101077.
90. Majdi, H., and J.A. Esfahani, *Energy consumption and drying time optimization of convective drying for performance improvement: Response surface methodology and lattice Boltzmann method.* Journal of Heat Transfer, 2018. **140**(10).
91. Hyuk, K.D., *Mathematical modeling and simulation of rapeseed drying on concurrent-flow dryer*, in *Current Drying Processes.* 2020: IntechOpen.
92. Gonzalez-Diaz, A. et al., *The potential of decarbonising rice and wheat by incorporating carbon capture, utilisation and storage into fertiliser production.* Green Chemistry, 2020. **22**(3): p. 882–894.
93. Atuonwu, J., and S. Tassou, *Decarbonisation of food manufacturing by the electrification of heat: A review of developments, technology options and future directions.* Trends in Food Science & Technology, 2021. **107**: p. 168–182.
94. Zhang, Z. et al., *Life cycle assessment of milk powder spray drying system in China.* Energy Sources, Part A: Recovery, Utilization, and Environmental Effects, 2021. **47**(2): p. 1–13.
95. Kumar, S., S.V. Jadhav, and B.N. Thorat, *Life cycle assessment of tomato drying in heat pump and microwave vacuum dryers.* Materials Today: Proceedings, 2022. **57**(4): p. 1700–1705.
96. Elhami, B., M. Ghasemi Nejad Raini, and F. Soheili-Fard, *Energy and environmental indices through life cycle assessment of raisin production: A case study (Kohgiluyeh and Boyer-Ahmad Province, Iran).* Renewable Energy, 2019. **141**: p. 507–515.
97. Merone, D. et al., *Energy and environmental analysis of ultrasound-assisted atmospheric freeze-drying of food.* Journal of Food Engineering, 2020. **283**: 110031.

2 Sustainable Drying Technology for Fresh Foods

Vishnuvardini Narayanan and Nagamaniammai Govindarajan
Department of Food Process Engineering, School of Bioengineering, College of Engineering and Technology, SRM Institute of Science and Technology, Kattankulathur, India

2.1 INTRODUCTION

Traditional Drying Methods: Traditional drying methods often rely on fossil fuel–based energy sources and inefficient processes, leading to significant energy waste and environmental pollution. Traditional drying methods for fruits such as sun drying or hot-air (HA) drying often result in prolonged drying time, loss of nutrients, and increased energy requirements.

Sustainable Drying Technology: Sustainable drying technology offers innovative solutions to address these challenges. Sustainable drying technology for fresh fruits refers to the use of energy-efficient and environment-friendly methods. It is to preserve the quality, nutritional value, and shelf life of fruits by removing moisture, while minimizing negative impacts on the environment. This approach aims to reduce energy consumption, greenhouse gas emissions, and waste generation, while maximizing product quality and minimizing overall costs. There are several key principles and techniques involved in sustainable drying technology.

Renewable Energy Sources: Sustainable drying technologies utilize renewable energy sources like solar, wind, biomass, and geothermal to reduce fossil fuel reliance and promote clean, abundant energy.

Energy-Efficient Systems: Sustainable drying technologies utilize energy-efficient designs, equipment, insulation, heat recovery, heat exchanger, and advanced control systems to optimize the drying process and minimize energy waste.

Waste Heat Recovery: Waste heat recovery systems use excess drying process heat for preheating air or other purposes, reducing energy consumption and enhancing the drying process.

Design of Drying System: Efficient drying system design considers airflow patterns, inlet temperature, heating source, material characteristics, and moisture content for uniform, energy-efficient drying.

Drying Technologies: Sustainable drying technologies like solar, microwave (MW), vacuum, freeze, and supercritical fluid drying cater to specific applications and materials, offering advantages and suitability.

Process Optimization: Continuous improvement and process optimization are crucial for sustainable drying, improving efficiency, reducing energy consumption, and enhancing product quality.

Material Handling and Preparation: Proper material handling and preparation before drying improve efficiency and sustainability, reducing drying time and energy requirements by removing excess moisture.

DOI: 10.1201/9781003344087-2

Life Cycle Assessment: Life cycle evaluation assesses environmental impact of drying processes, highlighting sustainable benefits for fresh fruit drying, including waste reduction, nutritional preservation, shelf life extension, and product quality improvement.

2.2 TRADITIONAL AND MECHANIZED DRYING METHODS

2.2.1 Traditional Drying Methods

2.2.1.1 Sun Drying

Sun drying is one of the oldest and most traditional methods of drying fruits (Petikirige et al., 2022). It involves spreading the fresh fruits in a single layer on trays or mats and placing them in direct sunlight. The fruits are left exposed to the sun for several days, regularly turning them to ensure even drying. Covering the fruits with a clean mesh or netting to protect them from insects, birds, or other pests while still allowing air circulation helps prevent contamination and ensures that the product is protected during the drying process (Al-Hooti et al., 1997). In hot, dry areas, this approach performs well. Sun drying is reliant on favorable weather conditions and can be time-consuming. Additionally, it may not be suitable for drying certain types of fruits or vegetables that are more sensitive to heat or require faster drying times. In such cases, alternative methods like shade drying, air drying, or using modern drying equipment may be more appropriate (Opara et al., 2009).

2.2.1.2 Shade Drying

Shade drying is a method of drying fruits using indirect sunlight and ambient airflow. It is a slower process compared to sun drying, but it can be useful in areas with high humidity or when direct sunlight is not available, and when you prefer a slower drying process (Radojčin et al., 2021). Traditional shadow drying requires patience and favorable environmental conditions. It is important to monitor the fruits regularly and protect them from rain, excess humidity, or pests (Yang et al., 2024).

2.2.1.3 Air Drying

Air drying involves hanging fruits in a well-ventilated area with good airflow. The fruits can be strung together with a needle and thread or placed on a wire rack. The natural airflow helps to dry the fruits slowly over time. It is important to consider the environmental conditions, as high humidity or rainy weather may prolong the drying time or affect the quality of the dried fruits (Boateng, 2023). This method is commonly used for smaller fruits such as berries, cherries, or figs.

2.2.1.4 Oven Drying

Oven drying is a traditional method used when sun drying is not feasible due to weather conditions or limited sunlight. It involves placing the fruits on baking sheets and drying them at a low temperature which will be around 130°–150°F, for several hours. The oven door is usually left slightly open to allow moisture to escape. This method requires careful monitoring to prevent overdrying or burning the fruits (Boateng, 2023). The impact of oven drying techniques on extraction has been studied (Kumar et al., 2022). This method is commonly used when environmental conditions are not suitable for sun drying.

2.2.1.5 Dehydrator Drying

The dehydrator uses a heating element, air circulation, and trays or shelves to evaporate moisture from food. It has an electric coil for heat generation, a fan for even distribution, and adjustable temperature settings for specific foods. The heated air evaporates moisture, effectively drying the food. The drying time varies based on factors like food type, thickness, moisture content, and temperature setting (Boateng, 2023).

Regardless of the method used, it is essential to prepare the fruits properly before drying. This may include washing, peeling, slicing, or pitting the fruits as needed. The drying time can vary depending on factors such as fruit size, moisture content, and the drying method employed. The fruits should be checked regularly during the drying process and removed when they are sufficiently dried but still pliable and leathery. Properly dried fruits should be stored in airtight containers to maintain their quality and prevent moisture absorption.

2.2.2 Mechanized Drying Methods

2.2.2.1 Tray Drying

Tray drying is a batch dryer with a drying mechanism that is similar to a cabinet dryer (Nwankwo et al., 2021). Sliced fruits are placed on trays or racks in a drying chamber or oven. The trays are then subjected to heated air circulation, either through natural convection or through forced convection using fans or blowers to blow air into the chamber (Figure 2.1). Uneven drying, which results from inadequate airflow circulation in the drying chamber, is the tray dryer's main issue. Typically, air inlet temperatures fall between 60°C and 80°C. To prevent entrainment of dry particles with the wind, air velocity must be adjusted to account for the size, shape, and density of the food particles.

2.2.2.2 Tunnel Drying

Tunnel drying involves passing a continuous stream of HA through a tunnel-like structure where fruits are transported on a conveyor belt. The HA dries the fruits as they move through the tunnel. The drying time can vary depending on the fruit type, thickness, moisture content, and desired moisture level in the final product. Despite the drying method's versatility, tunnel dryers need more effort to operate than continuous belt dryers, which limits its use. This method is suitable for large-scale fruit drying and offers efficient drying with consistent results.

2.2.2.3 Cabinet Drying

The fruits to be dried are placed on trays or racks inside the cabinets, and heated air is circulated to dry the fruits. Despite the low cost of the equipment, the process is batch-based, with high operational (labor) costs and poor performance (Nwankwo et al., 2021). Cabinet drying provides better control over the drying conditions and is commonly used for small- to medium-scale fruit drying.

2.2.2.4 Fluidized Bed Drying

In fluidized bed drying, a bed of hot air or gas is used to suspend and fluidize the fruit pieces. The fluidization temperature ranges from 70°C to 100°C, while the air velocity ranges from 0.98 m/s to 5.2 m/s. The final product should be dried fruit with a water content of 10–14% and an excellent

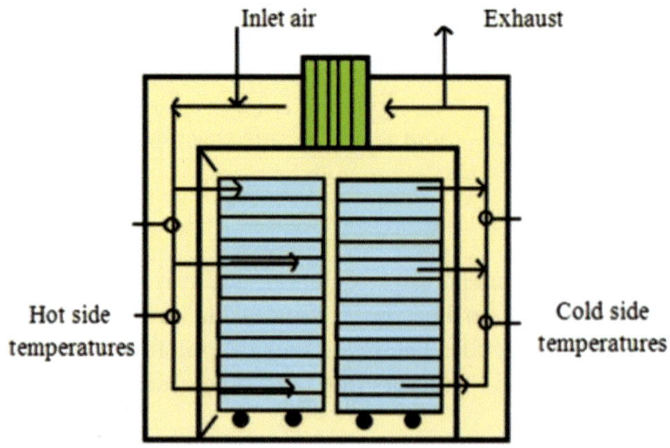

FIGURE 2.1 Tray dryer. (Misha et al., 2013.)

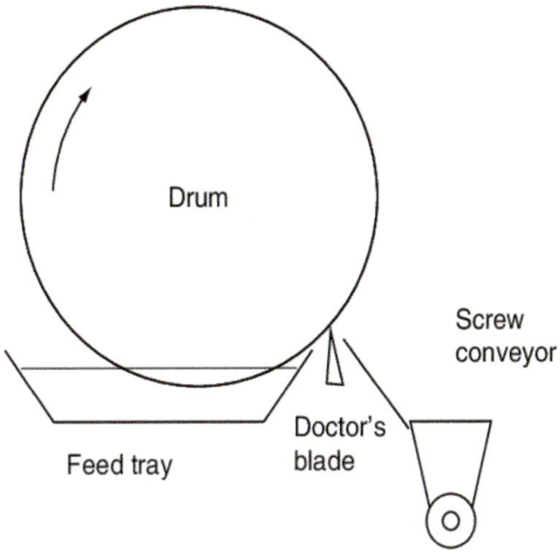

FIGURE 2.2 Drum dryer. (Ramli & Wan Daud, 2003)

rehydration quality ranging from 8 to 20 (Bauman et al., 2005). This method is suitable for delicate fruits and provides a shorter drying time compared to other methods.

2.2.2.5 Drum Drying

Drum drying is a mechanized method used for drying fruit purees or pulps. The fruit puree is spread in a thin layer on the surface of a heated rotating drum. As the drum rotates, the moisture evaporates, leaving behind dried fruit sheets that can be crushed or powdered (Figure 2.2). The temperature can range from 110°C to 160°C (Galaz et al., 2017).

2.2.2.6 Spray Drying

Spray drying is a specialized method used for drying liquid fruit extracts or concentrates. The fruit extract is atomized into fine droplets and sprayed into a hot drying chamber. The droplets quickly dry and solidify, forming dried particles that can be collected. The optimum conditions have been obtained with feed flow rate of 15 ml/min, inlet air temperature of 130°C, and outlet air temperature of 85°C (Chegini & Ghobadian, 2007).

These mechanized drying methods offer advantages such as faster drying times, better control over drying conditions, improved product quality, and higher production efficiency compared to traditional drying methods. The selection of a particular method depends on factors such as the nature of the material, desired drying rate, energy efficiency, and product specifications.

2.3 ADVANCE DRYING METHODS

There are several advanced methods of drying, namely, MW, vacuum, infrared, freeze, osmotic dehydration (OD), heat pump (HP) dryer, radio frequency (RF) drying, and various other technologies, used for fruits to offer best-quality dried products.

2.3.1 Microwave Drying

By absorbing electromagnetic energy through water, fat, and sugar molecules, MW drying is a way of heating food products. These molecules have a negative and positive charge on opposite sides, making them electrically dipoles. As a result, the molecules rub against one another due to the presence of fluctuating electric fields brought on by MWs. In the MW drying process, this heat energy

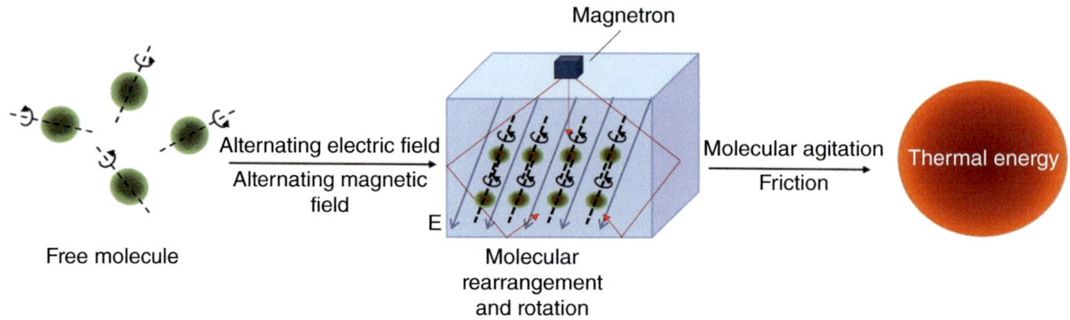

FIGURE 2.3 Microwave drying. (Hu et al., 2021.)

serves as a heater. When electromagnetic waves pass through a medium, they cause molecules to oscillate and produce thermal energy (Figure 2.3). This process is known as volumetric heating (Wardhani et al., 2022). Set the MW to a low power level, usually around 30–50% power: domestic MWs operate at a frequency of 2,450 MHz, whereas commercially large-scale MWs operate at a frequency of 900 MHz and a wavelength of 12.24 cm (Banerjee & Janghu, 2023). Start with shorter drying intervals, to monitor the progress and prevent overdrying. Some fruits with higher water content or delicate textures may not yield desirable results with this method.

2.3.2 Vacuum Drying

Vacuum drying (VD) involves placing the fruit in a vacuum chamber and reducing the pressure to lower the boiling point of water. Heat is then applied to the chamber in the range of 0.6–101 kPa (Nwankwo et al., 2021) causing the water in the fruit to vaporize. Gas that has been trapped in pores expands as a result of the pressure drop and escapes. The holes can be filled with the osmotic solution to increase the mass transfer surface area when the pressure is restored. The vacuum pressure is applied in the range of 50–100 mbar (Sagar & Kumar, 2010).

2.3.3 Infrared Drying

Infrared drying (IRD) utilizes infrared radiation in the wavelength range of 0.8–1,000 μm (Calin-Sanchez et al., 2020). The radiation enters the exposed material and increase internal temperature, causing the water within the fruit to evaporate (Figure 2.4) (Yadav & Ali, 2023). This method can be efficient and allows for better control of the drying process, resulting in improved product quality.

2.3.4 Freeze-Drying

Freeze-drying (FD), also known as lyophilization, is sublimation of ice fraction where water passes from a solid state to a gaseous state. It involves freezing the fruit at very low temperatures, typically around −30°C to −50°C and then placing it in a vacuum chamber ranging from 1 mbar to 0.5 mbar (Maulburg, 2020), which allows the frozen water in the fruit to sublimate. All degradation and microbial activity is prevented by the extremely low temperature, which improves the quality of the finished product (Sagar & Kumar, 2010).

2.3.5 Osmotic Dehydration

OD is a drying method used for fruits, particularly to remove moisture while preserving their texture, flavor, and nutritional properties (Figure 2.5). It involves immersing fruit pieces in a hypertonic

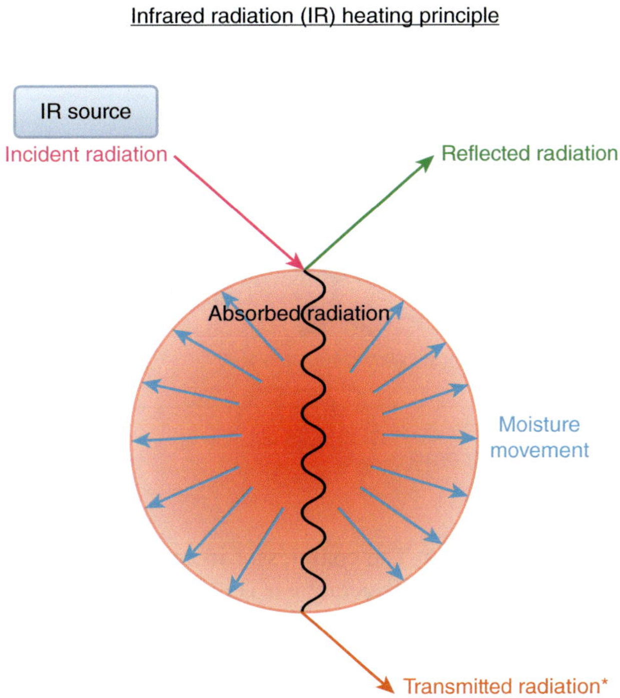

FIGURE 2.4 Infrared drying. (Delfiya et al., 2022.)

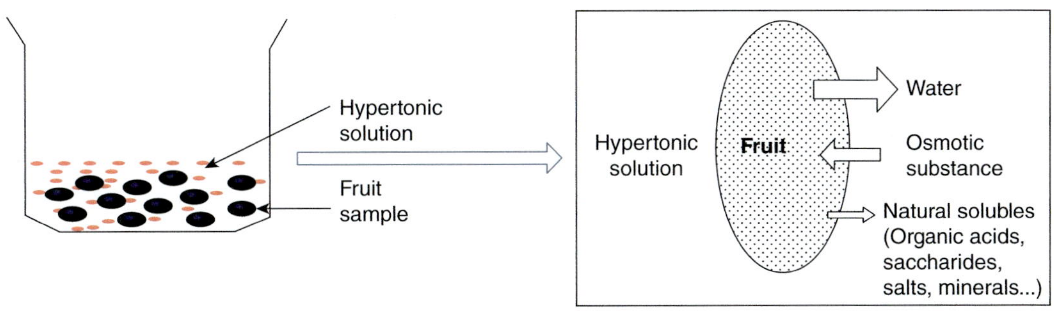

FIGURE 2.5 Osmotic dehydration. (Akharume et al., 2019.)

solution, that is, a solution that has a higher concentration of solute compared to another solution or the cytoplasm of a cell, which causes water to move out of the fruit through osmosis (Yadav & Singh, 2014). In this case, dissolved sugars (such as sucrose or glucose) or salts (such as sodium chloride) are in water. When food is placed in the hypertonic solution, solute from the solution diffuses into the tissue of the fruits. It is particularly suitable for fruits that are sensitive to heat or have delicate textures (Saleena et al., 2021).

2.3.6 Heat Pump Dryer

HP dryer is an energy-efficient drying method. It utilizes a HP system to remove moisture from the fruits by circulating HA within a closed drying chamber. The evaporator contains a refrigerant in a low-temperature state (40–70°C); as the air passes over by fan or blower, it absorbs heat from

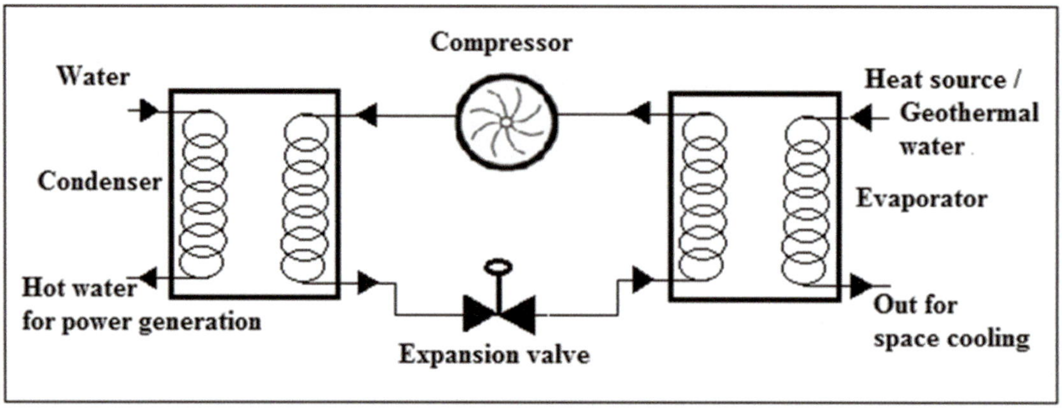

FIGURE 2.6 HP dryer. (Sircar et al., 2019.)

the air and changes from liquid to gas. The compressor increases the pressure and temperature of the low-pressure and low-temperature refrigerant gas (Figure 2.6). The hot and high-pressure (10–20 bar) refrigerant gas transfers heat to the surrounding environment in the condenser (air or water surrounding the condenser). The heated air is circulated over the material to be dried in the HP dryer to remove moisture. The HP system allows for efficient heat transfer and energy recycling. By using the heat from the condenser to reheat the air, it reduces the need for additional energy inputs and increases the overall energy efficiency of the drying process" (Fayose & Huan, 2016; Salehi, 2021; Loemba et al., 2023)

2.3.7 Radio Frequency Drying

RF drying is the interaction of an electromagnetic field created by a RF generator with the molecular species in a substance. The fruit slices are kept between two electrodes that are subjected to a 40,000,000 times per second changing electric field (Nwankwo et al., 2021). It causes the water molecules within the fruit to rapidly oscillate due to changing electric fields through molecular friction, leading to rapid and efficient drying.

2.4 HYBRID DRYING TECHNOLOGY

Hybrid drying technology refers to a combination of different drying methods or technologies to achieve optimal drying efficiency and quality for various products or materials. It typically involves the integration of two or more drying techniques, such as convective drying, MW drying, IRD, or VD (Preeti et al., 2023).

The purpose of hybrid drying is to take advantage of the strengths of different drying methods while minimizing their limitations. By combining multiple technologies, hybrid drying systems aim to improve drying rates, reduce energy consumption, enhance product quality, and achieve uniform drying results (Jha & Tripathy, 2021; Salehi, 2021).

2.4.1 Combined Drying with Microwave

Some disadvantages of microwave (MW) drying include inconsistent heating, the potential for textural damage (Orsat et al., 2006), and thermal runaway due to dimensional resonance phenomenon (Laguerre & Hamoud-Agha, 2018). To get around these issues, several drying techniques are combined.

2.4.1.1 Microwave with Convective Drying

Convective drying involves passing heated air over the material to remove moisture. In this process, HA acts as a medium for heat transfer and carries away the moisture evaporated from the material surface.

Due to its low initial investment, low operational costs, and simplicity of usage, conventional drying is the most widely used drying method. It comprises around 85% of all industrial drying processes. Value-added components degrade as a result of relatively lengthy drying durations (1–7 hours, depending on temperature, moisture content, and product structure) and high air temperatures (up to 180°C). Another aspect of the product that is significantly altered by HA drying is its structure. Foods that are HA-dried have volume retention values between 15% and 30%. Another drawback is the development of a hard crust (case hardening) on the product's exterior. Color and taste are evolving as well, and depending on the product and the consumer's preferences, they may be advantageous or harmful (Gaukel et al., 2017).

2.4.1.1.1 Working Principle

The HA reduces surface unbound moisture, while MW energy eliminates bound moisture from inside the product through volumetric heating (Calin-Sanchez et al., 2020). The moisture-carrying capacity of the airflow is determined by its velocity and temperature, which must be controlled considering the heat sensitivity of volatile components and phytochemicals. The process must be optimized for airflow, temperature, and MW power density. MW heating can be applied at the beginning, second, and end stages, reducing mass transfer and improving drying rate by removing bound water.

As an illustration, drying time was lowered by 64% when MW and HA drying were combined at the end stage as opposed to convective air drying (Laguerre & Hamoud-Agha, 2018). The air used is either hot (80°C, for example) to prevent surface cooling or cold (20°C, for example) to prevent surface overheating (Figure 2.7).

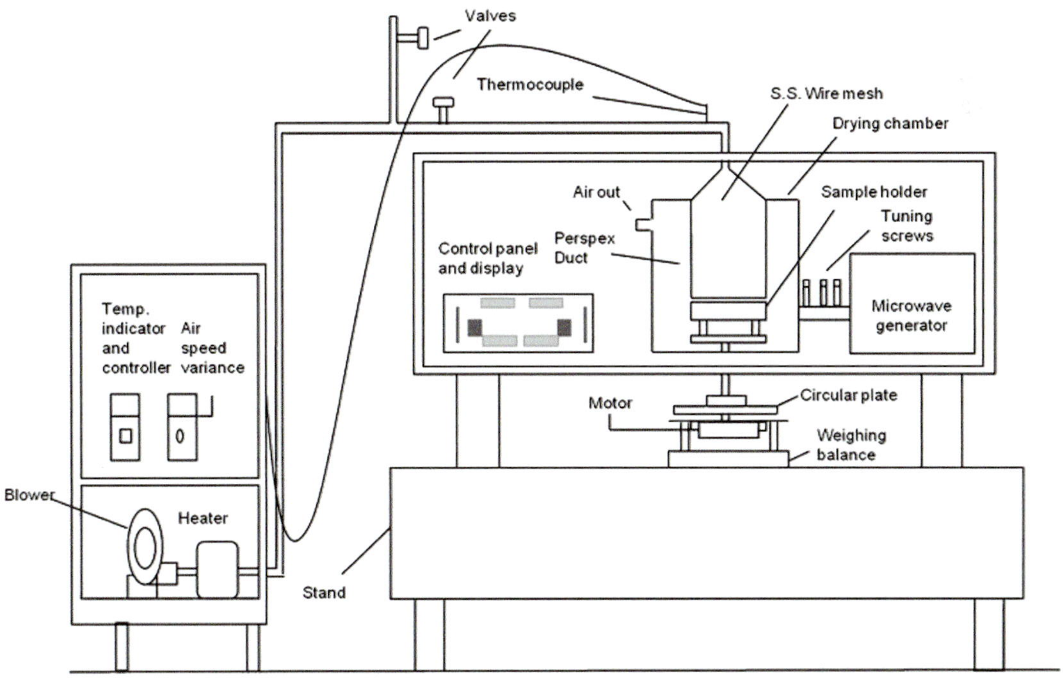

FIGURE 2.7 MW-assisted HA dryer. (Kumar et al., 2014.)

2.4.1.2 Microwave with Freeze-Drying

Lyophilization is a low-temperature dehydration process that freezes and sublimates heat-sensitive materials, preserving structure and minimizing loss of valuable compounds. However, it is limited by high costs and has a limited heat transfer rate due to no constant drying period. This process also results in a dry, porous structure in outer product parts, leading to a low heat transfer coefficient (Pansare & Patel, 2019; Bhambhani et al., 2023).

2.4.1.2.1 Working Principle

In MW FD, MW heating may be used simultaneously with sublimation to provide heat in a vacuum. Following a conventional lyophilization procedure, MW heating can also be utilized independently. Due to its volumetric heating, low processing temperature, and lack of oxygen in the processing environment, MW FD has various benefits, including a shorter drying time (Laguerre & Hamoud-Agha, 2018). Only the organic portion of frozen food is penetrated by MW energy, warming the entire product instead of heating it layer by layer as is the case with the traditional FD. However, thermal runaway could occur as MW energy does not mix with frozen water, which could lead to subpar product quality.

To ensure uniform food drying, FD can be done in a MW. There are two techniques: two-phase drying and simultaneous MW drying. The fruit is simultaneously exposed to MWs while it is in a vacuum to give energy for sublimation.

In two-phase drying, the product is first frozen before being heated with MW energy in a sealed space. This procedure raises the frozen food's core temperature, which causes the frozen water to thaw and turn gaseous. This transition establishes an interface between the frozen and dried components, which gradually disappears over time to produce a product that is uniformly dry (Banerjee & Janghu, 2023).

2.4.1.3 Microwave with Fluidized Bed Drying

The main issue with MW drying is the uneven drying of the product at a given period. Combining fluidized bed drying with the MW technique eliminates this sort of flaw (Khoshtaghaza et al., 2015). The unfavorable size reduction of food particles as a result of particle impact during fluidized bed drying is a drawback. As a result, the disadvantages of both types of drying procedures are reduced in the hybrid drying process (Chupawa et al., 2022).

2.4.1.3.1 Working Principle

In the fluidized drying process, the food to be dried is exposed to HA at high pressure through a porous bed, agitating the food particles. If a MW is then added, the radiation will be absorbed by each particle, resulting in uniform heating and a reduction in the drying process' diffusion time. When both approaches were used together with 2.1 and 3.5 W/g of MW energy in a fluidized bed drier as opposed to applying MW and fluidized bed drying separately, drying time was reported to be decreased by 50% (Banerjee & Janghu, 2023). Another element that determines whether a particular drying procedure will be successful is rehydration capability.

To achieve the best results for energy economy and consistent drying, fluidized bed drying requires careful consideration of the MW energy and HA temperature. When a high-power MW is used, the surface temperature of the particles is greater than the air temperature in the fluidized bed drier; as a result, airflow cools the particles rather than evaporating moisture.

2.4.1.4 Microwave with Vacuum Drying

By rapidly heating the food using MW radiation, the drying process known as MW drying speeds up the evaporation of water. MWs are used in the VD process to hasten the drying process. Additionally, this type of drying employs vacuum pressure to lower the boiling point of water, allowing drying to take place at low temperatures (Zhang et al., 2024). MW radiation technology may be appropriate for effective drying due to its special heating characteristics (Khodifad & Dhamsaniya, 2020).

2.4.1.4.1 Working Principle

A technique known as "microwave vacuum drying" combines the advantages of MW and VD, thereby improving both energy efficiency and product quality. The MW vacuum method combines the benefits of MW heating and VD. Water throughout the entire volume of the materials being dried absorbs MW radiation during the drying process. As a result, the material's core experiences high vapor pressure, which permits moisture to quickly diffuse into the surrounding vacuum and prevents structural collapse. This procedure, sometimes referred to as the puffing phenomenon, renders the substance porous, lowering its density (Wardhani et al., 2022).

MW VD reduces unwanted chemical changes, sensory transitions, and nutrient loss in food by removing water at lower temperatures. It also achieves a greater reduction in microbial count through shorter processing times and lower temperatures.

2.4.1.5 Microwave with Osmotic Dehydration

OD is a minor processing step that improves product quality, especially color, while having a minimal impact on texture and nutritional content. It involves the movement of solutes against the osmotic pressure over a semipermeable barrier. Along with water, other low-molecular-weight compounds, such as vitamins, minerals, or organic acids, can diffuse out of the cell. Given that the anthocyanin level of cranberries after OD was 10% lower than that of fresh berries, this dispersion may be significant in terms of quality. Because OD operates at moderate temperatures and doesn't need latent heat of vaporization, it is appropriate for industrial applications and is energy-efficient. Due to the lengthy process of traditional OD, strategies that may be used in conjunction with OD to speed up the process without compromising product quality have been developed (Wray & Ramaswamy, 2013).

2.4.1.5.1 Working Principle

Because water makes up the majority of the food product, the MW absorption causes a quick and uneven heating of the food product during MW-assisted dehydration. As a result, pressure builds inside the product, pushing moisture out of the product and promoting moisture loss. The higher rate of moisture loss under MW circumstances in the case of OD tends to lessen the inward flow of solids gain (Alharaty & Ramaswamy, 2023).

2.4.2 Combined Drying with Convective Hot Air

Convective hot-air drying (CHAD) is a widely used industry method for preserving and processing fruits and vegetables. It offers advantages such as uniform distribution of HA and temperature, reduced energy consumption, and better dried products under optimal conditions. However, the drying process conditions, such as temperature, air velocity, and relative humidity, can affect important product properties like texture, color, total carotenoid content, phenolic composition, antioxidant capacity, and other bioactive compounds (Nawi & Abdan, 2016). Despite attempts to improve the quality of dried products using low temperature, low relative humidity, and optimized drying parameters, these approaches still require high energy and drying time, leading to unacceptable total costs. Innovative methods combining non-thermal techniques and CHAD have gained increasing interest in the advancement of drying technology. Nonthermal technologies, such as ultrasound (US), pulsed electric fields, and ultraviolet (UV) radiation with convective dryer has a synergistic impact on the process of mass and heat transmission inside a product. These techniques offer enhanced quality attributes, reduced drying time, and increased processing efficiency (Cullen et al., 2012).

2.4.2.1 Hot Air with Ultrasound Drying

2.4.2.1.1 Ultrasound Drying

US drying utilizes high-frequency sound waves of 100 kHz and above (Demirdoven & Baysal., 2008) to promote the drying process. When applied to the material, US waves create vibrations that induce cavitation and microstreaming, leading to enhanced mass transfer and moisture removal from the material.

When combined, US is a type of sound energy that is conveyed by waves as pressure at frequencies of 20 kHz and above. Ultrasound can be divided into two categories: low frequency (high energy, high intensity) at frequencies between 20 kHz and 100 kHz and high frequency (low energy, low intensity) at frequencies >100 kHz. High-frequency low-intensity US technology is frequently used as an analytical method for nondestructive quality and process inspection and control, such as to identify different food properties, examine the product's quality, gauge the degree of spoilage, measure flow rate, and so on. Low frequency, high energy, or "power ultrasound" creates a number of events that lead to cavitation, which controls enzymes and accelerates tissue regeneration.

Power ultrasound also enhances the rate of heat and mass transfer, making it useful for various postharvest applications on crops. These applications include **drying, freezing, and OD** (Onwude et al., 2017).

2.4.2.1.2 Mechanism of Operation

The main drying concept used in the US relies more on mechanical than heating mechanisms, which shortens the drying process without considerably increasing product temperature. The generator, transducer, and application system are the three primary parts of an ultrasonic wave production system. In order to create ultrasonic fields, many forms of energy (such as electric, magnetic, kinetic, etc.) are transformed into acoustic energy. The system's energy is provided by the generator, and the transducer is in charge of converting that energy from the generator into acoustic energy (Figure 2.8).

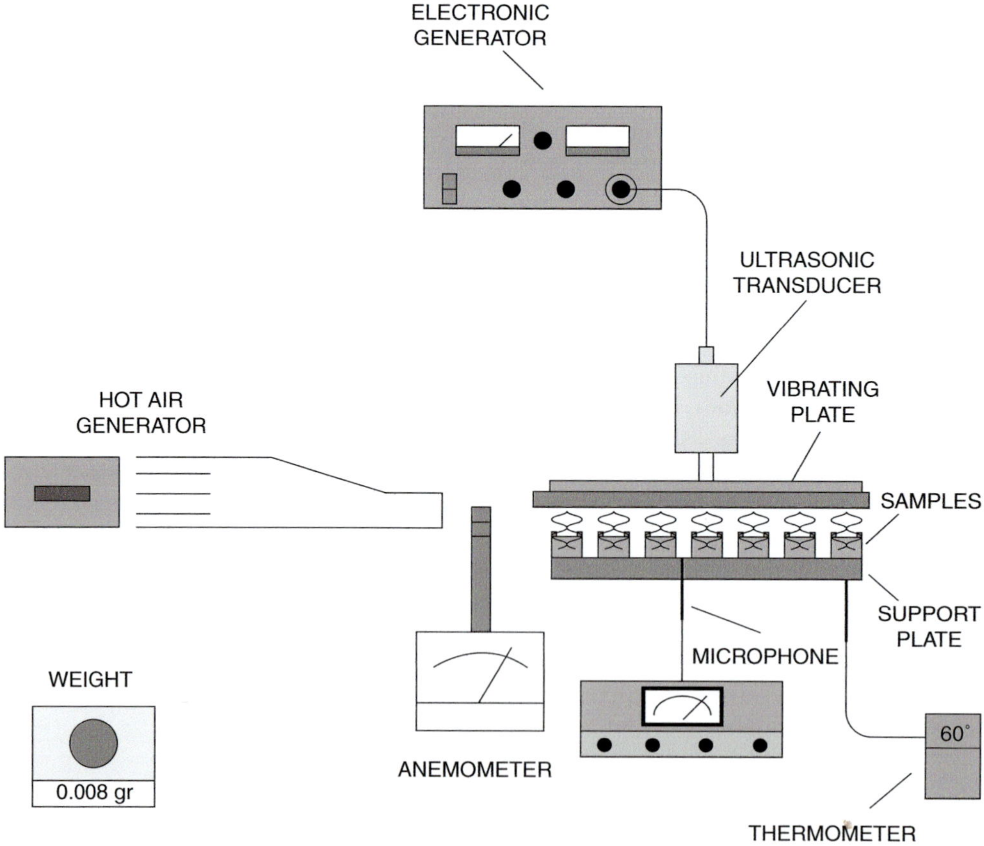

FIGURE 2.8 HA assisted with US dryer. (Pakbin et al., 2014.)

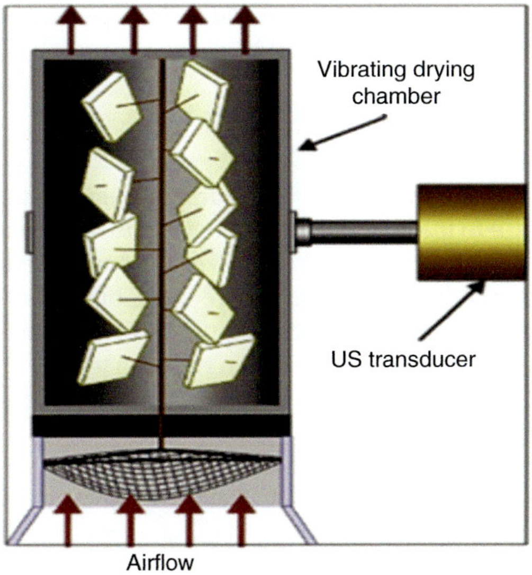

FIGURE 2.9 HA assisted with US. (Rojas et al., 2020.)

The acoustic energy produced by the US during operation causes periodic velocities and pulsations of the tissue particles at the interfaces, during which the water molecules' bonds to the product's surface may be disrupted. Cell cavitation takes place in parallel, causing diffusion from the product's inside to the boundary layers (Chavan et al., 2022; Madhu et al., 2019).

When ultrasonic waves pass through the series of oscillations, the product experiences alternate compressions and expansions. The cavitation required for eliminating strongly adhered moisture may be produced by ultrasonic waves (Onwude et al., 2017). High US power may result in an increase in alternating stress that is larger than the surface tension of the moisture within product capillaries. The capillaries lead to the formation of small porous channels within the product and an increase in the transport of water from the interior to the surface. Therefore, by using convective HA, it is simple to eliminate water vapor that has accumulated at the product's surface as a result of high-power US (Figure 2.9).

Advantages

1. Significantly shorter overall drying time and increasing drying rate, especially in the beginning of the drying process
2. Increased heat and mass transfer diffusion, increased energy savings, and improved product quality
3. It is efficient at low air velocity, low temperature, low US power, and high product moisture content (Khmelev et al., 2011)

Limitations

1. The first drying phase, when the drying rate is highest, is the best time to apply US for an industrial use; any additional US application could waste energy as sonic waves.
2. High capital cost of design and inadequate medium for power ultrasonic transmission.
3. If safety precautions are not applied, high US frequencies (above 16 kHz) might result in hearing damage among operators.
4. Free radicals may develop as a result of the US intensity needed to accomplish significant mass transfer augmentation.

5. Practical challenges for industrial-scale demand and complexity of equipment design; process conditions (Nowacka & Dadan, 2022)

2.4.2.2 Hot Air with Pulse Electric Field Drying

2.4.2.2.1 Pulse Electric Field Treatment

Pulse electric field (PEF) involves applying short-duration, high-voltage electrical pulses of 20–80 kV/cm (Zhang et al., 2010) to the material. Due to the significant distance in the treatment chamber and the pulse generator's power restriction, 1–8 kV/cm is employed for solid meals. The electric field may be applied as exponentially decaying, square wave, bipolar, or oscillatory pulses at ambient, sub-ambient, or slightly above-ambient temperatures. To treat the whole volume of the food sample, the pulses are administered at high repetition rates (up to 3,000 pulses per second) (Janahar et al., 2022). These pulses create an electric field that permeates the material, causing electroporation (formation of pores) in the cell membranes. This facilitates the release of intracellular moisture, promoting faster drying.

PEF is frequently applied to materials with low electrical conductivity, high electrical resistivity, and those free of bubbles. It can also be effectively applied on materials with high conductivity only for a short period of time (micropulses). The electrical properties must be determined before PEF is applied to agricultural crops. Due to their synergistic actions, PEF and CHAD may work together to enhance drying kinetics in a manner similar to heat drying. This hybrid method might serve as an alternative to traditional drying techniques.

2.4.2.2.2 Mechanism of Operation

Continual transmembrane pressure buildup presses the cell harder, thins the membrane, and eventually creates holes in the membrane. The membrane becomes weaker until it bursts as a result of the osmotic expansion of the cell. Electroporation in the lipid domains and protein channels is what causes this process. Even if the exact mechanism of PEF on food is unclear, it is generally accepted that the mechanism is either electroporation or electro-permeabilization. Poration theory may be used to combine PEF with CHAD. As previously noted, the creation of holes increases during electroporation of the product tissue (Figure 2.10). By applying HA, this lowers internal cell resistance and accelerates the process of mass and heat transfer (Onwade et al., 2017).

Advantages

1. The color of fruits and vegetables is retained by combined PEF and CHAD drying.
2. PEF and CHAD increase product temperature, benefiting mild temperatures which favor sensitive products' performance.

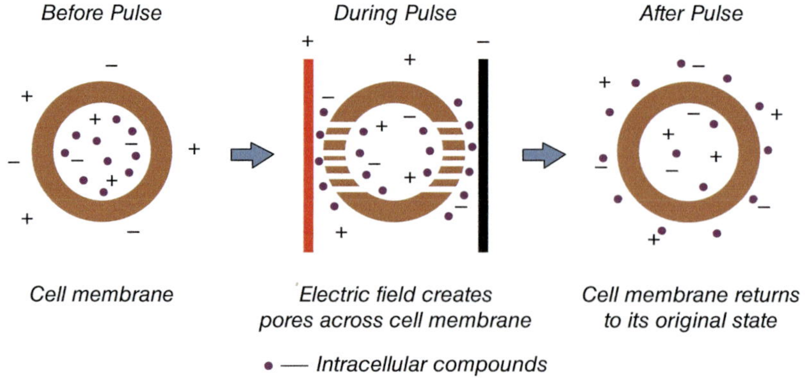

FIGURE 2.10 Electroporation in HA-assisted PEF. (Thulasidas et al., 2019.)

Limitations

1. Inadequate processing conditions and equipment design hinder optimal result.
2. Product suitability and electrical characteristics affect electrical damage, causing higher resistance (Ostermeier et al., 2020).

2.4.2.3 Hot Air with Ultraviolet

Ultraviolet (UV) drying utilizes UV radiation, typically from UV lamps, to facilitate the drying process. UV radiation interacts with the material's surface and promotes photochemical reactions, which result in the evaporation of moisture and drying (Patterson et al., 2018).

When combined, with wavelengths ranging from 100 nm to 400 nm, UV light is a part of the electromagnetic spectrum. Being nonthermal, chemical-free, and with no waste discharges, UV technology is a very environmentally benign energy source (Aksoy et al., 2019). Although the term "irradiation" has been used to refer to UV treatment, UV light is a nonionizing radiation and should not be confused with other types of IR radiation (such as gamma radiation). In the pharmaceutical, electrical, and aquaculture industries, UV laser radiation has been used to disinfect surfaces, render microbes in liquid food inert, and postharvest treat fruits and vegetables. However, UV light has gained popularity recently not only as a substitute for chemical sterilizing and traditional thermal heating techniques but also as a means of inactivating a variety of microorganisms (Raponi et al., 2009).

2.4.2.3.1 Mechanism of Operation

Because the maximal germicidal impact happens at a wavelength of 254 nm, UV-C (254 nm) light has been regarded as the best for treating food products. Microorganisms exposed to UV-C radiation typically experience DNA damage throughout the treatment procedure, which encourages microbial proliferation. Pyrimidine dimer production modifies the DNA helix's structure and prevents microbial cells from replicating. Cells die as a result of damaged reproductive systems that are unable to heal on their own. When other heat sources are applied, the process may eventually lead to greater pore development and initiate rapid moisture removal (Mshelia et al., 2023).

Advantages

1. Higher rate of drying when combined UV and CHAD
2. Decreased drying time and usage of energy is reduced when combined
3. Simple and inexpensive to develop and install

Limitation

1. The US and CHAD combined method of drying is yet unclear, the project is still in its early phases and requires more research.
2. There are no requirements for the process of US and CHAD to achieve the best results.

2.4.3 Convective Drying Followed by Vacuum–Microwave Drying

The procedure is divided into two stages: the first stage involves pre-drying the fresh product using convection, and the second stage involves finishing drying it using a vacuum MW. The convective pre-drying significantly reduces the fresh material's unbound moisture without harming their bioactive components. The moisture content is afterward finished dried in a vacuum–MW system to the necessary level. According to reports, these two drying techniques worked better together than they did when used independently (Calin-Sanchez et al., 2020).

2.4.4 Fluidized Bed Drying: Assisted by Microwaves, Far Infrared Rays, and Ultrasounds

2.4.4.1 Far Infrared Rays

Far infrared (FIR) rays are a form of electromagnetic radiation that can penetrate deep into the material, heating it from within. FIR rays have excellent heat absorption properties and promote efficient heat transfer. When combined with fluidized bed drying, FIR rays enhance the overall drying efficiency and contribute to uniform heat distribution throughout the material. When a fresh fruit or vegetable is exposed to electromagnetic radiation that has a wavelength range of 0.8–1,000 m, IRD takes place. The wavelength range of infrared is 0.75–1,000 m. Infrared radiation transfers the heat from the heating source to the food surface. On the other hand, the technique has no impact on the air in the vicinity. Due to the ease of equipment and energy savings, this method is one of the best for combining with conventional drying processes (Wen et al., 2024).

2.4.4.2 Synergistic Effect

FIR rays, located at the farthest side of visible rays, are used to raise food product temperatures to the above wavelength, allowing for better control of matter components. In contrast to MW-assisted fluidized bed dryers, this technique can be used at any step to regulate the quality of the matter components. For example, FIR rays–assisted fluidized bed was used in the first stage of drying to control the influence of garlic allicin. Both individual fluidized beds and techniques that combined FIR photons with fluidized beds did not have an impact on functional components such as amino acids. High-power US application can accelerate mass transfer and contribute to high-quality dried products in heat-sensitive horticultural products. This is due to lower temperatures and time needed for drying treatment. US methods can avoid negative effects like shrinkage, color darkening, cracking, or nutritional changes in fruits and vegetables. Additionally, this method is considered low-cost and energy-efficient (Chavan et al., 2022).

2.4.5 Intermittent Drying

Intermittent drying is a technique where material is subjected to alternating periods of drying and resting, achieving specific drying objectives. Controlling thermal energy sources and energy input mechanisms can achieve intermittency. This method reduces browning, hydrothermomechanical stress, and chemical reactions, while maintaining bioactive ingredients (Majumder et al., 2021).

2.4.5.1 Purpose

Intermittent drying improves efficiency, preserves product quality, and manages material characteristics, optimizing conditions for specific materials that benefit from intermittent exposure.

2.4.5.2 Drying–Resting Cycles

Intermittent drying involves alternating cycles of drying and resting periods, with material exposed to conditions and allowed to rest in static or controlled environments.

Benefits of Intermittent Drying

Efficiency Improvement: Intermittent drying enhances efficiency by reducing energy consumption and removing moisture, allowing for further drying cycles.

Quality Preservation: Intermittent drying preserves product quality by allowing resting periods for moisture redistribution and equilibration, reducing damage risk from excessive heat or prolonged exposure.

Control of Material Characteristics: Intermittent drying controls material characteristics like texture, structural integrity, and moisture content through alternating drying–resting cycles, allowing desired transformations.

Applications: Intermittent drying is utilized in food processing, pharmaceuticals, and materials science for delicate materials like fruits, vegetables, and grains, ensuring moisture content management and quality preservation (Kumar et al., 2014).

2.4.6 Smart Drying Systems

Smart drying systems for fresh fruits are intelligent and automated systems that utilize advanced technologies, monitoring, and control mechanisms to optimize the drying process, improve efficiency, and ensure consistent product quality. These systems use real-time data analysis, sensors, and automated controls to control temperature, humidity, and airflow, ensuring efficient drying and minimizing energy waste. IoT connectivity enables remote monitoring, control, and data analysis, while data analysis techniques identify patterns, trends, and correlations (Yao et al., 2023). Customization and preset programs are offered for different fruit varieties, ensuring optimal drying conditions. Remote monitoring and control enable operators to monitor the system's performance, make adjustments, and troubleshoot issues remotely. Energy optimization is achieved by integrating energy-efficient components, such as heat recovery systems or HPs, and analyzing real-time data to adjust energy usage and minimize waste. Quality assurance is maintained by closely monitoring parameters like moisture content, color, texture, and taste, preserving the nutritional value and sensory attributes of dried fruits. Integrating smart drying systems with traceability systems enhances product traceability, quality control, and compliance with food safety regulations. Overall, smart drying systems help fruit producers optimize operations, reduce waste, and ensure high-quality dried fruits for consumers (Petikirige et al., 2022).

2.5 SIGNIFICANCE, BENEFITS, AND LIMITATIONS

2.5.1 The Significance and Benefits of Sustainable Drying of Fresh Fruits

The sustainable drying of fresh fruits offers numerous benefits, including preserving nutritional value, extended shelf life, reducing food waste, energy efficiency, cost-savings, value-added product opportunities, and reducing environmental impact. These technologies use methods like HP drying, VD, or FD to minimize nutrient loss, ensuring the preservation of essential vitamins, minerals, and antioxidants. They also remove moisture content, preventing spoilage and extending shelf life, reducing food waste, and allowing fruits to be stored and enjoyed for an extended period (Sivamani et al., 2021). Fresh fruits are perishable and can spoil quickly if not preserved promptly. Sustainable drying technologies focus on energy-efficient methods and renewable energy sources, minimizing greenhouse gas emissions and environmental impact (Radojcin et al., 2021). By adopting sustainable drying practices, the overall carbon footprint of the fruit drying process can be significantly reduced. These technologies have positive economic implications for fruit producers, as they can extend the shelf life of fruits, reduce seasonal fluctuations, increase market opportunities, reduce inventory losses, and increase profitability. Dried fruits are lightweight, compact, and have a longer shelf life, making them convenient for storage, transportation, and consumption (Uthpala et al., 2020). They also contribute to a more sustainable and responsible agricultural sector, promoting efficient resource use, waste reduction, and environmental impact. Overall, sustainable drying is an essential aspect of fruit processing, benefiting both producers and consumers, supporting health and environmental sustainability (Patel & Kar, 2012).

2.5.2 The Limitations of Sustainable Drying of Fresh Fruits

Sustainable drying of fresh fruits offers numerous benefits but has certain limitations. It requires initial investment in equipment and infrastructure, which can be higher than traditional methods.

These technologies often require sophisticated process control and monitoring systems, which may require specialized knowledge and training. Longer drying times may impact production efficiency and require careful planning. Sustainable drying methods may also impact sensory attributes, such as texture, color, or flavor, making it difficult to balance nutritional value with desired sensory qualities. Sustainable drying technologies may have limitations in scale and capacity, with some suitable for small-scale operations or specific fruit varieties, while others require larger equipment and infrastructure (Kaimal et al., 2022). Renewable energy sources like solar or biomass may vary depending on geographic location and local infrastructure, making implementation more challenging in regions with limited renewable energy access. The optimal drying conditions and methods may vary for different fruit varieties, requiring additional research and development. Market demand and acceptance of dried fruits can vary depending on consumer preferences, cultural factors, and trends. Understanding and predicting demand and assessing consumer acceptance of sustainably dried products is essential for successful adoption and commercial viability (Sun & Liang, 2020). Despite these limitations, ongoing research, technological advancements, and knowledge-sharing can help overcome these challenges and optimize sustainable drying methods for efficient and environment-friendly fruit drying processes (Jha & Tripathy, 2021).

REFERENCES

Akharume F, Smith A, Sivanandan L, Tingi K. Recent progress on osmo-convective dehydration of fruits. 2019;4:956–69. doi: 10.25177/JFST.4.9.RA.613.

Aksoy A, Karasu S, Akcicek A, Kayacan S. Effects of different drying methods on drying kinetics, microstructure, color, and the rehydration ratio of minced meat. Foods. 2019;8(6):216. https://doi.org/10.3390/foods8060216

Alharaty G, Ramaswamy HS. Microwave-osmo-dehydro-freezing and storage of pineapple titbits: Quality advantage. Processes. 2023;11(2):494. https://doi.org/10.3390/pr11020494

Al-Hooti S, Sidhu JS, Qabazard H. Physicochemical characteristics of five date fruit cultivars grown in the United Arab Emirates. Plant Foods for Human Nutrition. 1997;50(2):101–13. doi: 10.1007/BF02436030.

Banerjee B, Janghu S. (2023). Role of food microwave drying in hybrid drying technology. In: Jovanovic, J.D. (ed.) A Comprehensive Review of the Versatile Dehydration Processes. InTechOpen. http://dx.doi.org/10.5772/intechopen.108894

Bauman I, Bobic Z, Dakovic Z, Ukrainczyk M. Time and speed of fruit drying on batch fluid-beds. Sadhana. 2005;30(5):687–98.

Bhambhani A, Stanbro J, Sethuraman A, Pande P. (2023). Principles and practice of lyophilization process and product development: Scale-Up and technology transfer. In: Jameel, F. (ed.) Principles and Practices of Lyophilization in Product Development and Manufacturing. AAPS Advances in the Pharmaceutical Sciences Series, vol 59. Springer, Cham. https://doi.org/10.1007/978-3-031-12634-5_25

Boateng ID. Thermal and nonthermal assisted drying of fruits and vegetables: Underlying principles and role in physicochemical properties and product quality. Food Engineering Reviews. 2023;15:113–55. https://doi.org/10.1007/s12393-022-09326-y

Calin-Sanchez A et al. Comparison of traditional and novel drying techniques and its effect on quality of fruits, vegetables and aromatic herbs. Foods. 2020;9:1261.

Chavan P, Sharma P, Sharma SR, Mittal TC, Jaiswal AK. Application of high-intensity ultrasound to improve food processing efficiency: A review. Foods. 2022;11(1):122. https://doi.org/10.3390/foods11010122

Chegini GR, Ghobadian B. Spray dryer parameters for fruit juice drying. World Journal of Agricultural Sciences. 2007;3(2):230–36.

Chupawa P, Sudathip Inchuen S, Jaisut D, Ronsse F, Duangkhamchan W. Effects of Stepwise Microwave Heating and Expanded Bed Height Control on the Performance of Combined Fluidized Bed/Microwave Drying for Preparing Instant Brown Rice. Food and Bioprocess Technolog . 2023. 16:199–215. https://doi.org/10.1007/s11947-022-02933-x

Cullen PJ et al. (2012) Status and trends of novel thermal and non-thermal technologies for fluid foods. Novel Thermal and Non-Thermal Technologies for Fluid Foods. Elsevier Inc., pp. 1–6.

Delfiya DSA, Prashob K, Murali S, Alfiya PV, Samuel MP, Pandiselvam R. Drying kinetics of food materials in infrared radiation drying: A review. Journal of Food Process Engineering. 2022;45(6):e13810. https://doi.org/10.1111/jfpe.13810

Demirdoven A, Baysal. T. The use of ultrasound and combined technologies in food preservation. Food Reviews International. 2008;25(1):1–11.
Fayose F, Huan Z. Heat pump drying of fruits and vegetables: Principles and potentials for sub-Saharan Africa. International Journal of Food Science. 2016;1–8. https://doi.org/10.1155/2016/9673029
Galaz P et al. Effect of drum drying temperature on drying kinetic and polyphenol contents in pomegranate peel. Journal of Food Engineering. 2017;208:19–27.
Gaukel V, Siebert T, Erle U. (2017). Microwave-assisted drying. The Microwave Processing of Foods (2nd edition, pp. 152–78). Karlsruhe Institute of Technology (KIT), Karlsruhe, Germany.
Hu Q, He Y, Wang F et al. Microwave technology: A novel approach to the transformation of natural metabolites. Chinese Medicine. 2021;16:87. https://doi.org/10.1186/s13020-021-00500-8
Janahar JJ, Jin ZT, Balasubramanium VM. Pulsed electric field processing applications in the food industry. Agriculture and Natural Resources. 2022. https://ohioline.osu.edu/factsheet/fst-fabe-1002
Jha A, Tripathy PP. Recent advancements in design, application, and simulation studies of hybrid solar drying technology. Food Engineering Reviews. 2021;13:375–410. https://doi.org/10.1007/s12393-020-09223-2
Kaimal AM, Tidke VB, Mujumdar AS et al. Food security and sustainability through solar drying technologies: A case study based on solar conduction dryer. Materials Circular Economy. 2022;4:7. https://doi.org/10.1007/s42824-022-00051-9
Kemp IC. (2012). Fundamentals of Energy Analysis of Dryers. Modern Drying Technology. Wiley.
Khmelev VN, Shalunov AV, Barsukov RV et al. Studies of ultrasonic dehydration efficiency. Journal of Zhejiang University Science A. 2011;12:247–254. https://doi.org/10.1631/jzus.A1000155
Khodifad BC, Dhamsaniya NK. Drying of food materials by microwave energy: A review. International Journal of Current Microbiology and Applied Sciences. 2020;9(5):1950–73. https://doi.org/10.20546/ijcmas.2020.905.223
Khoshtaghaza MH, Darvishi H, Minaei S. Effects of microwave-fluidized bed drying on quality, energy consumption and drying kinetics of soybean kernels. Journal of Food Science and Technology. 2015;52:4749–60. https://doi.org/10.1007/s13197-014-1557-6
Kumar C, Karim MA, Joardder MUH. Intermittent drying of food products: A critical review. Journal of Food Engineering. 2014;121:48–57.
Kumar D, Ladaniya MS, Gurjar M et al. Impact of drying methods on natural antioxidants, phenols and flavanones of immature dropped *Citrus sinensis* L. Osbeck fruits. Scientific Reports. 2022;12:6684. https://doi.org/10.1038/s41598-022-10661-7
Kumar D, Prasad S, Murthy GS. Optimization of microwave-assisted hot air drying conditions of okra using response surface methodology. Journal of Food Science and Technology. 2014;51(2):221–32. doi: 10.1007/s13197-011-0487-9.
Laguerre J-C, Hamoud-Agha MM. (2018). Food preservation and waste exploitation. Microwave Heating for Food Preservation. IntechOpen.
Lewicki PP. Design of hot air drying for better foods. Trends in Food Science and Technology. 2006;17(4): 153–63.
Loemba ABT, Kichonge B, Kivevele T. Comprehensive assessment of heat pump dryers for drying agricultural products. Energy Science & Engineering. 2023;11:2985–3014. doi: 10.1002/ese3.1326.
Madhu B, Srinivas MS, Srinivas G, Jain S. Ultrasonic technology and its applications in quality control, processing and preservation of food: A review. British Journal of Applied Science and Technology. 2019;32(5):1–11.
Majumder P, Sinha A, Gupta R et al. Drying of selected major spices: Characteristics and influencing parameters, drying technologies, quality retention and energy saving, and mathematical models. Food and Bioprocess Technology. 2021;14:1028–54. https://doi.org/10.1007/s11947-021-02646-7
Misha S, Sohif M, Ruslan M H, Salleh E. Review on the Application of a Tray Dryer System for Agricultural Products. world applied sciences 2013, Vol 22(3), pp 424-433.
Maulburg. The Role of Vacuum Technology in Freeze Drying. 2020. https://www.buschvacuum.com/th/en/success-stories/the-role-of-vacuum-technology-in-freeze-drying.html (Accessed 25 April 2025)
Mshelia R, Dz., Dibal NI, Chiroma S. Food irradiation: An effective but under-utilized technique for food preservations. Journal of Food Science and Technology. 2023;60:2517–25. https://doi.org/10.1007/s13197-022-05564-4
Mulet A, Carcel J, Garcia-Perez V, Riera E. (2010). Ultrasound-assisted hot air drying of foods. Ultrasound Technologies for Food and Bioprocessing. Springer. pp. 511–534. doi: 10.1007/978-1-4419-7472-3_19
Nawi J, Abdan. Modelling the convective drying process of pumpkin using an artificial neural network. International Food Research Journal. 2016;23(Suppl):S237–43.

Nowacka, D. (2022) Ultrasound-assisted drying of food: Methods and protocols in food science. Emerging Food Processing Technologies, Spinger. doi: 10.1007/978-1-0716-2136-3_7
Nwankwo C, Osinachi U, James M, Chigozie O, Dereje B, Irene C. Technological advancements in the drying of fruits and vegetables: A review. African Journal of Food Science. 2021;15:367–79. 10.5897/AJFS2021.2113.
Onwude D I, Hashim N, Janius R, Abdan K, Chen G , Oladejo A O. Non-thermal hybrid drying of fruits and vegetables: A review of current technologies. Innovative Food Science & Emerging Technologies. 2017. Volume 43, pp 223–238.
Opara LU, Al-Ani MR, Al-Shuaibi YS. Physico-chemical properties, vitamin c content, and antimicrobial properties of pomegranate fruit (*Punica granatum* L.). Food and Bioprocess Technology. 2009;2: 315–21.
Orsat V et al. Microwave drying of fruits and vegetables. Stewart Postharvest Review. 2006;6:4.
Ostermeier R, Parniakov O, Töpfl S, Jäger H. Applicability of pulsed electric field (PEF) pre-treatment for a convective two-step drying process. Foods. 2020;9(4):512. https://doi.org/10.3390/foods9040512
Pakbin B, Rezaei K, Haghighi M. (2014). An introductory review of applications of ultrasound in food drying processes. Journal of Food Processing and Technology. 6:1. 10.4172/2157-7110.1000410.
Pansare SK, Patel SM. Lyophilization process design and development: A single-step drying approach. Pharmaceutical Biotechnology. 2019;108(4):P1423–33.
Patel KK, Kar A. Heat pump assisted drying of agricultural produce: An overview. Journal of Food Science and Technology. 2012;49:142–60. https://doi.org/10.1007/s13197-011-0334-z
Patterson J, Kailasam S, Giftson H, Immaculate JK. Effect of drying technologies on the biochemical properties of *Stolephorus commersonnii*. Food Quality and Safety. 2018;2(3):153–58. https://doi.org/10.1093/fqsafe/fyy010
Petikirige J, Karim A, Millar G. Effect of drying techniques on quality and sensory properties of tropical fruits. International Journal of Food Science and Technology. 2022;57:6963–79. https://doi.org/10.1111/ijfs.16043
Preeti MAG, Khandetod YP, Dhande KG, Sawant PA. Solar hybrid drying technologies: A comprehensive review of recent trends and developments. The Pharma Innovation Journal. 2023;SP-12(12):1–6.
Radojčin M, Pavkov I, Bursać Kovačević D, Putnik P, Wiktor A, Stamenković Z, Kešelj K, Gere A. Effect of selected drying methods and emerging drying intensification technologies on the quality of dried fruit: A review. Processes. 2021;9(1):132. https://doi.org/10.3390/pr9010132
Ramli W, Wan Daud W. (2003). 9 Drum Dryers.
Raponi F, Moscetti R, Monarca D, Colantoni A, Massantini R. Monitoring and optimization of the process of drying fruits and vegetables using computer vision: A review. Sustainability. 2009;9(11). https://doi.org/10.3390/su9112009
Rojas ML, Augusto P, Cárcel ED. Ethanol pre-treatment to ultrasound-assisted convective drying of apple. Innovative Food Science & Emerging Technologies. 2020;61;102328. doi: 10.1016/j.ifset.2020.102328.
Sagar VR, Kumar PS. Recent advances in drying and dehydration of fruits and vegetables: A review. Journal of Food Science and Technology. 2010;47(1):15–26.
Saleena P, Jayashree E, Anees K. Recent developments in osmotic dehydration of fruits and vegetables: A review. The Pharma Innovation Journal. 2021;SP-11(2):40–50.
Salehi F. Recent applications of heat pump dryer for drying of fruit crops: A review. International Journal of Fruit Science. 2021;21(1):546–55. doi: 10.1080/15538362.2021.1911746.
Sircar A, Yadav K, Maiti S. (2019). Performance Evaluation of a Ground Source Heat Pump for the Coastal Climate. Proceedings, 44th Workshop on Geothermal Reservoir Engineering Stanford University, Stanford, CA, February 11–13, 2019.
Sivamani P, Mounika E, Sairam NH, Jagannadha Rao N. Applications of vacuum technology in food processing. The Pharma Innovation Journal. 2021;10(5):914–18.
Sun Y, Liang C. Factors determining consumers' purchase intentions towards dried fruits. International Journal of Fruit Science. 2020;20(Suppl 2):S1072–96. doi: 10.1080/15538362.2020.1774477.
Thulasidas JS, Varadarajan GS, Sundararajan R. Pulsed electric field for enhanced extraction of intracellular bioactive compounds from plant products: An overview. Novel Approaches in Drug Designing & Development. 2019;5(2). https://juniperpublishers.com/napdd/NAPDD.MS.ID.555657.php
Uthpala TGG, Navaratne S, Thibbotuwawa A. Review on low-temperature heat pump drying applications in food industry: Cooling with dehumidification drying method. Journal of Food Process Engineering. 2020;43(10):613502. https://doi.org/10.1111/jfpe.13502

Wardhani N, Amanda N, Sari A. (2022). Microwave vacuum drying of fruit: A review. Advances in Biological Sciences Research, vol 19. Atlantis Press International B.V. pp. 309–16.

Wray D and Ramaswamy H S. Microwave-Osmotic Dehydration of Cranberries under Continuous Flow Medium Spray Conditions. Journal of Microwave Science and TechnologyVolume 2013, Article ID 207308, 11. http://dx.doi.org/10.1155/2013/207308.

Wen J, Pan J, Ma J et al. Advances in far-infrared research: Therapeutic mechanisms of disease and application in cancer detection. Lasers in Medical Science. 2024;39:41. https://doi.org/10.1007/s10103-024-03994-4

Yadav A, Ali Z. Modern drying techniques in fruits and vegetables to reduce postharvest losses. Indian Farmer. 2023;10(03):101–104.

Yadav AK, Singh SV. Osmotic dehydration of fruits and vegetables: A review. Journal of Food Science and Technology. 2014;51:1654–73. https://doi.org/10.1007/s13197-012-0659-2

Yang Q, Hu C, Li J, Xiao H, Jia W, Wang X, Liu X, Tang Z, Chen B, Yi X et al. Effects of different natural drying methods on drying characteristics and quality of diaogan apricots. Agriculture. 2024;14(5):660. https://doi.org/10.3390/agriculture14050660

Yao J, Chen W, Fan K. Novel efficient physical technologies for enhancing freeze drying of fruits and vegetables: A review. Foods. 2023;12(23):4321. https://doi.org/10.3390/foods12234321

Zhang X, Chen M, Guan J. Regeneration behavior of solid desiccants with microwave drying. Journal of Thermal Analysis and Calorimetry. 2024;149:10927–40. https://doi.org/10.1007/s10973-024-13010-0

Zhang Y et al. Pulsed electric field processing effects on physiochemical properties, flavor compounds and microorganisms of longan juice. Journal of Food Processing and Preservation. 2010;34(6):1121–38.

3 Biomass Energy Applications for Drying Fresh Produce

Canan Ece Tamer
Department of Food Engineering, Faculty of Agriculture, Bursa Uludağ University, Bursa, Turkey

Azime Özkan Karabacak
Department of Food Processing, Gemlik Asım Kocabıyık Vocational School, Bursa Uludağ University, Bursa, Turkey

Özge Süfer
Food Engineering Department, Faculty of Engineering, Osmaniye Korkut Ata University, Osmaniye, Turkey

Senanur Durgut
Department of Food Engineering, Faculty of Agriculture, Bursa Uludağ University, Bursa, Turkey

3.1 TECHNOLOGIES USED IN DRYING FRESH PRODUCE

Drying, which prevents potential microbial activity by evaporating moisture, is a common method of food preservation used worldwide. This method allows for the extension of food shelf life, simpler transit and storage owing to volume reduction, access to nonseasonal items, and a greater selection of products for customers (Moses et al., 2014). Drying can also cause changes in the texture and amount of bioactive compounds in food products. These alterations in the food matrix are influenced by variables such as drying duration, drying temperature, and water activity of the material (Alaei et al., 2018). Both producers and customers place a great value on the product's quality after the drying process. Uncontrolled drying can result in flavor losses and undesired changes in color (Mousakhani-Ganjeh et al., 2021).

The equipment used in drying process should be selected specifically for the product (Sapmaz & Kılıçaslan, 2023). Additional classifications for dryers include industrial and traditional dryers based on price, solar, biomass, and fossil fuel dryers based on energy sources, open-sun, and indirect solar, and mixed-mode dryers based on drying model and forced air convection or natural air convection (Tchaya and Houdji, 2021). Figure 3.1 illustrates the classification of energy sources used in food drying.

Since ancient times, food has been dried using the conventional method of sun and shade drying in various parts of the world. Despite being a cheap and readily available method, it has some drawbacks. These disadvantages include the need for long drying times, the risk of contaminating the food, losing nutritional and medicinal compounds, and producing a poor quality product.

 DOI: 10.1201/9781003344087-3

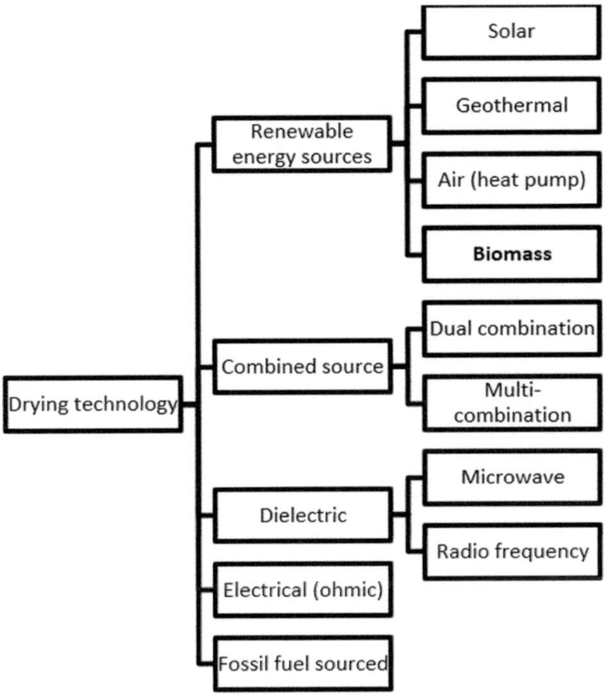

FIGURE 3.1 Classification of energy sources used in food drying. (Qu et al., 2021.).

Furthermore, the inability to provide controlled conditions is the most significant disadvantage of this method. In modern drying systems, controlled conditions can be achieved, leading to the production of higher quality products (Chua & Chou, 2003; Rathore & Panwar, 2010; Singh, 2011).

After the drying process, the product's nutritional and sensory qualities are of significance. In recent years, numerous researches have been done on pretreatments, equipment, and techniques in drying process to achieve high-quality final products. As a result of these studies, new systems have been developed.

The driving forces for the exploration and development of new drying methods can be listed as operational safety, better product quality, non-polluting nature, higher capacity, improved process control, better cost economics, benefits of hybrid drying, and shorter drying time (Moses et al., 2014).

Volumetric heating is a technique that is used in microwave-assisted drying. It ensures a homogeneous temperature increase in the material to be dried. This has a more favorable effect compared to the temperature increase from the surface to the core in traditional methods, and the product's drying process can be greatly accelerated. As a result, the dried product's sensory qualities are enhanced in comparison to conventional procedures (Zahoor et al., 2023). However, it has a significant initial investment cost and may cause the product's texture to alter unintentionally (Mahanti et al., 2021).

The method of freeze-drying, often referred to as lyophilization, involves freezing the product's moisture, before removing it by sublimation. According to reports, the sensory qualities of the finished products are better with freeze-drying method than with conventional drying techniques because of the low temperature and pressure environment in which drying takes place. Additionally, bioactive components of the food are better preserved in freeze-drying (Chumroenphat et al., 2021).

Despite these positive characteristics, freeze-drying involves expensive and energy-intensive systems (Chao et al., 2023).

Refractance window drying system uses radiation, convection, and all three types of heat transmission. It is used for drying heat-sensitive products. However, in terms of drying capacity, it is insufficient for industrial production and efforts are being made to improve this aspect (Mahanti et al., 2021).

Vacuum oven drying is a simple method; however, it can be costly for large-scale production (Correia et al., 2017). According to reports, vacuum oven drying requires more time than the conventional hot-air drying. The longer drying time is caused by the slower heat transmission rate. It can be used with infrared or microwave techniques to increase drying speed (Bao et al., 2023).

Spray drying is a widely used drying system that involves the dehydration of atomized liquid droplets in a hot convective environment. It is an economical and efficient method; however, some products require the use of a carrier substance during the drying procedure (Azhar et al., 2020).

Thermal drying methods need lots of energy to get the moisture out of the material. During the drying procedure, a significant quantity of latent heat from vaporization is released along with the moisture from the material. The moisture that is released during conventional hot-air drying operations is immediately discharged from the system. The heat in the discharged liquid can account for up to 55% of the overall energy intake. The latent heat in the air from the dryer is converted into perceptible heat using heat pump drying systems, which save electricity. This technique is being used more and more due to its potential to dry faster and save energy (Cheng et al., 2023; El Fil & Garimella, 2021).

Electrohydrodynamic (EHD) drying, employs a high voltage electric field to speed up water evaporation (Paul & Martynenko, 2021). In comparison to oven drying, EHD drying is less costly and results in food with better physiochemical attributes, including nutrient content, flavor, shrinkage, and color (Bajgai et al., 2006).

In food drying process, there are practical challenges due to the slow mass transfer depending on the food matrix and treatment conditions. In order to accelerate mass transfer and get over these challenges, an ultrasonic treatment might be used during drying. The rehydration capabilities of dried products can also be enhanced by utilizing ultrasonic waves to create microscopic channels that could facilitate losing and gaining of moisture. The procedure has been utilized to produce appropriate drying rates at lower temperatures or high drying rates at the same temperatures. It has generally been utilized in a variety of ways, including ultrasound-assisted dehydration (i.e., freeze, osmotic, convective, vacuum) (Başlar et al., 2016).

Superheated steam is defined as steam that has been heated over its saturation point at a specific pressure. Longer constant rate intervals and lower necessary moisture concentrations set the concept apart from conventional drying. Superheating intensity increases and appears to be the primary process variable, as operational demands decline. Moses et al. (2014) stated that increased microbicidal and insecticidal effects, decreased fire and explosion danger, decreased oxidation risk, no pollution, reduced equipment needs, and good drying performance are benefits of superheated steam drying. In addition to these methods, various combined food drying systems are being developed and utilized in the industry, such as pulsed electric field and assisted drying systems (Mousakhani-Ganjeh et al., 2021).

3.2 ENVIRONMENTAL IMPACTS OF ENERGY OPTIONS FOR DRYING FRESH PRODUCE

3.2.1 Fossil Fuels and Greenhouse Gases

Fossil fuels including coal, liquefied petroleum gas and electrical methods such as microwave radiation and infrared radiation are more common for drying of food products (Pardo & Zufía, 2012). Of the total energy used worldwide, 80% comes from fossil fuels, and half of the electricity produced stem from fossil fuel–powered plants (Clairand et al., 2020). In addition to these methods, various combined food drying systems are being developed and utilized in the industry, such as electrohydrodynamic, pulsed electric field, and ultrasound-assisted drying systems (Mousakhani-Ganjeh et al., 2021).

The world economy is significantly reliant on fossil fuels. Fossil fuel economy is influenced by fuel type, supply chain, and supply–demand dynamics (Edmonds & Reilly, 1983). Also, fossil fuels differ from other environmental pollutants due to their relationship with the climate crisis. The urgency to address climate issues is increasing both locally and internationally. In this regard, the development of policies for the supply chain of fossil fuels becomes crucial (Donaghy et al., 2023). Additionally, the use of these resources is limited due to the greenhouse gas emissions they release into the environment (Pardo & Zufía, 2012). The food industry also emits greenhouse gases (N_2O, CH_4, and CO_2). When fossil fuels are burned to create energy, CO_2 is emitted. Methane is produced in paddy, through fermentation in cattle, and in landfills from the breakdown of food waste, whereas nitrous oxide is produced when fertilizers are employed to boost crop yields (Clairand et al., 2020; Kling & Hough, 2010). According to projections from the International Energy Agency, worldwide energy consumption would rise by 70% and emissions of greenhouse gases will rise by 60% by 2050 compared to 2011. To restrict the rise in average global temperatures to 2°C by the year 2050, greenhouse gas emissions would need to be reduced by 50% from 2009 levels and the increase in energy consumption would need to be capped at a rate of 25% (Codina Gironès et al., 2017).

3.3 ENERGY SUSTAINABILITY CONSIDERATIONS IN DRYING FRESH PRODUCE

Sustainability is the ability of a product to reach future generations based on prevailing conditions. Activities carried out to meet current needs should not harm the ecological balance. Production conducted with environmental awareness not only preserves the nature of today, but also provides resources for future generations to meet the same needs (Arı & Ergin, 2018).

In the 1970s, various technologies were developed with the aim of preventing environmental pollution. These technologies focus on preventing the release of pollutants into the environment. By the 1980s, with the development of industrial ecology, producers began to reexamine every step of production to enhance environmental and energy performance. Up to the present day, these changes have evolved to improve waste management, increase energy efficiency, and enable the recyclability of materials (Çelik et al., 2016).

Energy is essential for the functioning of every system on earth, and sustainability is also a highly important issue. The world's energy sources could be categorized under two different groups: Coal, petroleum, and natural gas are examples of the first group of conventional fossil fuels that are finite. Solar, wind, hydro, and geothermal energy make up the second category of energy sources (Masud et al., 2020). The main factors driving the quest for alternative energy sources are the depleting supply of fossil fuels, as well as the damage that fossil fuels cause to atmospheric pollution and climate change (Shafizadeh et al., 2023). Furthermore, concerns about the depletion of fuel reserves by the end of this century have accelerated research into alternative, sustainable energy sources (Masud et al., 2020).

Each country uses a varied quantity of energy, but in developing nations, the food industry often uses lots of energy. African nations utilize approximately 65–75% of the energy needed to produce food inefficiently during cooking and food preparation. The energy needed to dry wet grain to the ideal storage moisture level is around 0.5–0.75 MJ/kg. Electricity, natural gas, or liquefied petroleum gas (LPG) all are potential sources of this energy (Clairand et al., 2020). The quality of the product that results from drying should be taken into consideration while making efforts to improve environmental concerns in the drying process. The sources employed in food drying systems should be dependable and accessible (Abunde Neba & Jiokap Nono, 2017).

According to Qu et al. (2021) and other sources, the energy consumption of industrial drying systems currently makes up around 25% of all industrial energy consumption. Additionally, according to Sapmaz and Kılıçaslan (2023), up to 70% of the total energy used in the production of some items is required for drying operations. This suggests that the several drying techniques used in the business are energy-intensive procedures. Additionally, the food industry uses a sizable amount (about 12%) of the overall power consumed in the industrial sector (Clairand et al., 2020).

3.4 BIOMASS ENERGY OPTIONS IN DRYING FRESH PRODUCE

In recent years, carbon-neutral biomass energy (bioenergy) has gained attention among renewable energy sources. This is because different solid, liquid, and gaseous forms of bioenergy may be created and used to replace different types of fossil fuels in a variety of applications. Bioenergy provides around 10% of the main energy needed by the globe. In most developing countries, around half of the world's biomass is utilized for heating and cooking. Modern energy sources, including biogas, biodiesel, bioethanol, and biopower, are effective with little detrimental environmental impact (Shafizadeh et al., 2023). Oils generated from biomass are transesterified using methanol to create biodiesel (Arias et al., 2023). In direct competition with their use as a source of food or feed, bioethanol (biodiesel) is mostly made from corn and sugarcane (soybean, sunflower, and rapeseed oil). The link between energy and agricultural food prices is elevated by this energy generation technology in terms of social and political interests (Yoon et al., 2022).

The global waste problem is increasingly growing due to population increase, urbanization, and economic development (Millati et al., 2019). Straw, sugarcane bagasse, corn cob, wood chips, sawdust, and other types of lignocellulosic biomass make up the majority of both agricultural and forestry wastes. It has been reported that coffee waste can also be utilized in biomass energy production. Particularly, the utilization of coffee husks as biomass is highly significant in minimizing waste generation (de Oliveira e Silva et al., 2023). Biowaste materials can be used to sustainably produce clean, renewable energy in the form of biofuels while also resolving the issue of waste disposal because of their biodegradable nature (Usmani et al., 2023). The natural polymers cellulose, hemicellulose, and lignin found in lignocellulosic biomass all are readily transformed into environmentally favorable biofuels. The use of fuels generated from biomass has been strongly marketed as a replacement for fossil fuels since it is believed that biomass is a carbon-neutral energy source (Fakudze & Chen, 2023; Khan et al., 2019).

A reliable source of thermal energy for drying for many years has been biomass (Masud et al., 2020). With 77.4% of all renewable energy resources and 10.4% of the world's total primary energy supply, biomass energy is by far the most prevalent renewable energy source (Carlos & Ba Khang, 2008). Thanks to biomass, the developing world might experience sustainable economic growth and development. For their primary energy needs, over 90% of rural families and 15% of urban households utilize biomass (Masud et al., 2020).

By-products from rice husk, bagasse from sugarcane, oil palm waste, and wood scraps are some of the most promising. The world's leading producers of agricultural goods, including rice, sugarcane, palm oil, coconut, and rubber, are several Southeast Asian countries. The region's quick economic growth has resulted in a rise in average annual energy consumption of 3.9% and average yearly carbon emissions of more than 5%. Therefore, it is anticipated that employing biomass energy will benefit these places (Tun et al., 2019).

In British Columbia, Canada, the biomass potential from leftover crops might cut greenhouse gas emissions by up to 2%, according to Wang et al. (2020). Additionally, Cyprus generates 13,000 tons of solid waste from the processing of olives using two separate techniques. Cyprus's projected annual energy potential is 38 GWh. Estimates show that olive waste pellets make up 0.9% of the energy required to heat homes, and that the energy demand in Cyprus is higher in this sector than in the industrial and agricultural sectors. The number of working days and the plant's capacity are crucial factors in determining if a process is feasible.

The most well-liked and accessible kind of renewable energy is solar power. Despite the low cost and wide availability of this technology, the product is frequently susceptible to uncontrolled extended drying times, contamination, and poor product quality. The drying process, however, was hampered by the lack of sunshine on overcast and during rainy days and nighttime. The main difficulty in using solar dryers is that sun radiation is intermittent and thus not always available. To enable a non-stop drying process in the event that the solar heat supply is inadequate, solar drying may be supplemented with an extra biomass-based heat producing unit (Mutlu et al., 2021).

3.5 BIOMASS ENERGY APPLICATIONS FOR DRYING SYSTEMS

The development of bacteria, plants, or animals results in the production of biomass, one of the main resources of renewable energy. Agricultural crop residue, forest residue, animal manure, solid municipal waste, and sewage are typical sources of biomass (Figure 3.2) (Huda et al., 2014; Newaj & Masud, 2014.; Masud et al., 2019).

Through thermochemical processes like combustion or gasification, biomass may be utilized as a source of renewable energy for the generation of both heat and electricity (Li et al., 2012). In the biomass-based drying process, the biomass is burnt to create hot air that is then blown through the dryer. Biomass dryers are frequently employed in underdeveloped nations, and also this type of resource used on an industrial basis. Low energy density is produced by combining solar and biomass energy, which is useful for drying crops and is ecologically friendly (Ndukwu et al., 2020).

Hot air that has been heated by burning biomass is pumped into the dryer, even though the usual dryer based on biomass has a low efficiency, which is problematic given the significant energy loss caused mostly by the poor design of these dryers. Traditional biomass dryers have significant shortcomings. However, newer designs may be able to overcome these shortcomings through low heat loss, improved heat and mass transfer rates, and excellent product quality. Improved biomass dryers such as fluidized bed dryers, flash dryers, and continuous direct-heat rotary dryers are some of the ones that are suggested (Liu et al., 2015; Masud et al., 2020).

Along with its several advantages, biomass food drying has certain drawbacks. The main drawback of this method is the inefficiency of traditional dryers made of biomass. Recent designs, with their minimal heat loss and improved heat and mass transfer rates, may be able to overcome the significant disadvantages of traditional biomass dryers. The difficulties show why biomass technology has to advance. With this technology, drying food has issues with CO_2 emissions, high upfront expenses, harmful combustion by-products that lower the quality of the dried food, and regular

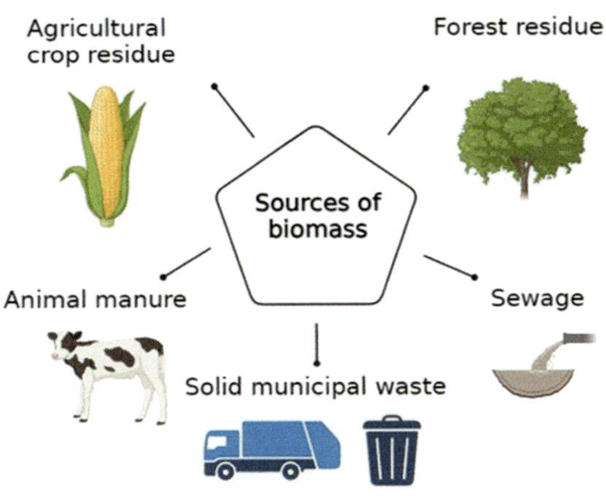

FIGURE 3.2 Sources of biomass.

maintenance needs. Some of the recommended improved biomass dryers are the fluidized bed dryer, flash dryer, and continuous direct-heat rotary dryer (Isaksson et al., 2013; Liu et al., 2015). Due to the accessibility and affordability of the feed materials that the dryer utilizes to create heat energy for drying, biomass food dryers continue to be a favored technique of drying in most developing countries in spite of the aforementioned disadvantages (Masud et al., 2020).

Both direct and indirect biomass dryer designs are commonly used. Direct dryers are highly efficient and can quickly dry low-moisture-content materials. Their simple design and operation make them cost-effective for specific applications. However, direct dryers may raise environmental concerns due to emissions and are not ideal for heat-sensitive biomass.

In contrast, indirect biomass dryers offer distinct advantages. They produce fewer emissions, are more environmentally friendly, and can handle heat-sensitive materials, which expands their application range. Despite these benefits, indirect dryers may be less efficient than direct dryers (Mishra et al., 2022; Murugan et al., 2021).

In a direct biomass dryer, the combustion waste is instantly pumped through the chamber containing the foods that need to be dried. Through convective heat transmission, the food gets dehydrated. The evaporation mass transfer hypothesis is used to remove the moisture from food's surface. The moisture level of the outer layer reduces as a result. Thus, the cell membrane allows water to move osmotically from the inner layer to the outer layer. The bulk is then moved once the water has evaporated. The aforementioned steps are continued till the food has dried to the necessary consistency. In the event of an indirect biomass food dryer, the heat from the combustion product is utilized to warm the secondary working fluid, which is commonly fresh air. Generally, a heat exchanger is employed for this. The working liquid is then put into the food-holding container. One of the primary problems biomass food dryers run into is the burning of the food's outer cells. Osmosis is a lengthy process that transfers moisture from an object's inside to outside. The outer layer cells of the meal finally burn since there is not enough time for the water to get from the inner layer to the outside layer. This issue has to be resolved, as it is inconvenient to continuously extinguish the burner's fire, and then relight it to flow warm air as needed. Designing hybrid dryers is currently a difficult process that typically requires expensive prototype systems and time-consuming research to determine the optimal materials to use and the ideal operating conditions for each connected unit (Dhanushkodi et al., 2015; Okoroigwe et al., 2015).

There are several types of biomass dryers used for drying food (Figure 3.3). One such dryer is the biomass rotary dryer. Any form of food can be dried using a biomass rotary dryer, with the exception of slurries, solutions, and pastes. The system's drying effectiveness is independent of load variations, because it may generate a significant quantity of heat energy. Biomass rotary dryer is very simple to use and is also strongly advised for businesses with little oversight. However, installing a

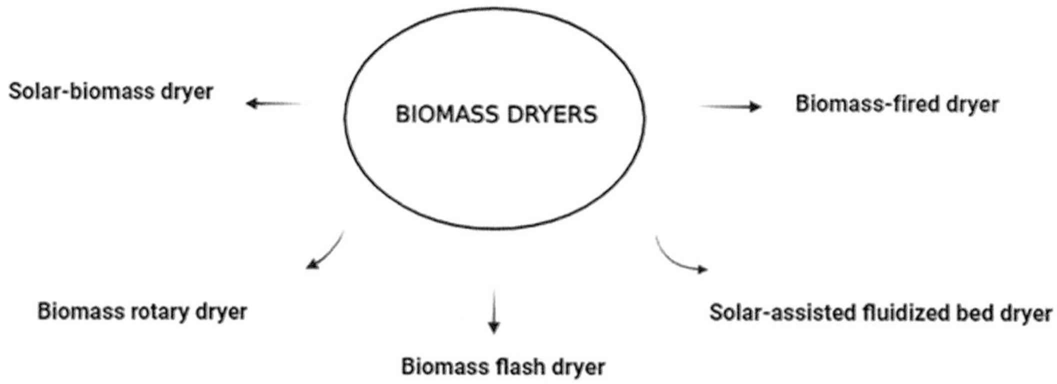

FIGURE 3.3 Biomass dryers.

biomass rotary dryer can be expensive, and annual maintenance costs may equal 10% of the initial cost. A complicated conduit for high-speed gas flow, a feeder that will distribute and add solid particles to the gas stream, a biomass combustion chamber, an airflow control system, and an enhanced biomass dryer based on the flash drying technique all should be included in the dryer. Particles from gases can be collected using a cyclone collector after being dried by solids. Particles as small as 1–2 mm may be dried by a biomass flash dryer. In contrast to a rotary-type dryer, the performance of a flash dryer greatly depends on the loading scenario. The flash drying system will not work, because overloaded materials cannot be fed into the dryer at high feed rates. Additionally, a biomass flash drying system would need more maintenance than standard drying systems. The two major components of a biomass continuous fluid bed dryer are a biomass combustion chamber that will provide the system with thermal energy and a continuous feeding system that maintains the fluidity of the moist food particle, until it comes into touch with the heated air. It is preferable if the input material has low moisture levels as fluidization decreases drying effectiveness (Van't Land, 2012).

3.5.1 Drying Kinetics on Biomass Energy Drying of Fresh Produce

Alternatively, interactive computer use, mathematical modeling, design simulation, and analysis software can improve the design and optimization of hybrid dryers for food process applications by taking a variety of design parameters into account and obtaining a suitable or an ideal design. Design software should consider both the solar irradiation process and the combustion of biomass when establishing the appropriate dryer design parameters in light of the growing interest in solar–biomass hybrid dryers. It would appear that engineers are required to create dryer design software that integrates drying process parameters with solar radiation and biomass combustion process characteristics (Abunde Neba & Jiokap Nono, 2017).

For the drying of agricultural items such as coconut, maize grain, bananas, herbs, pineapple, and ginger, several researchers have utilized sun drying systems with backup biomass heaters. Biomass is a possible energy source for drying agricultural goods. A mathematical modeling research of hot-air drying for several crops was carried out by Waewsak and Punlek (2006). Several agricultural materials, including lemon grass, red chili peppers, and leech lime leaves, were dried using a biomass dryer. The Midilli model, which was proven to be superior to the Wangh and Singh model for lemon grass, better represented the drying behavior of red chili peppers and leech lime leaves (Waewsak & Punlek, 2006).

3.5.2 Studies on Biomass Energy Applications in Drying of Fresh Produce

Studies have been carried out into the use of biomass energy in food drying (Table 3.1). In particular, biomass-assisted dryers have been designed and their effects on the drying of various fresh foods have been evaluated.

Madhlopa and Ngwalo (2007) used a solar biomass heater to dry pineapple: The design and evaluation of an indirect-type natural convection sun dryer with integrated collector–storage solar and biomass backup warmers. The dryer's primary elements are the collector–storage thermal mass and the biomass burner, which has a rectangular duct and the drying chamber with a standard solar chimney. Fresh pineapples were utilized to test the dryer in three distinct process modes (solar, biomass, and solar–biomass) under a variety of weather conditions. The findings demonstrate that the thermal mass helped to decrease energy loss and control temperature changes in the drying chamber by storing part of the heat produced by both solar and biomass air heaters.

Amer et al. (2010) suggested using a solar biomass dryer to dry banana slices. Moisture content of banana slices decreased from 82% to 18% in 8 hours.

For efficiently drying cassava, Sunday et al. (2012) created a solar dryer with a backup heater. It contained an incinerator, a drying chamber, and a solar collector. At night, drying trials were carried out. While solar drying dried materials the quickest, things dried more slowly utilizing incinerators and control drying.

TABLE 3.1
Studies on Biomass Energy Applications in Drying of Fresh Produce

Material	Equipment Type/Technology	References
Pineapple	Solar–biomass dryer	Madhlopa and Ngwalo (2007)
Banana	Solar–biomass dryer	Amer et al. (2010)
Cassava	Solar–biomass dryer	Sunday et al. (2012)
Fresh okra, fresh groundnuts, and fresh cassava chips	Solar–biomass dryer	Okoroigwe et al. (2015)
Cashew	Solar–biomass dryer	Dhanushkodi et al. (2014)
Paddy	Solar-assisted fluidized bed dryer	Yahya et al. (2017)
Cashew	Solar–biomass dryer	Dhanushkodi et al. (2017)
Fried glutinous rice cracker (*khaotan*)	Biomass-fired dryer	Thanompongchart et al. (2017)
Fish	Solar-powered biomass dryer	Hamdani et al. (2018)
Pepetak leiognatus spp.	Solar–biomass dryer	Yuwana et al. (2020)
Banana	Solar–biomass dryer	Kiburi et al. (2020)
Coriander leaves and Babul wood (*Acacia nilotica*)	Biomass-based drying system	Nalawade and Panwar (2021)
Tapioca	Biomass-based drying system	Murugan and Saji Raveendran (2021)

Numerous studies that deal with using a drying process along with solar and biomass technology have discussed a variety of aspects, such as the comparison of the operational models (solar, biomass, and biomass–solar) (Madhlopa & Ngwalo, 2007; Okoroigwe et al., 2015), the process design (Amer et al., 2010; Tonui et al., 2014), the modeling of the process with various software tools (such as computational fluid dynamics or Matlab) (Abunde Neba & Jiokap Nono, 2017; Sonthikun et al., 2016), or the process's economic viability (Dhanushkodi et al., 2014). Evidently, continuous, profitable, and concurrent use of solar-assisted biomass dryers is possible (Qu et al., 2021).

In accordance with Okoroigwe et al. (2015), the sun drying system's drying efficiency could be enhanced with the installation of a biomass backup heater. The original solar collector and biomass unit's drying efficiency is expected to improve with the addition of a rear pass solar collector, a biomass heating burner, and a gas to gas heat exchanger. Additionally, this will guarantee that only pure, smokeless, and ash-free hot air will reach the samples. The effectiveness of the dryer based on solar, biomass, and solar–biomass heating rose from 5.19% to 16.04% while drying fresh okra, fresh groundnuts, and fresh cassava chips, respectively.

For drying cashews, Dhanushkodi et al. (2014) created a solar–biomass dryer. According to the performance studies, it took 6 hours for active mode, 10 hours for passive mode, and 14 hours for sun drying to accomplish the required moisture reduction from 10% to 5%. Thermal efficiency of the solar collectors varied from 30% to 45% in passive mode to 65–70% in active mode. The results of the evaluation of the product's quality demonstrate that the suggested technique is suitable for cashew drying.

A hybrid solar–biomass fluidized bed dryer was created, assessed, and improved by Yahya et al. (2017). Using a solar-assisted fluidized bed dryer and an integrated biomass furnace, the drying kinetics of paddy were studied. The hybrid system used less specific energy and dried materials more quickly than solar dryers.

After analyzing the drying kinetics of cashew kernels in a hybrid solar–biomass dryer, Dhanushkodi et al. (2017) determined the best drying models for defining the cashew's drying behavior. For the study on drying, 40 kg of cashew kernels were distributed among 10 drying trays, each holding 4 kg. The experiment was conducted in solar, biomass, and hybrid modes for 7 hours. Also, Page and Modified Page models for the biomass and hybrid drying were best fitted.

Thanompongchart et al. (2017) developed a fried glutinous rice cracker (*khaotan*) dryer using biomass as fuel. The dry rice cracker's physical characteristics were deemed to be satisfactory.

The dryer powered by biomass outperformed older dryers powered by LPG in terms of energy efficiency. The dryer that was powered by biomass used less energy and cost less to operate. Overall, it was determined that the suggested biomass-fired dryer could deliver acceptable product quality, while lowering operational costs.

Abunde Neba and Jiokap Nono (2017) have created a brand-new model and piece of software (Dryer Design and Simulation Software, DryDSim) for a solar–biomass hybrid dryer. The main programming language for achieving this goal was Matlab. The design of the hybrid dryer was influenced by green pepper drying tests carried out in a lab. The objective of the design software is to determine the most efficient design and economical parameters for new solar–biomass hybrid dryers for a variety of agricultural food items. The software should also be improved to enable weather-based automatic switching between biomass and solar energy sources and customized system layouts. This will need creating a physical model of the system in DryDSim and running the mathematical simulations in Matlab's Simulink software. To see if the software produces predictions that are consistent with practice and maybe correct for any deviations, it was recommended a pilot unit of the hybrid dryer should be built and tested.

In a study performed by Hamdani et al. (2018) a solar-powered biomass fish dryer was utilized. Using hybrid fish dryers that combine solar and biomass energy, fish could be dried in as little as 15 hours. The system still works utilizing biomass energy when there is no sun. The dryer has a 2.6-year payback period, making it economically feasible.

Introducing a hybrid solar–biomass dryer and conducting energy-based thermal analysis and energy-sustainability analysis are Ndukwu et al.'s (2020) contributions. A biomass furnace and an indirect sun dryer were incorporated during construction. While sunshine was present, the indirect dryer-collected solar energy warming and the drying chamber's air are then circulated. The biomass furnace was used to create flue gases when solar radiation wasn't present, and those gases were subsequently routed via the copper tube's drying chamber. As a result, drying continues even when sunlight is not present. This hybrid dryer's improved functionality was verified by a thermal examination.

For the purpose of drying fish (*Pepetak Leiognatus* spp.), Yuwana et al. (2020) improved the functionality of a hybrid dryer powered by solar and biomass energies. The biomass used was coconut shell. Drying process was finished in 20.47 hours, which was quicker than solar drying (24.20 hours). The dryer needed to finish drying processes in 14.37 and 15.43 hours when operated by respectively biomass energy and solar energy continued with biomass energy. For every kilogram of wet fish, 0.633 kg of coconut shell was used as fuel.

A greenhouse dryer was designed by Kiburi et al. (2020). Banana slices were dried using three energy sources: sun, biomass, and solar–biomass. It has been demonstrated that the solar mode is less energy- and exergy-efficient than the biomass and solar–biomass models. The greenhouse dryer's solar, biomass, and solar–biomass models' respective average energy consumption ratios were determined to be 35.58±24.78, 40.60±10.52, and 33.46±13.45%. For the three energy modes employed, there was no statistically significant variation in the exergy efficiency of drying air.

In 2021, Nalawade and Panwar designed a biomass-based drying system, built it, and evaluated its effectiveness. Thermal efficiency, moisture removal efficiency, heat transfer efficiency, and drying efficiency all have been assessed for the proposed system. Coriander leaves (4 kg) with an initial moisture content of 89.40% (w.b.) were fed into the dryer to assess its effectiveness. They were dried to 9.54% (w.b.) final moisture content. Babul wood (*A. nilotica*) (6 kg) were used in the drying. The usual drying chamber temperature was 51.84°C. The dryer's efficiency in terms of thermal and heat transfer was 28.22% and 56%, respectively. The short payback period makes the designed solution technically and economically viable based on economic considerations. The payback period cost US$2.68, the net present value was 0.38 years, and the benefit-cost ratio was US$3398.

In order to dry tapioca, Murugan and Saji Raveendran (2021) designed a dryer and conducted periodical effectiveness tests during the drying process. Between 60°C and 70°C, and with a producer gas mass flow that was continuous, drying was performed. The tapioca's ultimate moisture content was found to range from 5% to 12%. The biomass-based dryer speeds up drying by 50%, with respect to open-air solar drying. The quality, flavor, or color of the tapioca was unaffected by the drying procedure while using a biomass-based dryer. According to experimental findings, the biomass-based dryer is more effective than the more widely used solar drying method. According to the researchers, this method may be used to dry a range of crops, including black pepper, cardamom and chilies, among others.

Tchaya and Houdji (2021) integrated a biomass energy source and storage system to enhance an indirect solar dryer. In the hybrid mode, guava dried in 19 hours as opposed to 26 hours with the indirect sun drying.

3.6 CONCLUSIVE REMARKS

Using biomass energy for drying foods offers both advantages and disadvantages. Biomass energy is renewable and can reduce reliance on fossil fuels, contributing to environmental sustainability. Additionally, it provides a reliable energy source crucial for food drying operations. On the other hand, biomass energy also poses challenges, such as emissions that can contribute to air pollution if not managed effectively.

Geographical regions vary in suitability for biomass production and energy generation. Areas with abundant arable land and favorable climates can support the cultivation of biomass feedstocks like corn and soybeans, benefiting from high plant production efficiency. Regions with access to biomass resources can capitalize on biomass energy production, fostering local economic development and enhancing energy security. Furthermore, with proper design and optimization, biomass power plants can boost overall efficiency, improving energy generation from biomass resources.

In summary, while biomass energy offers opportunities for sustainable food drying and energy generation, careful consideration of its pros and cons, along with regional advantages in biomass production, is essential for maximizing benefits and mitigating drawbacks. Despite its potential, challenges remain in the practical use of biomass drying, requiring further study in energy usage, investment costs, and environmental impacts. Scaling up experimental biomass dryers through extensive modeling and techno-economic assessment is crucial for their successful implementation.

REFERENCES

Abunde Neba, F., & Jiokap Nono, Y. (2017). Modeling and simulated design: A novel model and software of a solar–biomass hybrid dryer. *Computers and Chemical Engineering*, *104*, 128–140. https://doi.org/10.1016/j.compchemeng.2017.04.002

Alaei, B., Dibagar, N., Amiri Chayjan, R., Kaveh, M., & Taghinezhad, E. (2018). The effect of short and medium infrared radiation on some drying and quality characteristics of quince slices under vacuum condition. *Quality Assurance and Safety of Crops & Foods*, 10(4), 371–381. https://doi.org/10.3920/QAS2017.1252

Amer, B. M. A., Hossain, M. A., & Gottschalk, K. (2010). Design and performance evaluation of a new hybrid solar dryer for banana. *Energy Conversion and Management*, *51*(4), 813–820. https://doi.org/10.1016/J.ENCONMAN.2009.11.016

Arı, E. S., & Ergin, E. (2018). Sürdürülebilir İşletme olma yolunda Çözüm Önerileri. *BEÜ AİD*, *3*(4), 1–18.

Arias, D. M., García-Valladares, O., Besagni, G., & Markides, C. N. (2023). A vision of renewable thermal technologies for drying, biofuels production and industrial waste, gas or water recovery. *Applied Thermal Engineering*, *223*, 120022. https://doi.org/10.1016/j.applthermaleng.2023.120022

Azhar, M. D., Hashib, S. A., Ibrahim, U. K., & Rahman, N. A. (2020). Development of carrier material for food applications in spray drying technology: An overview. *Materials Today: Proceedings*, *47*, 1371–1375. https://doi.org/10.1016/j.matpr.2021.04.140

Bajgai, T. R., Raghavan, G. S. V., Hashinaga, F., & Ngadi, M. O. (2006). Electrohydrodynamic drying: A concise overview. *Drying Technology*, *24*(7), 905–910. https://doi.org/10.1080/07373930600734091

Bao, X., Min, R., Zhou, K., Traffano-Schiffo, M. V., Dong, Q., & Luo, W. (2023). Effects of vacuum drying assisted with condensation on drying characteristics and quality of apple slices. *Journal of Food Engineering*, *340*, 111286. https://doi.org/10.1016/J.JFOODENG.2022.111286

Başlar, M., Toker, ÖS., Karasu, S., Tekin, Z. H., & Yildirim, H. B. 2016. Ultrasonic applications for food dehydration. Handbook of Ultrasonics and Sonochemistry. Springer. pp 1–24.

Carlos, R. M., & Ba Khang, D. (2008). Characterization of biomass energy projects in Southeast Asia. *Biomass and Bioenergy*, *32*(6), 525–532. https://doi.org/10.1016/J.BIOMBIOE.2007.11.005

Çelik, İE., Akman Ceyhan, A., & Akman, V. (2016). Yeşil Pazarlamada Sürdürülebilirlik ve Dünya'dan Bir Örnek: Tchibo Sustainability in Green Marketing and an Example from the World: Tchibo. International Conference on Eurasian Economies.

Chao, E., Li, J., & Fan, L. (2023). Influence of combined freeze-drying and far-infrared drying technologies on physicochemical properties of seed-used pumpkin. *Food Chemistry*, *398*, 133849. https://doi.org/10.1016/j.foodchem.2022.133849

Cheng, J. H., Cao, X., Shao, L. L., & Zhang, C. L. (2023). Performance evaluation of a novel heat pump system for drying with EVI-compressor driven precooling and reheating. *Energy*, *278*, 127989. https://doi.org/10.1016/j.energy.2023.127989

Chua, K. J., & Chou, S. K. (2003). Low-cost drying methods for developing countries. *Trends in Food Science & Technology*, *14*(12), 519–528. https://doi.org/10.1016/J.TIFS.2003.07.003

Chumroenphat, T., Somboonwatthanakul, I., Saensouk, S., & Siriamornpun, S. (2021). Changes in curcuminoids and chemical components of turmeric (*Curcuma longa* L.) under freeze-drying and low-temperature drying methods. *Food Chemistry*, *339*, 128121. https://doi.org/10.1016/J.FOODCHEM.2020.128121

Clairand, J. M., Briceno-Leon, M., Escriva-Escriva, G., & Pantaleo, A. M. (2020). Review of energy efficiency technologies in the food industry: Trends, barriers, and opportunities. In *IEEE Access* (Vol. 8, pp. 48015–48029). Institute of Electrical and Electronics Engineers Inc. https://doi.org/10.1109/ACCESS.2020.2979077

Codina Gironès, V., Moret, S., Peduzzi, E., Nasato, M., & Maréchal, F. (2017). Optimal use of biomass in large-scale energy systems: Insights for energy policy. *Energy*, *137*, 789–797. https://doi.org/10.1016/j.energy.2017.05.027

Correia, R., Grace, M. H., Esposito, D., & Lila, M. A. (2017). Wild blueberry polyphenol-protein food ingredients produced by three drying methods: Comparative physico-chemical properties, phytochemical content, and stability during storage. *Food Chemistry*, *235*, 76–85. https://doi.org/10.1016/j.foodchem.2017.05.042

de Oliveira e Silva, A., Garcia, F. P., Perazzini, M. T. B., & Perazzini, H. (2023). Design and economic analysis of a pre-treatment process of coffee husks biomass for an integrated bioenergy plant. *Environmental Technology and Innovation*, *30*, 103131. https://doi.org/10.1016/j.eti.2023.103131

Dhanushkodi, S., Wilson, V. H., & Sudhakar, K. (2014). Thermal performance evaluation of indirect forced cabinet solar dryer for cashew drying. *Journal of Agricultural and Environmental Sciences*, *14*(11), 1248–1254. https://doi.org/10.5829/idosi.aejaes.2014.14.11.21871

Dhanushkodi, S., Wilson, V. H., & Sudhakar, K. (2015). Life cycle cost of solar biomass hybrid dryer systems for cashew drying of nuts in India. *Environmental and Climate Technologies*, *15*(1), 22–33. https://doi.org/10.1515/rtuect-2015-0003

Dhanushkodi, S., Wilson, V. H., & Sudhakar, K. (2017). Mathematical modeling of drying behavior of cashew in a solar biomass hybrid dryer. *Resource-Efficient Technologies*, *3*(4), 359–364. https://doi.org/10.1016/j.reffit.2016.12.002

Donaghy, T. Q., Healy, N., Jiang, C. Y., & Battle, C. P. (2023). Fossil fuel racism in the United States: How phasing out coal, oil, and gas can protect communities. *Energy Research and Social Science*, *100*, 103104. https://doi.org/10.1016/j.erss.2023.103104

Edmonds, J., & Reilly, J. (1983). A long-term global energy-economic model of carbon dioxide release from fossil fuel use. *Energy Economics*, *5*(2), 74–88. https://doi.org/10.1016/0140-9883(83)90014-2

El Fil, B., & Garimella, S. (2021). Waste heat recovery in commercial gas-fired tumble dryers. *Energy*, *218*, 119407. https://doi.org/10.1016/J.ENERGY.2020.119407

Fakudze, S., & Chen, J. (2023). A critical review on co-hydrothermal carbonization of biomass and fossil-based feedstocks for cleaner solid fuel production: Synergistic effects and environmental benefits. *Chemical Engineering Journal*, *457*, 141004. https://doi.org/10.1016/j.cej.2022.141004

Hamdani, Rizal, T. A., & Muhammad, Z. (2018). Fabrication and testing of hybrid solar–biomass dryer for drying fish. *Case Studies in Thermal Engineering*, *12*, 489–496. https://doi.org/10.1016/J.CSITE.2018.06.008

Huda, A. S. N., Mekhilef, S., & Ahsan, A. (2014). Biomass energy in Bangladesh: Current status and prospects. *Renewable and Sustainable Energy Reviews*, *30*, 504–517. https://doi.org/10.1016/J.RSER.2013.10.028

Isaksson, J., Åsblad, A., & Berntsson, T. (2013). Influence of dryer type on the performance of a biomass gasification combined cycle co-located with an integrated pulp and paper mill. *Biomass and Bioenergy*, *59*, 336–347. https://doi.org/10.1016/J.BIOMBIOE.2013.10.002

Khan, T. A., Saud, A. S., Jamari, S. S., Rahim, M. H. A., Park, J. W., & Kim, H. J. (2019). Hydrothermal carbonization of lignocellulosic biomass for carbon rich material preparation: A review. *Biomass and Bioenergy*, *130*, 105384. https://doi.org/10.1016/j.biombioe.2019.105384

Kiburi, F. G., Kanali, C. L., Kituu, G. M., Ajwang, P. O., & Ronoh, E. K. (2020). Performance evaluation and economic feasibility of a solar–biomass hybrid greenhouse dryer for drying banana slices. *Renewable Energy Focus*, *34*, 60–68. https://doi.org/10.1016/j.ref.2020.06.009

Kling, M., Hough, I. (2010). The American carbon foodprint: Understanding and reducing your food's impact on climate change. Brighter Planet, Inc. http://brighterplanet.com.

Li, H., Chen, Q., Zhang, X., Finney, K. N., Sharifi, V. N., & Swithenbank, J. (2012). Evaluation of a biomass drying process using waste heat from process industries: A case study. *Applied Thermal Engineering*, *35*(1), 71–80. https://doi.org/10.1016/j.applthermaleng.2011.10.009

Liu, Y., Kansha, Y., Ishizuka, M., Fu, Q., & Tsutsumi, A. (2015). Experimental and simulation investigations on self-heat recuperative fluidized bed dryer for biomass drying with superheated steam. *Fuel Processing Technology*, *136*, 79–86. https://doi.org/10.1016/J.FUPROC.2014.10.005

Madhlopa, A., & Ngwalo, G. (2007). Solar dryer with thermal storage and biomass backup heater. *Solar Energy*, *81*(4), 449–462.

Mahanti, N. K., Chakraborty, S. K., Sudhakar, A., Verma, D. K., Shankar, S., Thakur, M., Singh, S., Tripathy, S., Gupta, A. K., & Srivastav, P. P. (2021). Refractance WindowTM-drying vs. other drying methods and effect of different process parameters on quality of foods: A comprehensive review of trends and technological developments. *Future Foods*, *3*, 100024. https://doi.org/10.1016/J.FUFO.2021.100024

Masud, M. H., Ahamed, R., Joardder, M. U. H., & Hasan, M. (2019). Mathematical model of heat transfer and feasibility test of improved cooking stoves in Bangladesh. *International Journal of Ambient Energy*, *40*(3), 317–328. https://doi.org/10.1080/01430750.2017.1392354

Masud, M. H., Karim, A., Ananno, A. A., & Ahmed, A. (2020). Sustainable Food Drying Techniques in Developing Countries: Prospects and Challenges. Springer International Publishing. https://doi.org/10.1007/978-3-030-42476-3

Millati, R., Cahyono, R. B., Ariyanto, T., Azzahrani, I. N., Putri, R. U., & Taherzadeh, M. J. (2019). Agricultural, industrial, municipal, and forest wastes: An overview. *Sustainable Resource Recovery and Zero Waste Approaches*, 1–22. https://doi.org/10.1016/B978-0-444-64200-4.00001-3

Mishra, L., Hauchhum, L., & Gupta, R. (2022). Development and performance investigation of a novel solar–biomass hybrid dryer. *Applied Thermal Engineering*, *211*, 118492.

Moses, J. A., Norton, T., Alagusundaram, K., & Tiwari, B. K. (2014). Novel drying techniques for the food industry. *Food Engineering Reviews*, *6*(3), 43–55. https://www.academia.edu/15199399/Novel_Drying_Techniques_for_the_Food_Industry

Mousakhani-Ganjeh, A., Amiri, A., Nasrollahzadeh, F., Wiktor, A., Nilghaz, A., Pratap-Singh, A., & Mousavi Khaneghah, A. (2021). Electro-based technologies in food drying: A comprehensive review. *LWT*, *145*, 111315. https://doi.org/10.1016/J.LWT.2021.111315

Murugan, P. C., & Saji Raveendran, P. (2021). Experimental studies on the application of biomass gasifier for drying tapioca in remote areas. *IOP Conference Series: Materials Science and Engineering*, *1084*(1), 012107. https://doi.org/10.1088/1757-899x/1084/1/012107

Mutlu, ÖÇ, Büchner, D., Theurich, S., & Zeng, T. (2021). Combined use of solar and biomass energy for sustainable and cost-effective low-temperature drying of food processing residues on industrial-scale. *Energies*, *14*(3), 561. https://doi.org/10.3390/en14030561

Nalawade, R., & Panwar, N. L. (2021). Experimental investigation on biomass fired dryer for drying of agricultural products. *International Journal of Ambient Energy*, *42*(15), 1765–1768. https://doi.org/10.1080/01430750.2019.1614990

Ndukwu, M. C., Simo-Tagne, M., Abam, F. I., Onwuka, O. S., Prince, S., & Bennamoun, L. (2020). Exergetic sustainability and economic analysis of hybrid solar–biomass dryer integrated with copper tubing as heat exchanger. *Heliyon*, *6*(2). https://doi.org/10.1016/j.heliyon.2020.e03401

Newaj, N., & Masud, M. H. (2014). Utilization of Waste Plastic to Save the Environment. International Conference on Mechanical, Industrial and Energy Engineering, 25–26 December 2014, Khulna, Bangladesh.

Okoroigwe, E. C., Ndu, E. C., & Okoroigwe, F. C. (2015). Comparative evaluation of the performance of an improved solar–biomass hybrid dryer. *Journal of Energy in Southern Africa*, *26*(4), 38–51. https://doi.org/10.17159/2413-3051/2016/V26I4A2092

Pardo, G., & Zufía, J. (2012). Life cycle assessment of food-preservation technologies. *Journal of Cleaner Production*, *28*, 198–207. https://doi.org/10.1016/J.JCLEPRO.2011.10.016

Paul, A., & Martynenko, A. (2021). Electrohydrodynamic drying: Effects on food quality. *Drying Technology*, *39*(11), 1745–1761. https://doi.org/10.1080/07373937.2021.1906694

Qu, H., Masud, M. H., Islam, M., Khan, M. I. H., Ananno, A. A., & Karim, A. (2021). Sustainable food drying technologies based on renewable energy sources. *Critical Reviews in Food Science and Nutrition*, *62*(25), 6872–6886. https://doi.org/10.1080/10408398.2021.1907529

Rathore, N. S., & Panwar, N. L. (2010). Experimental studies on hemi cylindrical walk-in type solar tunnel dryer for grape drying. *Applied Energy*, *87*(8), 2764–2767. https://doi.org/10.1016/J.APENERGY.2010.03.014

Sapmaz, S., & Kılıçaslan, İ (2023). Study on increasing the energy efficiency of a dryer used for the conversion of sewage sludge to biofuel. *Biomass Conversion and Biorefinery*, *13*(3), 2459–2468. https://doi.org/10.1007/s13399-022-02990-4

Shafizadeh, A., Shahbeik, H., Rafiee, S., Moradi, A., Shahbaz, M., Madadi, M., Li, C., Peng, W., Tabatabaei, M., & Aghbashlo, M. (2023). Machine learning-based characterization of hydrochar from biomass: Implications for sustainable energy and material production. *Fuel*, *347*. https://doi.org/10.1016/j.fuel.2023.128467

Singh, P. L. (2011). Silk cocoon drying in forced convection type solar dryer. *Applied Energy*, *88*(5), 1720–1726. https://doi.org/10.1016/J.APENERGY.2010.11.016

Sonthikun, S., Chairat, P., Fardsin, K., Kirirat, P., Kumar, A., & Tekasakul, P. (2016). Computational fluid dynamic analysis of innovative design of solar–biomass hybrid dryer: An experimental validation. *Renewable Energy*, *92*, 185–191. https://doi.org/10.1016/J.RENENE.2016.01.095

Sunday, I., Eloka-Eboka, A. C., Igbum, O. G., Eloka-Eboka, A. C., & Nwadinigwe, C. A. (2012). Performance evaluation of an efficient solar dryer with a backup incinerator for grated cassava under Makurdi humid climate. *International Journal of Environment and Bioenergy*, *1*(2), 119–130. www.ModernScientificPress.com/Journals/IJEE.aspx

Tchaya, G. B., & Houdji, E. T. (2021). Performance improvement of a hybrid biomass/indirect solar dryer (HBISD) for drying guava (*Psidium guajava*). *International Journal of Innovative Science & Technology*, *8*(8). https://www.researchgate.net/publication/357118659

Thanompongchart, P., Pintana, P., Phimphilai, K., & Tippayawong, N. (2017). Utilization of biomass energy in drying of glutinous rice crackers. *Energy Procedia*, *138*, 331–336. https://doi.org/10.1016/j.egypro.2017.10.131

Tonui, K. S., Mutai, E. B. K., Mutuli, D. A., Mbuge, D. O., & Too, K. V. (2014). Design and evaluation of solar grain dryer with a back-up heater. *Research Journal of Applied Sciences, Engineering and Technology*, *7*(15), 3036–3043. https://doi.org/10.19026/RJASET.7.639

Tun, M. M., Juchelkova, D., Win, M. M., Thu, A. M., & Puchor, T. (2019). Biomass energy: An overview of biomass sources, energy potential, and management in southeast Asian countries. *Resources*, *8*(2). https://doi.org/10.3390/resources8020081

Usmani, Z., Sharma, M., Tripathi, M., Nizami, A. S., Gong, L., Nguyen, Q. D., Reddy, M. S., Thakur, V. K., & Gupta, V. K. (2023). Converting biowaste streams into energy-leveraging microwave assisted valorization technologies for enhanced conversion. *Journal of the Energy Institute*, *107*. https://doi.org/10.1016/j.joei.2022.101161

Van't Land, C. M. (2012). Drying in the Process Industry. Wiley. https://doi.org/10.1002/9781118105818

Waewsak, J., & Punlek, C. (2006). A mathematical modeling study of hot air drying for some agricultural products. In *Thammasat International Journal of Science & Technology*, *11*(1), 14–20.

Wang, H., Zhang, S., Bi, X., & Clift, R. (2020). Greenhouse gas emission reduction potential and cost of bioenergy in British Columbia, Canada. *Energy Policy*, *138*, 111285. https://doi.org/10.1016/J.ENPOL.2020.111285

Yahya, M., Fudholi, A., & Sopian, K. (2017). Energy and exergy analyses of solar-assisted fluidized bed drying integrated with biomass furnace. *Renewable Energy*, *105*, 22–29. https://doi.org/10.1016/j.renene.2016.12.049

Yoon, P. R., Lee, S.-H., Choi, J.-Y., Yoo, S.-H., & Hur, S.-O. (2022). Analysis of climate change impact on resource intensity and carbon emission in protected farming systems using water–energy–food-carbon nexus. Resources, Conservation & Recycling, 184, 106394. https://doi.org/10.2139/SSRN.4054485

Yuwana, Y., Silvia, E., & Sidebang, B. (2020). Observed performances of the hybrid solar–biomass dryer for fish drying. *IOP Conference Series: Earth and Environmental Science*, *583*(1). https://doi.org/10.1088/1755-1315/583/1/012032

Zahoor, I., Mir, T. A., Ayoub, W. S., Farooq, S., & Ganaie, T. A. (2023). Recent applications of microwave technology as novel drying of food: Review. *Food and Humanity*, *1*, 92–103. https://doi.org/10.1016/j.foohum.2023.05.001

4 Waste Heat-Based Drying Systems

Ashok Prabhakar and Navneet Kaur
Department of Chemical Engineering, Chandigarh University, Mohali, Punjab, India

Deep Raj
Department of Environmental Science and Engineering, SRM University, Guntur, Andhra Pradesh, India

4.1 INTRODUCTION

Amid energy crisis, the proper utilization of heat energy with minimal heat loss is one of the great challenges. The researchers are contributing their knowledge to minimize the heat loss during the various processes occurring in industry or in day-to-day life. The optimization of heat recovery is a continuous ongoing process and it gets better with the new technological development. This recovery process and its proper utilization will really help to save energy and, consequently, it will benefit the living being and the environment. In a recent study in the European Union, it is estimated that 70% energy in the industry is used to produce heat for furnaces, reactor boilers, and dryers, but at the same time one-third of the energy is wasted (Tanasić et al., 2021). There are a few industrial sectors where the waste heat generation is largest and heat recovery can be done. These sectors are the food and tobacco (8.64%), pulp and paper (10.58%), metal (11.4%), chemical (11%) and nonmetallic minerals (11.4%) industries (Agathokleous et al., 2019). In this study, several aspects of heat generation, recovery, and utilization will be discussed. The heat generation that occurs during various processes may be direct combustion or the exothermic chemical reaction. The phrase waste heat is coined for the heat which is not being reused despite its potential energy content and is simply thrown to the nearby surroundings. The waste heat can be utilized in a variety of ways in industry and other places. This also impacts the cost of product-making process, which can be reduced up to a significant level by reusing this heat. Some examples of heat generation are boilers in power plant, kilns, ovens, and furnaces. The recovery of unused heat can minimize the consumption of the primary fuel during the entire process resulting in energy saving as well as economic benefits.

Waste heat generation can be classified as follows:

i. Waste heat in flue gas
ii. Unused heat in vapor steam
iii. Losses of heat during cooling of water
iv. Convective and radiant low-grade heat

There are various equipment which can be used to recover the waste heat from different sources. The post-recovery waste may be utilized in a number of different segments, ranging from typical industrial use to day-to-day life requirements. The present discussion is based on the utilization of waste heat in the food industry. Despite the malnutrition and starvation of the people around the world in the 21st century, the food wastage are reported to be 130 crore tons per year (Chen et al., 2017; Kummu et al., 2012). Moreover, the increment in the food waste causes damage to

DOI: 10.1201/9781003344087-4

the environment, and has negative impact on the economy and on social conditions (Joardder and Masud, 2019a). Several efforts are in line by the developed countries together with the developing nations to provide a sustainable solution to mitigate the food waste. The reduction in the rate of food waste can be done by the cumulative effort of developed and developing worlds and significant steps are being taken (Fujii and Kondo, 2018; Joardder and Masud, 2019b). The preservation of food materials post-harvesting is done by food drying technique and it is adopted by several industries and storage houses. The fresh raw foods obtained from farmland are high in water content and are more prone to be affected by the rapid growth of microorganism (Joardder et al., 2017). The heat and mass transport are the prime mechanism for removal of water from food and that is done by means of heat application (Joardder and Masud, 2019b; Joardder and Masud, 2019c). The removal of water molecules from food items via food drying process needs significant amount of heat energy, which may be equivalent to the latent heat of vaporization of water. Several drying techniques are being used across developing and developed countries and they use electrical and mechanical energy as input (Chua and Chou, 2003; Masud et al., 2020). The huge energy requirement in food drying process causes environmental pollution, emerging as a great concern for the civilization. The current drying technology in industry uses 20–25% of total energy requirement in food processing; approximately 15–90 MJ of energy is required for drying of 1 kg raw food material (Ananno et al., 2020; Joardder et al., 2017; Erbay and Hepbasli, 2017; Mohanraj, 2014).

The utilization of fossil fuel for energy generation in drying process creates huge problem to the environment due to release of waste heat in the atmosphere. It is of great concern and need to be addressed with alternative energy source (Arefin et al., 2017; Masud et al., 2017a, Masud et al., 2017b). Moreover, the newly developed more economical drying process would lead to less energy need along with its good impact on food readiness. The application of waste heat across the global industries are outlined in Table 4.1 (Karvonen et al., 2016; Ismail et al., 2012). The discussion of interest in this study is to analyze the waste heat generation source, its recovery using various techniques, and the last and most important the application of waste heat in the

TABLE 4.1
Temperature Distribution of Waste Heat Source

S. No.	Industry Type	Waste Heat Temperature (°C)
1. High temperature source	Iron casting and steel industry	1,450–1,550
	Nickel refining	1,370–1,650
	Copper refining furnace	760–820
	Hydrogen production industry	650–980
	Incinerator	650–1430
	Glass manufacturing unit	1,300–1,540
	Coke oven	650–1,000
2. Medium temperature source (230–650°C)	Exhaust from steam boiler	70–230
	Gas exhaust from turbine	370–540
	Heating furnace	430–650
	Baking ovens	220–549
	Cement-making units	450–620
3. Low <230°C	Condensate of steam	50–90
	Drying, baking, and curing ovens	90–230
	Hot processed liquids	30–230
	Cooling water	30–50

US Department of Energy (2008).

food industry. The phrase food industry includes all the relevant segments of food, it may range from food packaging, transportation, storage, and its distribution. Hence, it must be successfully applied to each and every segment so that it can be used to its fullest potential. The subsequent sections will deal with all the prospective of waste heat generation to its application in various segments, particularly to the food industry.

4.2 SOURCE OF WASTE HEAT PRODUCTION

There are various sources of waste heat generation, which must be identified and properly handled to minimize the energy loss by means of its maximum reutilization. We will discuss the various sources in the following sections.

4.2.1 Waste Heat from Industry

There are a number of industries which produce significant amount of waste heat in bulk quantity: food industry, paper and pulp, tobacco, basic metals, chemical industry, non-metallic minerals industry (Brückner et al., 2015). The range of waste heat from industry varies from low temperature to very high temperature depending upon the source of rejection. This heat can be utilized in various other operations based on the temperature scale. The waste heat having higher temperature has more potential to be reutilized in other operational work. Some of the waste heat sources based on the degree of temperature scales are provided in Table 4.1. The heat utilization in industrial sector is about 26%, which is the total energy consumption in the European Union (Eurostat, 2015). The more the waste heat generation, the more the heat utilization in the industry. The distribution of sector-wise waste heat generation is depicted in Figure 4.1. A study in the United States concluded that 20–30% of the energy in industry is lost in the form of heat energy and in the form of losses to run the pump, motors, and other electromechanically equipment (Papapetrou et al., 2018).

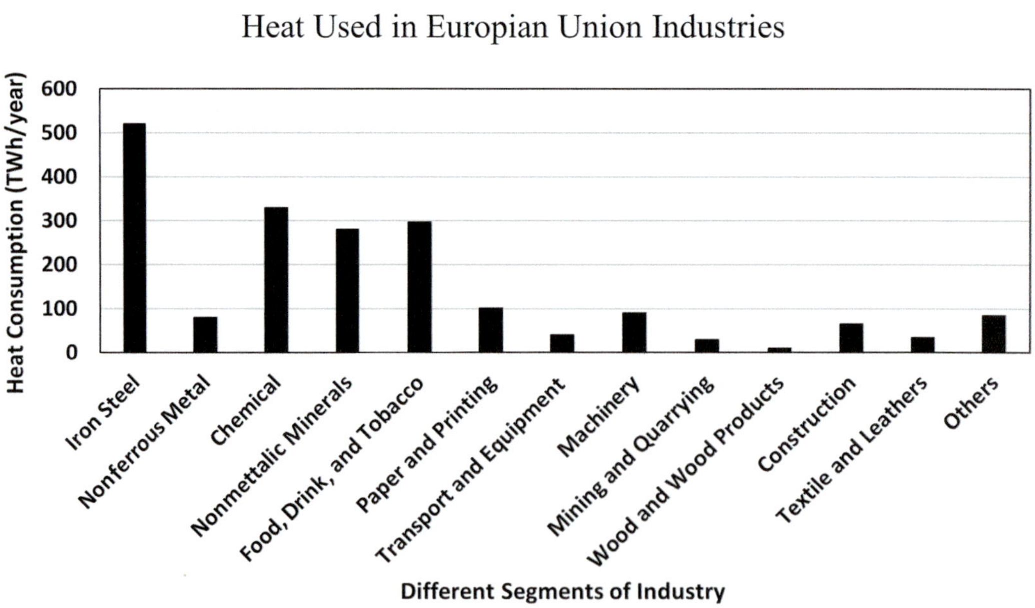

FIGURE 4.1 Heat consumption in European Union by industrial sector. (Eurostat, 2015.)

4.2.2 Waste Heat from Food Industry

According to estimates, the food industry uses roughly 26% of the energy used in the EU overall and ranks as the UK's fourth-largest industrial energy user (Carbon Trust, 2017a, Monforti-Ferrario et al., 2015). Low-medium temperature is the most common classification for waste heat generated in the food business (Thekdi and Nimbalkar, 2014). Despite this, the quantity of waste heat that is accessible in the food business greatly varies with industry type and depends mainly on the sort of procedure in issue. This is primarily because different sectors employ different production processes, which suggests that the only way to accurately estimate the quantity of beneficial waste heat is to undertake an extensive audit of the processes' energy consumption.

However, it is asserted that potential for waste heat recovery in the food business is high as per Feldman's study (Feldman, 1984). It is believed that energy waste ranges from 10% to 45%, depending on the operation. Potentially, heat from production processes, hot streams of air or water utilized in manufacturing, and heating and cooling systems are the main sources of waste heat (Featherstone, 2015). For instance, the source of waste heat in the red meat processing sector can be divided into by-product rendering, heat recovered from refrigeration unit, and meat processing (Pollock and Mason, 2015). For instance, the most energy-intensive process in an abattoir is the cooling of carcasses. On the other hand, rendering of by-products can be a significant energy user. Additionally, activities like scalding, singeing, and hair removal can take a significant amount of energy (Alfen, 2014). The cleaning process that requires a lot of hot water can also be considered a substantial energy consumer.

Major heat loss takes place in the atmosphere during hog singeing processes, where heat is required to dry off the hog corpses. Waste heat recovery can be heavily utilized in this process to increase production efficiency by providing the energy needed for the dehairing and scalding processes. The waste heat from sintering operations having temperature nearly 800°C may be useful for the boiler operation along with preheating the feedwater via economizer (Ashrafi et al., 2015, Environment Agency, 2009). On the other hand, it is also possible to extract heat from the hot water overflow in scalding processes by using automatically operated scalding chambers (Alfen, 2014; Environmental Protection Agency, 2000).

Since the manufacture of processed meat entails processes like cooking, cooling, smoking, etc., it requires more energy than the actual slaughter of animals. According to Fritzson and Berntsson (2006), refrigeration and product curing account for significant amount of heat loss in food industry. Waste heat can be recovered from condenser, wastewater, smoking vents, and cooking exhaust depending on the type of operation (Wang, 2008). Again, recovery from these sources needs to be evaluated based on specific circumstances because recovery from unused/waste heat sources, including waste heat from water source and cooker exhaust, may be challenging and uneconomical due to the grease and food waste items in the exhaust.

On the other hand, it has been demonstrated that in the poultry processing industry, where bird meat is prepared and processed, the processes for scalding, cooling, and freezing account for the greatest amount of energy use and energy loss.

According to Shupe and Whitehead (1979), heat recovery can be carried out, for example, when scalder and chiller overflow occurs, by gathering and transferring the energy back to the scalder or chiller systems. The boiler that is used to process wash water can be preheated using heat that can be recovered from refrigeration condenser systems (Mattsson and Sonesson, 2003). A de-superheater can also be used to recover heat from the temperature zone of 60–90°C in refringent condenser as the same setup is used in the middle of compressor and condenser to recover heat (Carbon Trust, 2017b).

The dairy plant rejected the waste heat coming out from the pasteurization process and refrigeration condensers (Niamsuwan et al., 2013, Sapali et al., 2014). Any waste heat source from dairy industry can be utilized to its optimum level for the spray drying process by heating the air supply (Atkins et al., 2011). Having said that, by using heat exchangers such as economizers or CO_2 heat pumps, chilled milk can be heated in the regenerator by recovering the heat from the pasteurization along with milk chilling operations (Singh and Dasgupta, 2017). The heat from refrigeration condensers is utilized to heat culture tanks for various activities, preheat boiler feedwater, and provide hot water for cleanup (Sapali et al., 2014; Singh and Dasgupta, 2017). Singh and Dasgupta (2017) have shown that the total fuel cost for production may be decreased by over 46% with a payback period of roughly 40 months when using a heat pump with an internal heat exchanger for combined heat recovery and hot water generation.

In order to heat culture tanks for various purposes, preheat boiler input water, and provide hot water for cleaning, refrigeration condensers are used (Sapali et al., 2014, Singh and Dasgupta, 2017). When employing a heat pump with an internal heat exchanger for combined heat recovery and hot water generation, Singh and Dasgupta (2017).

Similar to the dairy business, surplus heat from the pasteurization process can be recovered in the egg processing industry and utilized to warm the cold egg product through a regenerator. In order to preheat the boiler feedwater, heat the egg wash water and preheat the dryer air; waste heat from the refrigeration systems, hot waste water from washing eggs, and exhaust air from drying eggs can also be recovered (Jouhara et al., 2017b). Fruit and vegetable processing constitutes the bulk of the procedures involved in freezing and canning processes (Feldman, 1984). The refrigeration system condenser is where most of the waste heat is released during freezing operations. The waste heat is easily recovered and comes from heated refrigerant (Reindl and Jekel, 2007). On the other hand, it has been claimed that wastewater and retort vents are the main sources of waste heat in canning operations. Waste heat recovered from these fruit and vegetable processing processes can be used to heat water, wash cans, create blancher makeup water, clean up the factory, and heat boiler feedwater (Featherstone, 2015).

The main waste heat sources in bakeries and biscuit factories are flue gases from cooking ovens, fryers, pan washers, and boilers. The recovered waste heat can then be used to generate hot water for cleaning (Atkins et al., 2011; Sapali et al., 2014). For instance, Mumith et al. (2014) recovered low-grade waste heat from the exhaust gas of cooking ovens used to make biscuits using a thermoacoustic heat engine (TAHE). The engine of the technology, which comprises two heat exchangers and a stack of parallel plates housed in a cylindrical case and converts thermal energy to acoustic energy, works without any mechanical parts. According to the study, it is possible to produce 1,030 W of acoustic power with a thermal engine efficiency of 5.4% by recovering waste heat with a temperature of 150°C.

In a different study, Mukherjee et al. (2017) showed that the use of air-preheaters can reduce oven fuel use by 4%. In an experiment, the primary air supply of the burner was raised to 105°C, and heat was recovered from the exhaust flow of an industrial baking oven. According to this study, raising the primary air temperature can result in operating cost-savings of at least £4,200.

4.3 RECOVERY OF WASTE HEAT

In industry the heat generated must be recovered to minimize the costing of plant operation and to avoid the energy crisis and environmental safety. Hence, waste heat must be recovered by using utmost effort and technology to maximize the recovery and minimize the loss. A number of equipment are used for waste heat recovery. The only condition is the device must be selected as per their applicability to the relevant domain so that the heat recovery is optimized to its fullest capacity.

4.3.1 Advantage of Waste Heat Recovery

Waste heat recovery can benefit us directly or indirectly.

4.3.1.1 Direct Benefits

The direct benefits of the heat recovery is reflected in the efficiency of the process. It is also reflected in the utility consumption along with the costing.

4.3.1.2 Indirect Benefits

4.3.1.2.1 Effects on Pollution

When burned in incinerators, several toxic combustible wastes, including CO gas, sour gas, carbon black off-gases, oil sludge, acrylonitrile, and other plastic chemicals, are generated.

4.3.1.2.2 Effects on Equipment Size

Recovering waste heat lowers the amount of fuel needed, which lowers the amount of flue gas produced. This causes all flue gas handling equipment, including fans, stacks, ducts, and burners, to have smaller equipment sizes.

4.3.2 Equipment Is for Waste Heat Recovery

4.3.2.1 Recuperator

Through metallic or ceramic walls, a recuperator transfers heat from the air to the flue gases. The waste heat stream is contained on the other side of ducts or tubes that convey the air for combustion that has been preheated (Figure 4.2).

The metallic radiation recuperator is the most basic recuperator configuration in Figure 4.2.

The external annulus transports combustion air from the atmosphere to the air inlets of the furnace burners, while the inner tube transports hot exhaust gases. The incoming combustion air, which is now carrying more energy into the combustion chamber, cools the hot gases. Because this energy does not need to come from the fuel, less fuel is used for a given furnace loading.

Because burning less fuel also results in using less combustion air, stack losses are reduced by both reducing the stack gas temperatures and releasing fewer exhaust gas emissions. Because a sizable amount of heat transfer from the heated gases to the surface of the inner tube occurs by radiative heat transfer, the radiation recuperator is so named. However, only convection heat transfer occurs to the entering air because the cold air in the annuals is almost opaque to infrared radiation. The two gas flows in the diagram are typically parallel, but counter flowing the flows would make the setup simpler and increase the efficiency of heat transmission. The use of parallel flow is necessary because recuperators frequently have the additional benefit of cooling the exhaust gas removal duct, hence increasing the life of the duct. The tube type or convective recuperator is a

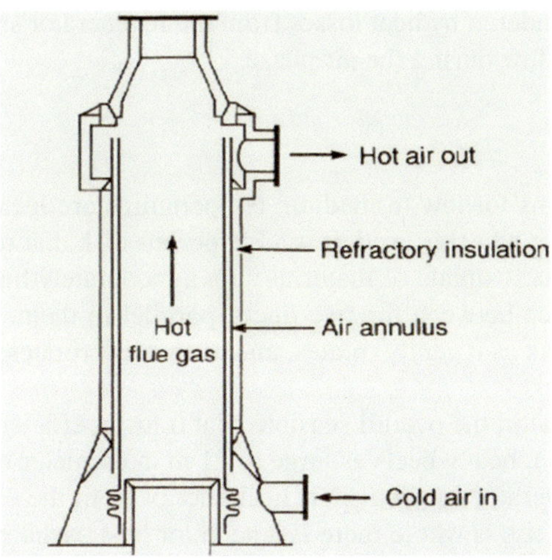

FIGURE 4.2 Recuperator.

second typical arrangement for recuperators. The heated air enters a shell that surrounds the hot tubes and passes over them once or more in a direction normal to their axes. The heated gases are conveyed through a number of parallel, small-diameter tubes. The heat exchanger is referred to as a two-pass recuperator if gases are allowed to pass over the tubes twice; a three-pass recuperator if two baffles are utilized; and so on. Baffling boosts the efficiency of heat exchange even if it raises the exchanger's price and the pressure drop in the combustion airstream. Due to the increased heat transfer surface made possible using several tubes and many passages of the gases, shell and tube type recuperators are typically more efficient and compact than radiation recuperators.

4.3.2.1.1 Radiation/Convective Hybrid Recuperator

Combinations of radiation and convective designs are employed for optimal heat transfer efficiency, with the high-temperature radiation recuperator being used first and then the convection type. Although more expensive than straightforward metallic radiation recuperators, these are more compact.

4.3.2.1.2 Ceramic Recuperator

The shortened life of the liner at intake temperatures above 1,100°C is the main factor limiting the heat recovery of metal recuperators. Ceramic tube recuperators, whose materials enable operation on the gas side to 1,550°C and on the preheated air side to 815°C on a more or less realistic basis, have been developed to overcome the temperature constraints of metal recuperators. Early ceramic recuperators were made of tile and bonded together with furnace cement, but thermal cycling quickly weakened the tubes and caused the joints to shatter. In later advances, several short silicon carbide tube types that can be connected by flexible seals found in the air headers were added. Leakage rates in earlier systems ranged from 8% to 60%. According to reports, the new designs have substantially reduced leakage rates and can withstand air preheat temperatures of up to 700°C for two years.

4.3.2.2 Regenerator

In glass and steel melting furnaces, regeneration, which is preferred for large capacities, has been employed extensively. The regenerator size is significantly correlated with the thickness and conductivity of the bricks along with the heat storage ratio.

In the regenerator, the intervals in reversal play a significant role. More heat storage is proportional to the longer duration and it will ultimately lead to increased cost. Additionally, prolonged periods of reverse lower the preheats average temperature, which lowers fuel efficiency (Figure 4.3).

As a furnace ages, dust and sludge buildup on the surfaces decreases the efficiency of heat transfer. Heat transfer is also hindered by heat losses from the regenerator's walls, air infiltration during the gas period, and air outflow during the air phase.

4.3.3 Heat Wheels

Waste heat recovery systems for low to medium temperatures are increasingly using heat wheels. Figure 4.4 shows how a heat wheel is used. It is a big porous disk that rotates between two adjacent ducts—cold and hot gas duct—made of material with a moderately high heat capacity. The disk's axis is situated on the space between the two ducts, parallel to them. Sensible heat is transferred along with hot air to the disk as it gently rotates, and as the disk rotates, sensible heat is transmitted from the disk to the cold air.

For this type of regenerator, the overall sensible heat transfer efficiency can reach 85%. With air capacity up to 1,130 m^3/min, heat wheels as large as 21 m in diameter have been constructed.

A rotary gas heat regenerator can transport heat energy from the exhaust point to the entering point of gases. Its primary use is where there is a need for heat exchange between huge air masses with small temperature differences. Common applications include heating and ventilation systems and heat recovery from dryer exhaust air.

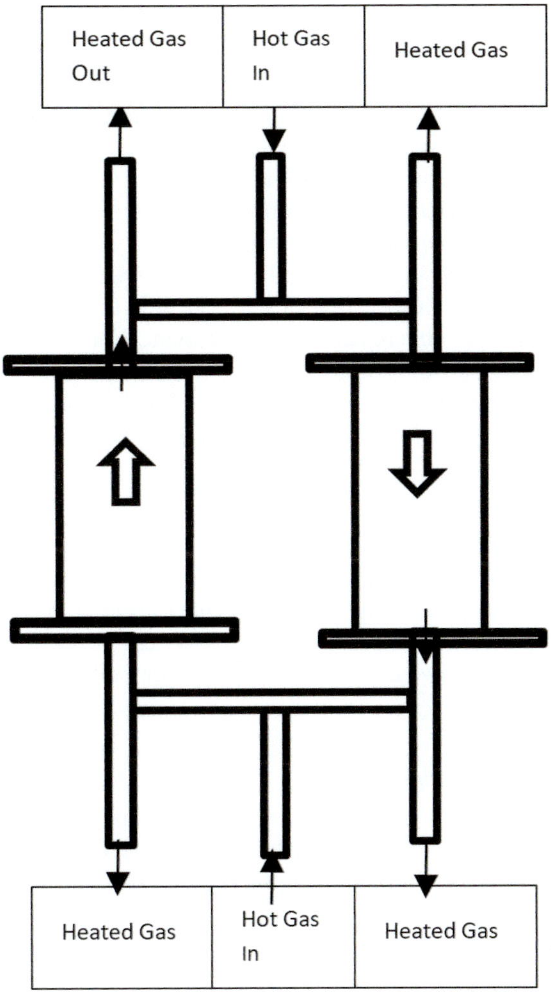

FIGURE 4.3 Fixed bed regenerator. (Adapted from https://www.thermopedia.com/content/1087/.)

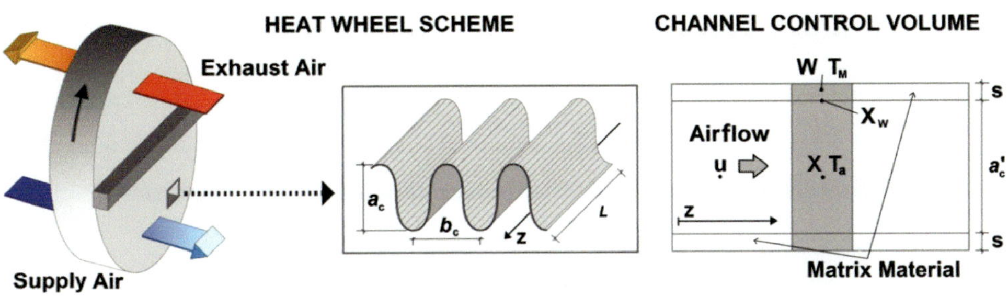

FIGURE 4.4 Heat wheel.

4.3.3.1 Heat Pipe

As compared to copper, the most common conductor, a heat pipe can transfer up to 100 times more thermal energy. In other words, a heat pipe is a device for absorbing and transporting thermal energy. Because there are no moving parts, maintenance is minimal. Three components make up the heat pipe—a capillary wick structure, a sealed container, and a working fluid. The internal surface of the container tube has the capillary wick structure integrally built into it, which is then vacuum-sealed. As the container tube is vacuum-sealed, thermal energy transferred to the heat pipe's exterior surface is in balance with its own vapor.

The working fluid close to the surface instantly evaporates when thermal energy is delivered to the heat pipe's exterior surface. This portion of the heat pipe transforms into an evaporator zone as the resulting vapor absorbs the latent heat of vaporization. When the thermal energy is removed at the other end of the pipe, the vapor is forced to condense back into liquid, releasing the latent heat of condensation. The condenser area functions as this section of the heat pipe. The liquid then returns to the area where it had previously evaporated. Figure 4.5 depicts a heat pipe.

4.3.3.1.1 Features of Heat Pipe

Heat pipe heat exchangers (HPHEs) are portable, small systems for recovering heat. Since there are no moving parts to break down, it essentially requires no mechanical maintenance. It doesn't require input power to operate, and it doesn't require lubrication or cooling systems. Additionally, it reduces the need for fan horsepower and improves the system's overall thermal performance. The heat pipe heat recovery systems have a 315°C operational temperature range and an efficiency range of 60–80%.

4.3.3.1.2 Application of Heat Pipe

4.3.3.1.2.1 Industrial Process to Space Heating The thermal energy from process exhaust is transferred through the HPHE for use in heating buildings. If necessary, the warm air can be mixed. There is a significant reduction or elimination of the need for extra heating equipment to deliver hot makeup air.

4.3.3.1.3 One Process to Other Process

The waste thermal energy from the process exhaust is recovered and transferred to the entering process air via the HPHEs. As a result, the entering air warms up, can be used for the same operation or other processes, and uses less energy overall.

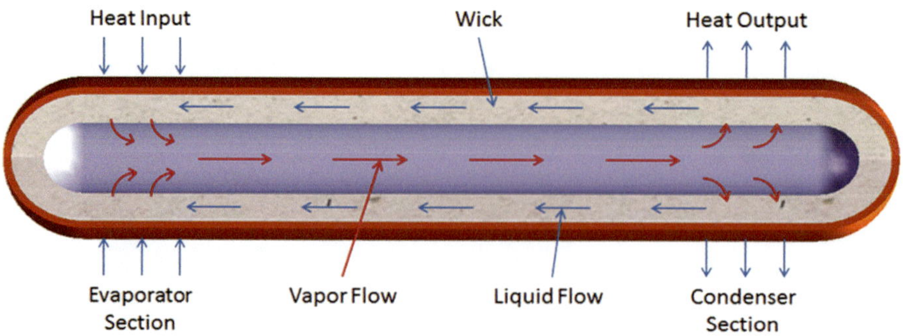

FIGURE 4.5 Heat pipe.

4.3.3.1.4 HVAC Applications

4.3.3.1.4.1 Cooling Aside from the operating savings of the cooling system, HPHEs precool the building's makeup air in the summer and so lower the total tons of refrigeration. The hot supply makeup air receives the thermal energy supply recovered from the cold exhaust.

4.3.3.1.4.2 Heating During the winter, the procedure is reversed to preheat the makeup air.

4.3.3.2 Economizers

The recovery of low to medium waste heat is done with the help of economizer or heat exchanger and transferred to liquids. The coated metallic fins over tubes are used to optimize the heat transport rate and available area for the transportation of heat and its absorption (Thermtech, 2014).

The liquid is passed through the circular tube and the mounted fins over the tube absorb heat. The thermal efficiency is then maximized and improved by feeding the heated liquid back into the system (Sen, 2006).

According to the research by Spirax Sarco (2011), using an economizer for a boiler system can lead to boosting efficiency by 1% for every 5°C decrease in flue gas temperature. This suggests that the system's fuel usage can be decreased by 5–10% with a payback period of less than two years (Advanced Manufacturing Office, 2012). The system efficiency increased by preheating of entering fluid (i.e., demineralized feedwater in boiler) by utilizing the waste heat through the economizer. This leads to the less energy requirement to reach the boiling point. Moreover, it is reported by a researcher that a single degree reduction in flue gas temperature will lead to the reduction of 7% of fuel consumption irrespective of system design (Maxxtec GmbH, 2014a).

There are multiple different types of economizers available for various applications, but it has been found that they all perform similarly (Electrical4u, 2017). Finned tubes, coiled tubes, noncondensing economizers, and condensing economizers are examples of these designs. The condensing and noncondensing types are typically used plants and big processing units for waste heat recovery from discarded flue gas, while the other types are mostly employed to improve boiler efficiency. Considering this, Vandagriff (2001) investigates the availability of deep economizers, used to recover heat from the low-temperature zone and are constructed of cutting-edge materials which withstand the deposition of acidic condensate on the heat exchanger's surface. On the other hand, glass-tubed economizers are utilized for low- to medium-temperature applications and for gas-to-gas heat recovery (Dabhoya, 2015).

4.3.3.3 Waste Heat Boilers

Waste heat boiler consists of parallel water tubes in multiple numbers in the vicinity of waste heat released from the boilers. The absorbed waste heat is used to produce steam and that too from medium- to high-temperature exhaust gases. The steam can be further used as an additional advantage for the production of electricity from steam-based power plant (Worrell et al., 2008).

For instance, J+G (2017) noted that the heat produced during the coal combustion reaches the temperature of 1,000°C and is used to produce steam. But in the end, post-steam generation and running of turbine cause plenty of heat energy left with steam which can be recovered by using waste heat boiler to produce steam and run boilers to run turbines and generators. The waste heat temperature has a significant impact on both the pressure and the pace of steam production. An additional burner unit can be added to make up the deficient heat to the process of steam production by using waste heat (Ganapathy, 2015).

According to Turner (2009), waste heat boilers can also be connected to other waste heat recovery devices like afterburners, preheaters, and finned-tubed evaporators to increase efficiency by warming the feedwater and, if necessary, producing superheated steam (Figure 4.6).

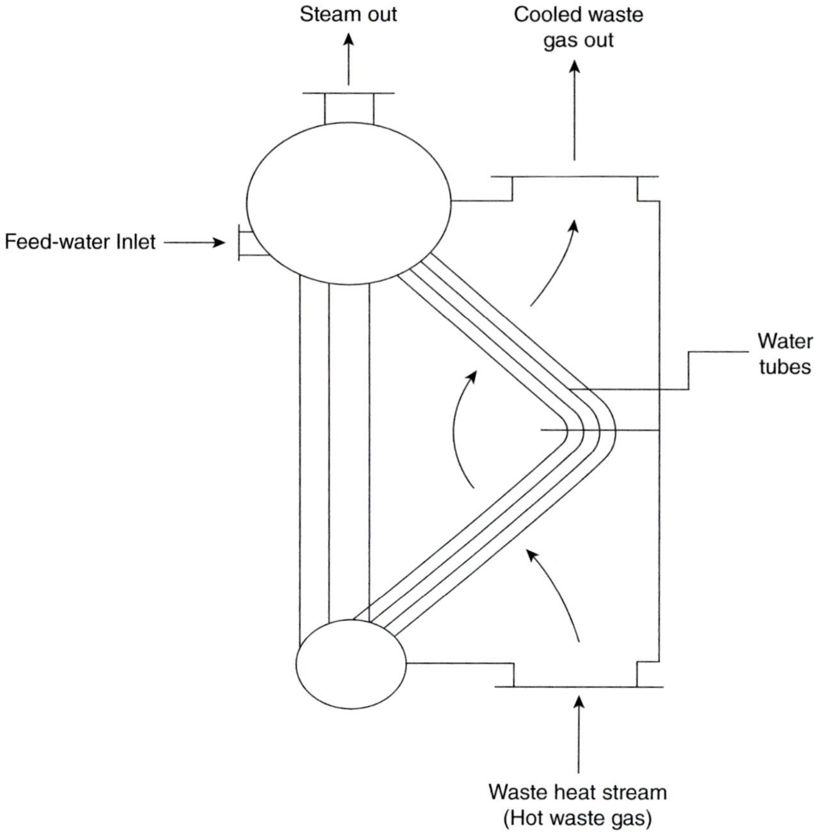

FIGURE 4.6 Waste heat boiler.

4.3.3.4 Air Preheaters

The air preheater is used at low to medium heat content in any waste stream (Figure 4.7). This application must be avoided when the waste stream consists of significant presence of contaminations. Many such examples are waste heat from furnaces, ovens, and steam boilers along with exhaust from gas turbines (Maxxtec GmbH, 2014b).

The design of preheater may be of several types, including plate and heat pipe type. The incoming cold air encountered with the bunch of plates positioned perpendicular to the entrance. The hot air passes through the wise space between plates resulting in the plate heating and these plates ultimately used to heat the cold air. The heat pipe, on the other hand, is made up of a collection of numerous parallel arrangements of pipes in a suitable vessel. The hot and cold stream have the separate provision to enter inside and exit out from the container.

The container pipe having suitable fluid get evaporated after receiving heat energy from waste gases at one end and after heat up the fluid reaches to other end to heat up the cold passing air (Yodrak et al., 2010). The rise in temperature of the cold air occurs due to its flow over the heated section (due to heat absorption) of the pipe, and this cycle is repeated (Yodrak et al., 2010).

According to Nicholson (1983), there are primarily three types of air preheaters that are used frequently and are categorized as regenerators: rotary regenerators, run-around coil (RAC), and recuperative. Although they all use the same basic idea as air preheaters, these technologies come in a variety of designs and serve a variety of functions.

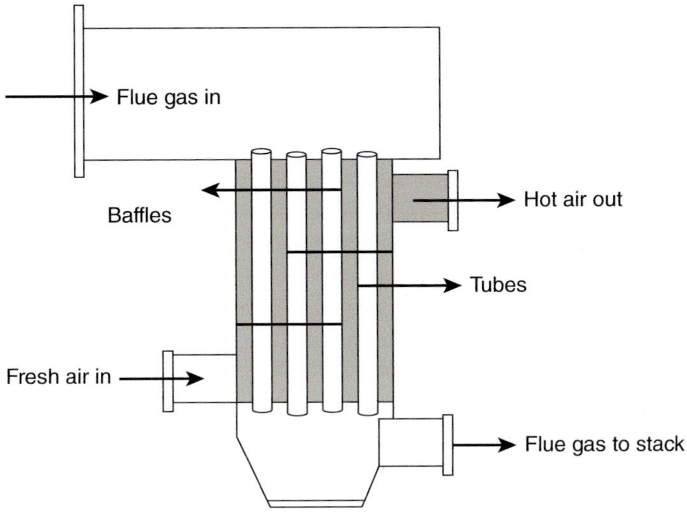

FIGURE 4.7 Air preheater. (*Source:* http://www.egcfe.ewg.apec.org/Documents/Costs&EffectivenessofUpgradingOlderCoal-FiredPowerPlantsFina.pdf)

4.3.3.5 Run-around Coil

This system, as shown in Figure 4.8 and as described by Toolan (2012), is composed of two coil-style heat exchangers connected to one another by RAC, filled with either water, glycol, or a combination of the two (Wallin et al., 2010). The waste heat from exhaust gases are recovered and transferred to the liquid in the coil, which is then combined with supply air in the secondary recuperator. The liquid round coil system, which is connected to the two airstreams by pumped pipework, allows heat exchange between them. RAC shows less effectiveness as compared to the direct recuperator and it needs a pump to function which leads to more energy input and maintenance (HK EE Net, 2017). The more effective and efficient utilization of this technique can be availed by using an additional heat source (Carbon Trust, 2011).

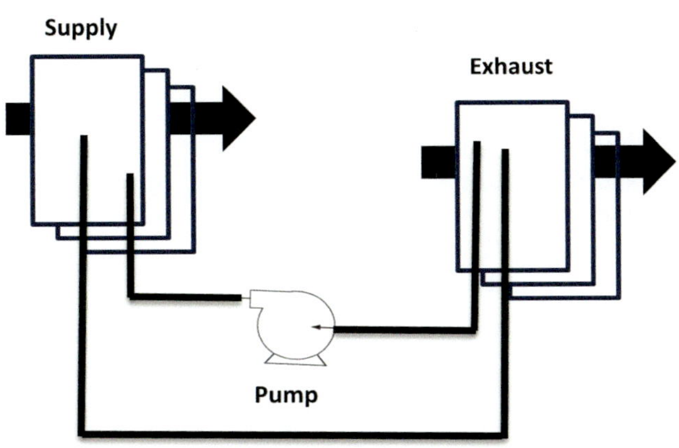

FIGURE 4.8 Schematic of run-around coil system.

4.3.3.6 Plate Heat Exchanger

To address the issue of cross-contamination, plate heat exchangers are used for the transportation of heat from one heat source to another fluid at low temperature. The parallel arrangement of the metallic plates stacked together to form a hollow metallic shell is named as plate heat exchanger.

In the fluid flow measurement and turbulence creation for heat transfer improvement, every single plate consists of pressed patterns encircled by gaskets (Sreejith et al., 2014). One type of fluid passes through one gap created by the gaskets while the other fluid is forced through the gap next to it (Sondex, 2012). The flowing across the front and back of the plates, the hot and cold fluids move through each segment of the heat exchanger, exchanging heat while avoiding cross-contamination. The plate heat exchanger provides the higher surface exposure per unit volume which results in increased heat transfer coefficient and this gives an extra edge over other heat exchangers, i.e., shell and tube and others (Aghayari et al., 2015).

According to reports, three different types of plate heat exchangers are available (Wang et al., 2007, Srihari and Das, 2008). The plates of plate heat exchangers can be joined via welding, brazing, or gasketing (Anusha and Kishore, 2017; Kelvion, 2017). A gasket, often constructed of a polymer substance, is positioned between the plates of a gasket plate heat exchanger and serves as a seal and separator around the edges of the plates (Laval Alfa, 2017). Thicker pressure plates and tightening bolts help position the plates in the frame permanently. Because of this, the heat exchanger can be disassembled for cleaning or modified to have a variable capacity by plates variation (Fernandes et al., 2008). The use of gaskets in this design has the benefit of resistance to heat and sudden pressure variation, while allowing the plate pack to be flexible. This is appropriate for applications that undergo continuous thermal cycling because of exposure to temperature fluctuations (Spirax Sarco, 2017). Nevertheless, the cycle's working temperature and pressure limit the usage of gaskets, which is a drawback (Webbusterz Engineering, 2015).

However, the heat exchanger having gasket plate shows efficient heat recovery rate of approximately 90% (Javanjal and Parande, 2017). The plates of brazed plate heat exchangers are brazed together using copper/nickel metal in vacuum furnace (Thulukkanam, 2013). The design of brazed plate heat exchanger offers high resistance towards high pressure and high temperature with respect to gasket exchanger and also this is inexpensive to maintain (Brazetek, 2012). Its brazing quality leads to problems if cleaning or size adjustments are necessary.

The brazed plate heat exchanger is more prone to thermal stress as compared to the gasket-type exchangers and this leads to the failure of structures when an abrupt change in temperature takes place. As per the report from IITD, the brazed exchangers are more suitable for slow heating and gradual temperature variation such as those involving thermal oils (IITD, 2012).

The welded plate exchangers are more suitable as compared to brazed heat exchangers as these are more sustainable to temperature cycling and pressure variation (Spirax Sarco, 2017). This benefit is attained by using laser welding methods, which form welded seams to hold the plate pack together (Reppich, 1999). This kind of heat exchangers has been demonstrated to have higher operating temperature and pressure limits, making them appropriate for heavy-duty applications (Valutech, 2016). Nevertheless, they are unable to be disassembled or altered in size, just as brazed heat exchangers. In an experiment, Cipollone et al. (2015) showed that a plate heat exchanger can be utilized as a component to superheat the working fluid of a system, which can enhance the evaporator duty when used for heat recovery.

4.3.3.7 Heat Pipe System

A heat pipe is used to relocate the heat by means of the condensation and vaporization mechanism of a fluid within it. A heat pipe consists of sealed container along with a wick structure used for heat transport using fluids, such as sodium, water, acetone, methanol, or ammonia, which is in equilibrium with its own vapor (Faghri, 2014). The evaporator section, the adiabatic transport segment, and the condenser section are the three distinct parts of a heat pipe.

The working fluid inside the heat pipe receives heat at hot end and is converted to vapor, which leads to increased vapor pressure and due to this it is traveled from hot end to cold end where it leaves the latent heat and condenses itself to liquid form and this is how heat is transported from hot end to cold end. To repeat the cycle, the menisci in the wick generates capillary pressure which helps to transport back the working fluid from the cold end to the hot end of the pipe and the process continues (Jouhara et al., 2017a). The thermal conductivity of heat pipe is very high (from 5,000 W/m·K to 200,000 W/m·K) as compared to the solid conductors (from 250 W/m·K to 1,500 W/m·K) (AAVID, 2017). The material of construction of heat pipes are Al, Cu, Monel, stainless steel, Inconel, and W. Moreover, the requirement of temperature and the working fluid material suitability play an important role in the material construction of heat pipes (ACT, 2017a, 2017b).

As already indicated, the working liquid is transported from the condenser back to the evaporator with the help of the heat pipe wick structure. The heat pipe wick structure is built using a variety of materials and methods according to PSC (2006). The heat pipes known as thermosyphons are available; they operate only by gravity and lack a wick structure. These heat pipes must be installed vertically because they cannot be used in a horizontal position. Having said that, heat pipes with a wick construction do not have this restriction and can function in both horizontal and vertical orientations.

Wick structures are often created by expanding copper or stainless-steel wicks against the pipe wall (Figure 4.9). This structure allows for the creation of heat pipes that can move the working fluid vertically, against gravity, and at a very modest inclination to the horizontal (PSC, 2006). Contrarily, grooved wick structures are constructed of elevated dents that are threaded or extruded, typically from copper or aluminum materials, perpendicular to the pipe surface. Similar to screen wick structures, heat pipes created with this kind of structure may convey liquid at a modest incline from the horizontal and operate in gravity-aided and horizontal orientations (CAPRI, 2017).

The temperature range of the application for which the heat pipe is being utilized heavily influences the kind of working fluid used in the heat pipe. For instance, ammonia, acetone, "Freon" refrigerants, and water are utilized for low-temperature applications in the temperature range of 200–550 K, according to Faghri (2014) and ACT (2017). The majority of heat pipe applications are said to be in this temperature range, and water is the most frequently employed fluid since it is affordable, has good thermophysical characteristics, and is also secure to handle (Faghri, 1995). In general, heat pipes have higher thermal conductivities than other types of heat exchangers, which results in a small temperature drop when transferring heat over long distances. They also have longer lifespans that do not require maintenance because they use passive operation and do not have any moving parts that could break down.

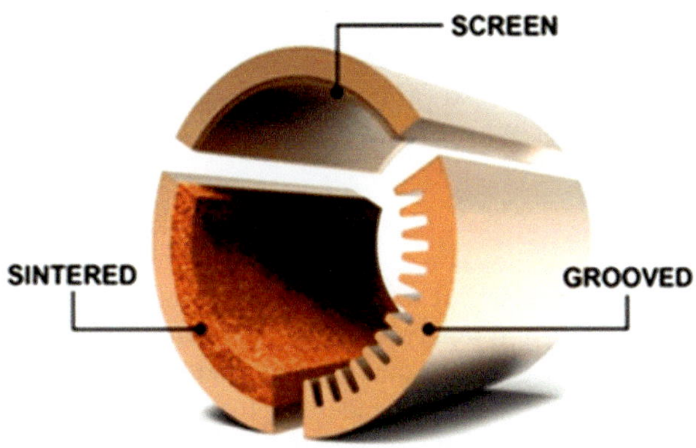

FIGURE 4.9 Common wick types of heat pipe.

4.4 UTILIZATION OF WASTE HEAT

The utilization of waste heat can be done by two different pathways: its internal and external uses.

The internal uses of the waste heat can be implemented in three different steps:

i. The very first step of heat loss recovery is to minimize the heat loss by means of good thermal insulation along with the optimization of the entire process to identify and maximize the process efficiency which ultimately helps to reduce the heat loss and will provide the economic benefit to the organization.
ii. After implementation of first step, if any waste heat is still there, then this can be utilized by reintegrating the process by using several heat recovery equipment.
iii. The third step is to utilize these waste heat to generate another form of energy like electricity by means of organic Rankine cycle (ORC) or in thermal cooling.

The spatial proximity of the waste heat source and demand is currently required for the most economically viable utilization of waste heat. Electricity that can be exported to the utility company or is portable seems to be the most adaptable kind of energy conversion. However, insufficient technological efficiency in the production of energy from waste heat remains a challenge that requires more attention.

4.4.1 Application of Waste Heat in Food Industry

Drying is one of the most widely used techniques in the food processing industry. The waste heat recovered from other processes can be used to dry the food stuffs. The dryer may be of various types, the details are discussed in the subsequent sections.

4.4.1.1 Fluidized Bed Dryer

This is the low-cost drying system available at ease. Its wide application range from food industry to agriculture and pharma industry (Mujumdar and Devahastin, 2000). This drying system is reported to be of high drying rate, with high thermal efficiency and ease of control. A simple schematic is shown in Figure 4.10.

4.4.1.2 Infrared Drying

Using energy from infrared light, infrared drying heats the polymer granule or flake's bulk material directly. There is no other transfer medium used; the supplied energy is applied straight to the granule. The granule's bulk material and any inside moisture are heated by the supplied energy, which also induces molecular oscillation and internal heating. The granule is surrounded by a stream of colder ambient air, and the heat from within pushes the moisture out into the stream of cooler air, eliminating it from the process.

The temperature differential between the vaporized water and the surrounding airstream drives infrared drying because it produces a partial pressure gradient from the granule's inside to its outside, which is a powerful driver of moisture.

4.5 CONCLUSION

Industrial heat loss is known as industrial waste heat. In the industrial sector, waste heat recovery refers to techniques for gathering and repurposing the heat that is wasted during production. It can subsequently be utilized to produce energy that is useful and lower energy usage overall. Heat loss is primarily divided into three temperature grades: high, medium, and low. Waste heat recovery systems are introduced in accordance with each range of waste heat. The choice of heat recovery strategies and tactics is primarily influenced by important variables like the kind, quantity, and

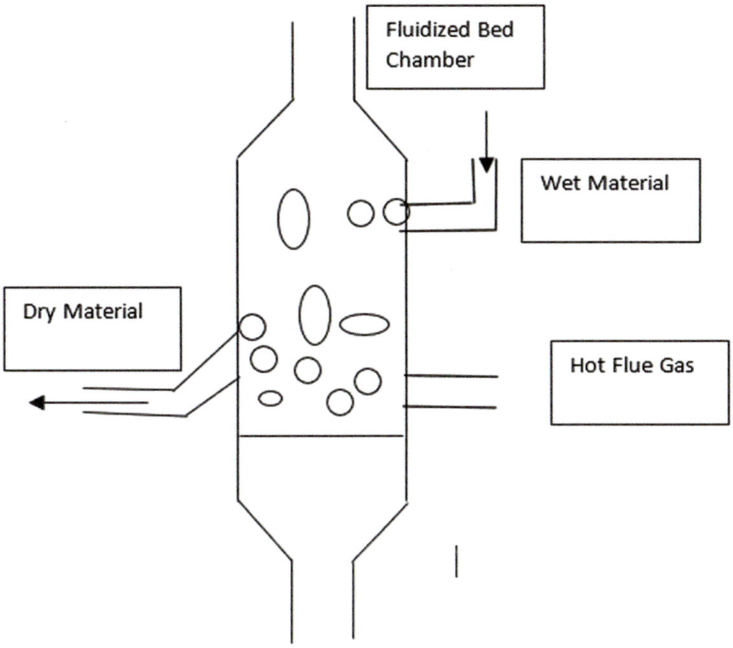

FIGURE 4.10 Fluidized bed dryer.

suitability of the heat source. A waste heat recovery unit, which is essentially an energy recovery heat exchanger, is the primary component of the various heat recovery technologies that have been found to be available for capturing waste heat. These units, which mostly consist of standard waste heat recovery systems, function according to the same principle of capturing, recovering, and exchanging heat that has the potential to contain energy during a process.

We must acknowledge that a significant quantity of waste heat from human activity has entered and is still entering the climate system, contributing to the current global warming. Approximately 80% of the energy consumed worldwide is waste heat, of which 44% is contributed by industry, 36% by business and residential buildings, and 20% by transportation. Everyone unintentionally and consciously contributes to it. The current global warming is not the fault of greenhouse gases.

The three components of the climate system—the air, land, and oceans—can have their past temperature changes simulated and future warmings reliably and with high certainty by using the Equivalent Climate Change Model and allocating the waste heat stream to each based on its specific heat capacity. The present analysis of waste heat recovery procedures will help to minimize the energy loss and ultimately the entire environment will get benefited.

REFERENCES

AAVID, "Heat Pipe Technology: Passive Heat Transfer for Greater Efficiency," 2017. Available at: https://www.thermacore.com/thermal-basics/heat-pipe-technology.aspx (accessed November 22, 2017).

ACT, "Compatible Fluids and Materials," Lancaster, PA: Advanced Cooling Technologies, Inc., 2017a.

ACT, "Everything You Need to Know about Heat Pipes," Lancaster, PA: Advanced Cooling Technologies, Inc., 2017b.

ACT, "Operating Temperature Range," Lancaster, PA: Advanced Cooling Technologies, Inc., 2017c.

Advanced Manufacturing Office, "Use Feedwater Economizers for Waste Heat Recovery," Washington, DC: Energy Efficiency and Renewable Energy, 2012.

Agathokleous R. et al., 2019. "Waste heat recovery in the EU industry and proposed new technologies," *Energy Procedia*, vol. 161, pp. 489–496.

Aghayari R., Maddah H., Ashori F., Hakiminejad A. and Aghili M., 2015. "Effect of nano-particles on heat transfer in mini double-pipe heat exchangers in turbulent flow," *Heat Mass Transfer*, vol. 51, pp. 301–306.

Alfen N. K. V., 2014. Encyclopedia of Agriculture and Food Systems, 1st ed., Davis: Elsevier Academic Press.

Ananno A. A., Masud M. H., Dabnichki P. and Ahmed A. et al., 2020. "Design and numerical analysis of a hybrid geothermal PCM flat plate solar collector dryer for developing countries," *Solar Energy*, vol. 196, pp. 270–286. https://doi.org/10.1016/j.solener.2019.11.069. In this issue. https://www.sciencedirect.com/science/article/abs/pii/S0038092X19311697 https://www.sciencedirect.com/science/article/abs/pii/S0038092X19311697

Anusha G. and Kishore P., 2017. "Heat transfer analysis of gasketed plate heat exchanger," *International Journal of Engineering Research & Technology*, vol. 5, no. 12, pp. 943–947.

Arefin A. M. E., Masud M. H., Joardder M. U. H., Mourshed M. and Naim-Ul-Hasan F. R., 2017. "Waste heat recovery systems for internal combustion engines: A review." In: 1st International Conference on Mechanical Engineering and Applied Science, Military Institute of Science and Technology, pp. 1–4.

Ashrafi O., Bédard S., Bakhtiari B. and Poulin B., 2015. "Heat recovery and heat pumping opportunities in a slaughterhouse," *Energy*, vol. 89, pp. 1–13.

Atkins M. J., Walmsley M. R. and Neale J. R., 2011. "Integrating heat recovery from milk powder spray dryer exhausts in the dairy industry," *Applied Thermal Engineering*, vol. 31, no. 13, pp. 2101–2106.

ATS, How Wicks and Orientation Affect Heat Pipe Performance, Norwood, MA: Advanced Thermal Solutions, Inc., 2009.

Brazetek, "Everything About Heat Exchanger," 2012. Available at: http://www.brazetek.com/blog/advantages-of-brazed-plate-heat-exchangers/ (accessed October 11, 2017).

Brückner S., Liu S., Miró L., Radspieler M., Cabeza L. F. and Lävemann E., 2015. "Industrial waste heat recovery technologies: An economic analysis of heat transformation technologies," *Applied Energy*, vol. 151, pp. 157–167. ISSN 0306-2619.

CAPRI, Metal Mesh Structured Heat Pipes, Hyderabad, India: Capri Cables Pvt. Ltd., 2017.

Carbon Trust, "Heat Recovery," London, UK: Carbon Trust Ltd., 2011.

Carbon Trust, "Food and Drink: Energy Efficiency and Carbon Saving Advice for the Food and Drink Sector," London, UK: Carbon Trust, 2017a.

Carbon Trust, "How to Implement Heat Recovery in Refrigeration," London, UK: Carbon Trust, 2017b.

Celsia, "Heat Pipe & Vapor Chamber Technology Overview," 2017. Available at: http://celsiainc.com/heat-pipe-and-vapor-chamber-technology-over-view/ (accessed November 4, 2017).

Chen J., Lau S. K., Chen L., Wang S. and Subbiah J. et al., 2017. "Modeling radio frequency heating of food moving on a conveyor belt," *Food and Bioprocess Technology*, vol. 102, pp. 307–319. https://doi.org/10.1016/j.fbp.2017.01.009

Chua K. J. and Chou S. K., 2003. "Low-cost drying methods for developing countries," *Trends in Food Science and Technology*, vol. 14, pp. 519–528. https://doi.org/10.1016/j.tifs.2003.07.003

Cipollone R., Bianchi G., Battista D. D. and Fatigati F., 2015. "Experimental and numerical analyses on a plate heat exchanger with phase change for waste heat recovery at off-design conditions," *Journal of Physics: Conference Series*, vol. 655, 012038.

Dabhoya S., "How to Reduce Corrosion in Air Preheater or Economiser," Gujarat, India: Altret Performance Chemicals Gujarat Pvt. Ltd., 2015.

Electrical4u, "Economiser in Thermal Power Plant: Economiser," 2017. https://www.electrical4u.com/economiser-in-thermal-power-plant-economiser/ (accessed October 23, 2017).

Environment Agency, The Red Meat Processing (Cattle, Sheep and Pigs) Sector (EPR 6.12), Bristol: Environment Agency, 2009.

Environmental Protection Agency, Cleaner Production Assessment in Meat Processing, 1st ed., United Nations Environment Programme, 2000.

Erbay Z. and Hepbasli A., 2017. "Exergoeconomic evaluation of a ground-source heat pump food dryer at varying dead state temperatures," *Journal of Cleaner Production*, vol. 142, pp. 1425–1435.

Eurostat, "Consumption of Energy," 2015. http://ec.europa.eu/eurostat/statisticsexplained/index.php/Consumption_of_energy

Faghri A., 1995. Heat Pipe Science and Technology, 1st ed., Washington, DC: Taylor & Francis.

Faghri A., 2014. "Heat pipes: Review, opportunities and challenges," *Frontiers in Heat Pipes*, vol. 5, no. 1.

Featherstone S., 2015. A Complete Course in Canning and Related Processes, 14th ed., Oxford: Elsevier.

Feldman M., "Heat Recovery for Canadian Food and Beverage Industries," Ottawa, Canada: Ministry of Supply and Services, 1984.

Fernandes C. S., Dias R. P. and Maia J. M., 2008. "New plates for different types of plate heat exchangers," *Recent Patents on Mechanical Engineering*, vol. 1, pp. 198–205.

Fritzson A. and Berntsson T., 2006. "Energy efficiency in the slaughter and meat processing industry: Opportunities for improvements in future energy markets," *Journal of Food Engineering*, vol. 77, pp. 792–802.
Fujii H. and Kondo Y., 2018. "Decomposition analysis of food waste management with explicit consideration of priority of alternative management options and its application to the Japanese food industry from 2008 to 2015," *Journal of Cleaner Production*, vol. 188, pp. 568–574.
Ganapathy V., 2015. Steam Generators and Waste Heat Boilers for Process and Plant Engineers, 1st ed., Boca Raton, FL: CRC Press, Taylor & Francis Group.
Gernert N. J., "Heat Pipes Get the Heat Out of Multikilowatt Electronic Devices," Lancaster, PA: AAVID Thermacore Inc., 2010.
HK EE Net, "Air-Side System/Equipment," Energy Efficient Technologies, 2017. Available at: http://ee.emsd.gov.hk/english/air/air_technology/air_tech_air.html (accessed October 15, 2017).
IITD, Classification of Heat Exchangers, Delhi: Indian Institute of Technology, 2012.
Ismail Y., Durrieu D., Menegazzi P., Chesse P. and Chalet D., 2012. Potential of Exhaust Heat Recovery by Turbocompounding. SAE Technical Paper.
J + G, "Waste Heat Boilers: New Energy from Old Energy," Jünger+Gräter GmbH, Schwetzingen, 2017.
Javanjal J. K. and Parande M., 2017. "Experimental studies on heat transfer using plate heat exchanger," *Journal of Chemical, Biological and Physical Sciences*, vol. 3, no. 4, pp. 2806–2817.
JHCSS, "Heat Pipes & Heat Exchangers," 2017. Available at: http://www.jhcss.com.au/products-1/thermal-management/heat-pipes-heat-exchangers (accessed November 20, 2017).
Joardder M. U. H. and Masud M. H., 2019a. "Foods and developing countries." In: Food Preservation in Developing Countries: Challenges and Solutions. Springer, pp. 1–25.
Joardder M. U. H. and Masud M. H., 2019b. "Feasibility of advance technologies." In: Food Preservation in Developing Countries: Challenges and Solutions. Springer, pp. 219–236.
Joardder M. U. H. and Masud M. H., 2019c. "Food preservation techniques in developing countries." In: Food Preservation in Developing Countries: Challenges and Solutions. Springer, pp. 67–125.
Joardder M. U. H., Masud M. H. and Azharul M., 2017. "Relationship between intermittency of drying, microstructural changes, and food quality," In: Law, C. L. and Karim, A. (Eds.), Intermittent and Nonstationary Drying Technologies: Principles and Applications, Advances in Drying Science and Technology, CRC Press, pp. 123–137.
Jouhara H., Chauhan A., Nannou T., Almahmoud S., Delpecha B. and Wrobel C., 2017a. "Heatpipe based systems: Advances and applications," *Energy*, vol. 128, no. 1, pp. 729–754.
Jouhara H., Tassou S., Ge Y., Bianchi G., Holgate A., Boocock M., Leroux A., de Miol M., Pailette B., Schweiger H., Vannoni C., Woolass S., Bennett S. and Godley F., "D 2.3 report on estimation of energy, environmental and economic potential for heat recovery in EU28," I-ThERM, London: Brunel University London, 2017b.
Jouhara H. et al. "Waste heat recovery technologies and applications," *Thermal Science and Engineering Progress*, vol. 6, pp. 268–289. Available from: https://www.researchgate.net/publication/324811679_Waste_Heat_Recovery_Technologies_and_Applications (accessed February 7, 2024).
Karvonen M., Kapoor R., Uusitalo A. and Ojanen V., 2016. "Technology competition in the internal combustion engine waste heat recovery: A patent landscape analysis," *Journal of Cleaner Production*, vol. 112, pp. 3735–3743.
Kelvion H., "Quality and Flexibility Plate Heat Exchangers," 2017. Available at: https://www.kelvion.com/products/category/plate-heat-exchangers/ (accessed November 17, 2017).
Kummu M., de Moel H., Porkka M., Siebert S., Varis O. and Ward P. J., 2012. "Lost food, wasted resources: Global food supply chain losses and their impacts on freshwater, cropland, and fertiliser use," *Science of the Total Environment*, vol. 438, pp. 477–489.
Laval Alfa, "Genuine Gaskets and Plates," 2017. Available at: http://www.alfalaval.co.uk/gasketed-plate-heat-exchangers/service/spare-parts/ (accessed November 4, 2017).
Masud M., Joardder M. U. H., Islam M. T., Hasan M. M. and Ahmed M. M., 2017a. "Feasibility of utilizing waste heat in drying of plant-based food materials." In: International Conference on Mechanical, Industrial and Materials Engineering, p. 500.
Masud, M.H., Akhter, M.S., Islam, S., Parvej, A.M., and Mahmud, S., 2017b. Design, construction and performance study of a solar assisted tri-cycle. Period. Polytech. - Mech. Eng. 61 https://doi.org/10.3311/PPme.10240.
Masud M. H., Karim A., Ananno A. A. and Ahmed A., et al., 2020. Insights of Drying. Sustainable Food Drying Techniques in Developing Countries: Prospects and Challenges, 1st ed. Springer International Publishing, Switzerland. https://www.springer.com/gp/book/9783030424756?fbclid¼IwAR2leJNpdPb8pDdXzbUdU_UweO6IRszBTaEaHU5HZJo1VRkrolukdP1ms84#aboutBook

Mattsson B. and Sonesson U., 2003. Environmentally-Friendly Food Processing, 1st ed., Boca Raton: CRC Press LLC.
Maxxtec GmbH, "Air Preheater," 2014a. Available at: http://www.maxxtec.com/en/product/automatisch-aus-titel-generieren/heat-exchanger/air-preheater. Maxxtec GmbH, "Economizer," Sinsheim, 2014b.
Mohanraj M., 2014. "Performance of a solar-ambient hybrid source heat pump drier for copra drying under hot-humid weather conditions," *Energy for Sustainable Development*, vol. 23, pp. 165–169.
Monforti-Ferrario F., Dallemand J.-F., Pinedo Pascua I., Motola V., Banja M., Scarlat N., Medarac H., Castellazzi L., Labanca N., Bertoldi P., Pennington D., Goralczyk M., Schau E. M., Saouter E., Sala S., Notarnicola B., Tassielli G. and Renzulli P., "Energy use in the EU food sector: State of play and opportunities for improvement," Brussels: European Commission, Joint Research Centre, Institute for Energy and Transport and Institute for Environment and Sustainability, 2015.
Mujumdar, A. S. and Devahastin S., 2000. "Fluidized bed drying," *Trends in Food Science and Technology*, vol. 1, 59–111.
Mukherjee S., Asthana, Howarth A. M. and Mcniell R., 2017. "Waste heat recovery from industrial baking ovens," *Energy Procedia*, vol. 123, pp. 321–328.
Mumith J.-A., Makatsoris C. and Karayiannis T., 2014. "Design of a thermoacoustic heat engine for low temperature waste heat recovery in food manufacturing: A thermoacoustic device for heat recovery," *Applied Thermal Engineering*, vol. 65, no. 1–2, pp. 588–596.
Niamsuwan S., Kittisupakorn P. and Mujtaba I. M., 2013. "A newly designed economizer to improve waste heat recovery: A case study in a pasteurized milk plant," *Applied Thermal Engineering*, vol. 60, no. 1–2, pp. 188–199.
Nicholson R., 1983. "Recuperative and regenerative techniques," *Heat Recovery Systems*, vol. 3, no. 5, pp. 385–404.
Papapetrou M., Kosmadakis G., Cipollina A., La Commare U. and Micale G., 2018. "Industrial waste heat: Estimation of the technically available resource in the EU per industrial sector, temperature level and country," *Applied Thermal Engineering*, vol. 138, pp. 207–216. ISSN 1359-4311,
Pollock G. and Mason C., "Options to Maximise Process Heat Recovery at Red Meat Processing Facilities," Australian Meat Processor Corporation, Sydney, 2015.
PSC, "Heat Pipe Design," 2006. Available at: http://www.pscmf.com/pro-duct/showproduct.php?id=192&lang=en (accessed November 3, 2017).
Reindl D. T. and Jekel T. B., 2007. "Heat recovery in industrial refrigeration," *ASHRAE Journal*, vol. 49, pp. 22–28.
Reppich M., 1999. "Use of high-performance plate heat exchangers in chemical and process industries," *International Journal of Thermal Sciences*, vol. 38, pp. 999–1008.
Sapali S., Pise S., Pise A., and Ghewade D. "Investigations of waste heat recovery from bulk milk cooler," *Case Studies in Thermal Engineering*, vol. 4, pp. 136–143, 2014.
SEC Heat Exchangers, "SEC Brazed Plate Heat Exchangers (BPHE)," 2017. Available at: http://www.brazed-plate.com/ (accessed November 16, 2017).
Sen M., Basic Mechanical Engineering, New Delhi, India: Laxmi Publications (P) Ltd., 2006.
Shupe W. and Whitehead W., 1979. "Heat recovery from poultry processing scald water," *Journal of Agricultural Engineering Research*, vol. 24, no. 3, pp. 325–330.
Singh S. and Dasgupta M. S., 2017. "CO_2 heat pump for waste heat recovery and utilization in dairy industry with ammonia-based refrigeration," *International Journal of Refrigeration*, vol. 78, pp. 108–120.
Sondex, "Plate Heat Exchanger: Working Principle," 2012. Available at: http://www.sondex.net/Sondex-Global/Products/Heat-Exchangers/Working-Principle.aspx (accessed November 9, 2017). https://beeindia.gov.in/sites/default/files/2Ch8.pdf
Spirax Sarco, "An Explanation of Specialist Boiler Types and Other Specialist Features," Spirax Sarco, Cheltenham, 2011.
Spirax Sarco, "Steam Consumption of Heat Exchangers," 2017. Available at: http://www.spiraxsarco.com/Resources/Pages/Steam-Engineering-Tutorials/steam-engineering-principles-and-heat-transfer/steam-consumption-of-heat-ex-changers.aspx (accessed November 16, 2017).
Sreejith K., Varghese B., Das D., Devassy D., Harikrishna K., and Sharath G. K., "Design and cost optimization of plate heat exchanger," *Research Inventy: International Journal of Engineering and Science*, vol. 4, no. 10, pp. 43–48.
Srihari N. and Das S. K., 2008. "Transient response of multi-pass plate heat exchangers considering the effect of flow maldistribution," *Chemical Engineering and Processing: Process Intensification*, vol. 47, no. 4, pp. 695–707.

Tanasić N., Ivošević M. and Simonović T. "Optimisation of waste heat recovery system in paper machine dryer section," 2021, 6th International Symposium on Environment-Friendly Energies and Applications (EFEA), IEEE. doi: 10.1109/EFEA49713.2021.9406269

Thekdi A. C. and Nimbalkar S. U., "Industrial Waste Heat Recovery: Potential Applications, Available Technologies and Crosscutting R&D Opportunities," Oak Ridge, TN: Oak Ridge National Laboratory, 2014.

Thermtech, "Reducing Energy Costs with Economisers," 2014. Available at: http://www.thermtech.co.uk/reducing-energy-costs-with-economisers/ (accessed October 16, 2017).

Thulukkanam K., 2013. Heat Exchanger Design Handbook, 2nd ed., Boca Raton, FL: CRC Press.

Toolan S., 2012. "Heat Recovery for Commercial Buildings," Dublin: Dublin Institute of Technology.

Turner W. C., 2009. Energy Management Handbook, 7th ed., London: Taylor & Francis Ltd.

US Department of Energy, 2008. "Energy Use, Loss and Opportunities Analysis: US Manufacturing and Mining."

Valutech, "Welded Plate Heat Exchangers," 2016. Available at: http://www.valutechinc.com/products/heat-exchangers/welded-plate-heat-exchangers/ (accessed November 14, 2017).

Vandagriff R., Practical Guide to Industrial Boiler Systems, New York: Marcel Dekker, Inc., 2001.

Wallin J., Madani H. and Claesson J., "Run-around coil ventilation heat recovery system: A comparative study." In: *International Conference on Applied Energy (ICAE 2010)*, Singapore, 2010.

Wang L., 2008. Energy Efficiency and Management in Food Processing Facilities, 1st ed., Boca Raton, FL: CRC Press.

Wang L., Sundén B. and Manglik R. M., 2007. Plate Heat Exchangers: Design, Applications and Performance, 1st ed., Southampton, UK: WIT Press.

Webbusterz Engineering, "What Is a Plate Heat Exchanger," 2015. Available at: http://www.webbusterz.org/what-is-a-plate-heat-exchanger/ (accessed November 14, 2017).

Worrell E., Galitsky C., Masanet E. and Graus W., *Managing Your Energy: An Energy Star Guide for Energy and Plant Managers*, Berkeley, CA: Energy Analysis Department, Lawrence Berkeley National Laboratory University, 2008, p. 60.

Yodrak L., Rittidech S., Poomsa-ad N. and Meena P., 2010. "Waste heat recovery by heat pipe air preheater to energy thrift from the furnace in a hot forging process," *American Journal of Applied Sciences*, vol. 7, no. 5, pp. 675–681.

5 Geothermal Energy for Drying Fresh Produce

Özge Süfer
Food Engineering Department, Faculty of Engineering, Osmaniye Korkut Ata University, Osmaniye, Türkiye

Azime Özkan Karabacak
Department of Food Processing, Gemlik Asım Kocabıyık Vocational School, Bursa Uludag University, Bursa, Türkiye

Canan Ece Tamer and Senanur Durgut
Department of Food Engineering, Faculty of Agriculture, Bursa Uludag University, Bursa, Türkiye

5.1 INTRODUCTION

The constant rise in generation of energy has been one of the aspects that has significantly contributed to world progress. The ability of a country to consume energy is correlated with its gross domestic product, according to economic research (1). It is known that almost 80% of the world's electricity is supplied by fossil fuel power plants (2). Global economic, societal, and environmental risks are raised by an overreliance on fossil fuels. Additionally, the current system of producing electricity is unable to satisfy the rising demand. Therefore, energy systems that are efficient and environment-friendly are needed to meet the constantly expanding demand for energy (3). Moreover, since a nation's development relies on its consistent, high-efficiency energy production, a promising solution is needed to combine continuous power with national energy production from renewable resources. These days, more and more attention is being drawn to the use of clean, renewable energy sources as a substitute for fossil fuel systems (4). The generation gap can be reduced by including renewable energy sources like geothermal, biomass, and solar energy as prospective energy sources. However, basing electricity generation on solar, wind, water, and biomass are inefficient, because they depend on weather conditions. Contrarily, geothermal energy has a very high generation capacity and is not entirely dependent on the weather (1). Additionally, geothermal energy is less expensive than fossil fuels and alternative renewable sources. It also has the highest availability factor in all renewable energy sources. Therefore, the addition of geothermal energy to the traditional grid system lowers the cost of purchasing energy per unit area, which ultimately boosts a nation's gross domestic product (5).

The earth's core has created and emitted heat for billions of years. The temperature of the earth's core, which is identical to that of the sun's surface, is 5,500 °C (6). The primary source of geothermal energy is the heat that has been generated and stored naturally throughout the earth's development for millions of years (5). The thermal energy found beneath the earth's crust, known as geothermal energy, appears to be a safe and sustainable way to slow down global warming. Its sources have been employed to produce electricity for more than a century (7). The majority of geothermal power plants were constructed worldwide in the 1970s and 1980s, as a result of the 1973 oil crisis (8). Geothermal installed capacity is rising steadily worldwide, according to a report from the international renewable energy agency (9). The geothermal energy resource can be used in one

 DOI: 10.1201/9781003344087-5

of two major ways: directly or indirectly. For space cooling, heating, and industrial applications involve employing the heat directly; indirect use includes turning the heat into electricity, which is then utilized (10). Depending on the geological and geophysical factors, the concentration of the world's geothermal energy varies. The first is the circum-Pacific belt's western coast, which encompasses Australia, West Indonesia, the Philippines, Japan, China, and Taiwan. The Mid-Atlantic Ridge, the second zone, reaches the northern tip of Iceland and is mostly ocean; the Mediterranean to Himalayan region, the third zone, comprises Tibet in China and Italy (11). The ten countries with the highest installed geothermal power generation efficiency as of 2020 are stated. Among them, the United States, Indonesia, the Philippines, Turkey, Kenya, Mexico, New Zealand, Italy, Japan, and Iceland have installed 3,700, 2,289, 1,918, 1,549, 1,193, 1,105, 1,604, 916, 550, and 755 MWe, respectively (5).

Geothermal energy has a wide range of household and commercial applications, including space heating, greenhouse and open-air heating, heating for aquaculture ponds and races, heating for industrial processes, agricultural drying, snow melting, and cooling (12). Since many technologies require more or less concentrated heat energy sources, the agricultural and food processing industries are interesting areas for the utilization of geothermal energy (13).

In order to prevent waste and ensure that healthy food is available all year long especially during droughts, drying agricultural products is an essential procedure (14). The main goal of drying agricultural products is to offer greater quality, higher prices, earlier harvesting that lowers field losses, and long-term storage without deterioration. Industrial drying procedures involve the use of heat and energy for powering the auxiliary machinery. Basic energy requirements include the energy needed to heat the product to the proper temperature to initiate the evaporation process as well as the energy needed to evaporate a particular percentage of the moisture. Most of the time, the procedure involves direct contact between the material (such as grain, vegetables, fruits etc.) and warm air that is relatively low in temperature (between 35 °C and 80 °C). In this context, geothermal fluids with low temperatures can be used as a source of energy to heat the air for agricultural products to dehydrate (15, 16). Geothermal drying reduces the moisture content of raw materials by passing hot air over or through trays or belts containing the raw materials. During geothermal drying, electric power is employed to run fans and pumps (17).

This chapter aims to present a comprehensive review on geothermal drying theory/principles, applications, and dryer types for agricultural products, as well future perspectives of geothermal drying.

5.2 GOVERNING EQUATIONS

Energy, exergy, and economic analysis procedures of geothermal energy are discussed in this section.

5.2.1 Energy Balance

Mass and energy conservation equations are valid for analyzing geothermal energy as such in other energy types or dryers, and some assumptions are considered like steady-state process and steady flow (7). Regarding these claims, Equation 5.1 (mass balance) and Equation 5.2 (energy balance) are useful for calculating energy and/or efficiencies (18, 19):

$$\sum \dot{m}_{input} = \sum \dot{m}_{output} \tag{5.1}$$

$$\dot{E}_{\text{input}} - \dot{E}_{\text{output}} = \Delta \dot{E} \tag{5.2}$$

where $\dot{m}$ is the mass flow rate and $\dot{E}$ is the total energy.

If the second mathematical expression is expanded, Equation 5.3 is obtained (20):

$$Q+\Sigma\dot{m}_{input}h_{input}=\dot{W}+\Sigma\dot{m}_{output}h_{output} \tag{5.3}$$

Q: heat input rate (net)
$\dot{W}$: network output rate
h: enthalpy

Furthermore, system efficiency can be determined by Equation 5.4 (21):

$$\eta=\frac{\dot{W}}{Q} \tag{5.4}$$

5.2.2 Exergy Balance

The exergetic content refers to the amount of productive work that can be produced by a process in a designated state of inactivity. The exchange of heat energy between the process and its reference environment happens at a definite temperature disparity, which plays a crucial role in determining the rate of irreversibility for the process (11).

General form of exergy (ex) balance equation is able to be written as follows (22):

$$\Sigma\left(\dot{m}_{inlet}ex_{input}+\dot{W}_{input}+\dot{E}x_{Q,input}\right)=\Sigma\left(\dot{m}_{outlet}ex_{output}+\dot{W}_{output}+\dot{E}x_{Q,output}\right)+\frac{dE_x}{dt}+\dot{E}x_d \tag{5.5}$$

Ex refers the Equation 5.6:

$$Ex=ex_{ph}+ex_{kn}+ex_{pt}+ex_{ch} \tag{5.6}$$

where ex_{ph}, ex_{kn}, ex_{pt}, and ex_{ch} indicate physical, kinetic, potential, and chemical exergy, respectively. They can be expressed with the aid of Equations 5.7–5.10:

$$ex_{ph}=(h-h_0)+T_0(s-s_0) \tag{5.7}$$

T: temperature
s: specific entropy
0: denotes reference or ambient conditions

$$ex_{kn}=\frac{1}{2}\upsilon^2 \tag{5.8}$$

υ: process velocity

$$ex_{pt}=g(z-z_0) \tag{5.9}$$

g: gravitational acceleration
z: elevation

$$ex_{ch}=\Delta_f G^0+\sum_{elements}\nu ex_{element}^{ch} \tag{5.10}$$

$\Delta_f G^0$: Gibbs free energy (formation)
ν: moles of element in chemical formation's equation

Considering same assumptions declared in energy balance and other equations mentioned in Dincer and Ozturk's book (11) about both Gibbs free energy and enthalpy of formation, as well as entropy generation and exergetic variables because of heat transfer, Equation 5.5 is reduced to Equation 5.11.

$$\Sigma\left(\dot{m}_{inlet}\,ex_{input} + \dot{W}_{input} + \dot{E}x_{Q,input}\right) = \Sigma\left(\dot{m}_{outlet}\,ex_{output} + \dot{W}_{output} + \dot{E}x_{Q,output}\right) + \dot{E}x_{d} \tag{5.11}$$

Moreover, ex efficiency (ψ) can be determined by Equation 5.12 (11):

$$\Psi = \frac{\dot{E}x_{output}}{\dot{E}x_{input}} \tag{5.12}$$

5.2.3 Economic Analysis

Bejan et al. (23) defined total cost on hour basis as follows (7):

$$\hat{Z}_{total} = \frac{CRF \times \Phi \times \Sigma k Z_k}{t_{hour}} \tag{5.13}$$

$\hat{Z}$: initial equipment cost
Φ: annual costs for maintenance and operating
CRF: capital recovery factor

Equation 5.14 is useful for computing CRF (7, 24):

$$CRF = \frac{i\,(1+i)^N}{(1+i)^N - 1} \tag{5.14}$$

i: interest rate
N: system life

On the other hand, levelized cost of hydrogen production (LCHP, \$/kg) and electricity (LCE, cent/kWh) are specified by Equations 5.15 and 5.16, respectively (7, 25, 26):

$$LCHP = \frac{\hat{Z}_{total}}{\dot{m}_{H_2} \times 3{,}600} \tag{5.15}$$

$$LCE = \frac{\hat{Z}_{total}}{W_{net,total}} \tag{5.16}$$

5.3 GEOTHERMAL DRYERS

Geothermal energy usage can be conceptually divided into two categories: at the surface and deep underground. The deep geothermal is also divided into high- and low-enthalpy reservoirs (27). Drying applications, which take the first place in terms of importance and economy among the areas where geothermal energy can be used in the agricultural area, can be carried out with geothermal energy drying systems under controlled conditions. Fruits and vegetables can be dried using geothermal resources with low enthalpy at temperatures below 150 °C. Geothermal wells are used to extract heat in the form of steam or hot water. Geothermal drying can make good use of the

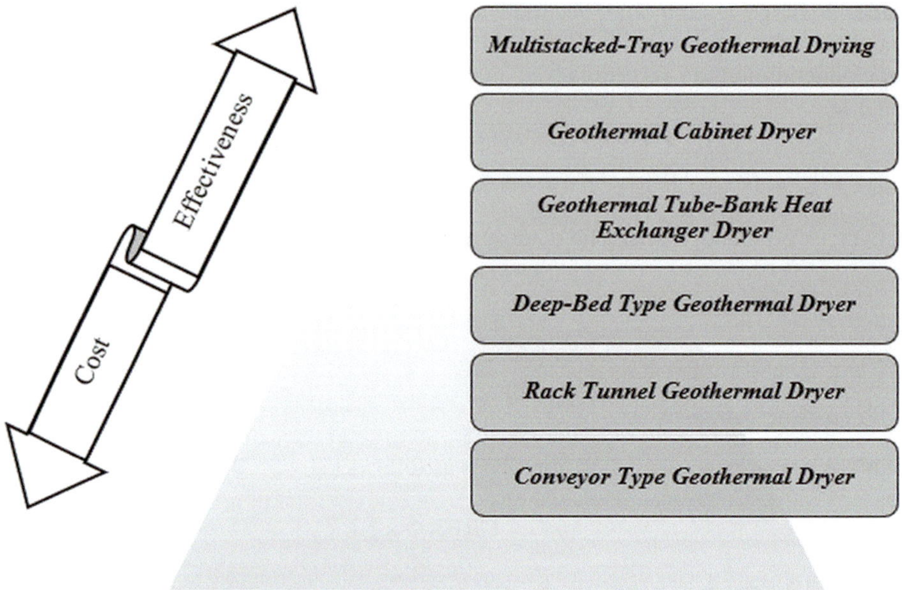

FIGURE 5.1 Comparison of the different geothermal drying methods [Adapted from Ref. (31)]

large quantity of waste heat produced by geothermal power facilities (28). Geothermal energy is a very cost-effective way to dry food because it is much cheaper to use hot water or steam in food processing than oil or electricity (29). Due to their proximity to plate borders, many developing nations have abundant geothermal resources. Therefore, inexpensive geothermal energy would be a great source for environment-friendly drying technologies in these countries (30). Food drying requires a tremendous amount of energy, which can be obtained from geothermal sources (14). By utilizing the plentiful geothermal energy sources that certain developing nations have access to, low-cost sustainable drying solutions can be utilized. Geothermal dryers are simple to install, since geothermal resources provide direct thermal energy (31). Suitable drying conditions can be used to generate better quality dried foods for each unique type of agricultural material. Different dryers designed for drying foods using geothermal energy are discussed below. Comparison of the different geothermal drying methods based on their effectiveness and cost was also illustrated in Figure 5.1.

5.3.1 Multistacked-Tray Geothermal Drying

One of the most common geothermal drying mechanisms is a multistacked-tray system. This system involves pumping geothermal hot water out of the ground and putting it through a cross-flow heat exchanger. The thermal energy of hot water is used for heating the air. The drying chamber, which is constructed with multistacked trays on which food stuffs are placed, is then circulated with the heated air (31).

In 2001, Nea Kessani, Xanthi, Greece, began adopting geothermal drying that included a multistacked tray system. Their drying system consists of 14 rectangular dryers measuring 2 m in height and 1 m in width. They used hot geothermal water that was 59 °C to dry the tomato slices (14).

A multistacked-tray geothermal drier for drying fruits was designed by Lund and Rangel (32). The dryer could dry 1 ton every cycle at a temperature of 60 °C. They stated that fruit samples were dried in 24 hours, from an initial moisture level of 80% to a final moisture content of 20%.

5.3.2 Geothermal Cabinet Dryer

The principal thermal energy source for a geothermal cabinet drier is recovered waste heat from a geothermal power station. Usually, waste heat is recovered as hot water, which is then used to heat the air using a counter-flow heat exchanger. The hot air is then circulated through a closed cabinet containing the food products. The cabinet system is useful because it gives the user control over the humidity and temperature in the drying chamber. The geothermal cabinet dryer system also has a short drying time. Additionally, it uses less energy than multistacked-tray geothermal drying (Figure 5.1). This type of dryer was used to dry olive leaves in Balcova-Narlidere, Turkey (33), garlic and chilies in Thailand (31), wheat and maize in Menengai, Kenya, and tomato in Turkey (34).

5.3.3 Geothermal Tube-Bank Heat Exchanger Dryer

The technology for drying geothermal tube-bank heat exchangers was developed in Indonesia due to the country's huge opportunity for geothermal resources. Through the use of this technology, a variety of food goods, such as tea, beans, coffee berries, fish, and rough rice, was dried (35, 36). Geothermal fluid is split into brine and steam in this dryer's separator. Steam is used to power a turbine, which then produces electricity. The brine that exits the evaporator moves to the re-injection wells. A steam generator is another name for the evaporator. Drying procedures utilize the superheated vapor of the working fluid that is produced by the evaporator. In the following step, an air-blown heat exchanger is used to condense the superheated working fluid vapor from the evaporator. The working fluid loop continues as the condensed working fluid is then circulated using a feed pump to the evaporator and so forth. Hot air is used as the cooling medium instead of fresh air. The drying operations are aided by the hot air flowing through the drying radiators (36). When the heated air reaches the desired temperature, it is blown into the four-tray drying chamber. This system's geothermal tube-bank heat exchanger has a 1,000 W heat transfer rate. This dryer's optimal air velocity and temperature ranges are 4–9 m/s and 45–60 °C, respectively. The amount of moisture in the source material has a considerable impact on how quickly the system dries (31).

5.3.4 Deep-Bed and Conveyor-Type Geothermal Dryer

Two distinct drying systems that are included in the category of an advanced geothermal dryer are the deep-bed (batch dryer) and conveyor-type geothermal dryer. Although these two systems are pricey, their efficiency and capacity are very good (31). Conveyor dryers continuously operate by placing the drying material in a chamber from which it is transported to the loading strip. The air (or gas) used for drying is required to be between 70 °C and 170 °C in temperature. The use of axial fans makes it possible to circulate the drying air. In a basket or other form of transport, dried material is gathered from the strips (16). These dryers are designed with lines that are 40 m long and 2–2.2 m wide. Negative effects of this dryer include uneven drying over the height of the sheet, which is significantly lessened by mixing while moving from one strip to another. The dirtiness that results from particles slipping past the protective screen and landing on the heat exchanger is another drawback.

In western Nevada, the United States, a large-scale conveyor geothermal dryer was constructed for drying onions and garlic. It had the capacity to dry 3,000–4,300 kg of wet onions per hour (37). Geothermal energy is used in deep-bed drying, a method that effectively dries grains. This device comprises a blower or fan that gives the air the required kinetic energy to move through the geothermal heat exchanger. In this procedure, the air is heated to an appropriate drying temperature and then diffused evenly across the dryer's perforated floor. By adjusting the hot water flow through the geothermal heat exchanger, this system's temperature can be controlled. While some grains can be dried at temperatures as high as 90 °C, the system performs best at a medium temperature of 50–60 °C with a relative humidity of 40% (35).

5.3.5 Rack Tunnel Geothermal Dryer

A rack tunnel geothermal dryer looks similar to a multistacked-tray geothermal dryer in its most basic form, with the exception that it contains a hot air recirculation system. Rack tunnel geothermal dryers are used to dry large quantities of material, so hot air is circulated to conserve energy and effectively recover the thermal energy contained therein. In order to provide constant air circulation throughout the dryer, chimneys are placed inside this device. The main heat source used to raise the temperature of the air around it is geothermal hot water and steam. In this system, along the chamber's rails, the connected wagons are slowly traveling. Hermetic doors are present at the tunnel's entry and exit. They are simultaneously and infrequently opened to allow material to be loaded and unloaded into the dryer. Depending on how the dried material is moving, drying air is going in the same or opposing directions. The movement of the air can be forced or natural, although countercurrent forced movement produces the best results (16). The capacity of a standard rack tunnel geothermal dryer, compared to other geothermal dryers, is 2,000–4,000 tons per batch (14). Rack tunnel geothermal dryer is usually used for drying fish products, especially in Iceland. In this dryer, two-step idea of drying is used to reduce the moisture of fish from 80 to 15% (wb) (29).

In Greece, the geothermal field of Neo Erasmio has been home to the world's first rack tunnel "geothermal" tomato dryer since 2001. The system heats ambient air to 55–58 °C in finned-tube, air-heater coils using low-salinity geothermal water (60 °C). Once the trays of fresh tomatoes are placed in the drying tunnels, heated air moves through. In this system, each tunnel is 1.24 m wide, 13.5 m long, and 2.2 m high. Twenty-five series of trays can fit along the tunnel's length. On each tray, approximately 8–9 kg of raw tomatoes are inserted (38).

5.3.6 Hybrid Geothermal–Solar Dryer

Solar drying has long been a method of significant interest for stabilizing and preserving agricultural products like fruits, vegetables, and plants. Studies regarding the sun drying systems have covered a wide range of processing methods based on the dried product, the environment of the operating region, and the application (39–43). Products are exposed to rain, wind, and dust during open-sun drying operations. They come into contact with birds, insects, and rodents. During this drying process, there are significant losses, and the food quality could be severely harmed (44, 45). Using a direct sun drier, some of such difficulties have been solved. Crops are directly exposed to sunlight in direct solar drying systems, which causes them to become dehydrated. In systems that use direct solar energy to dry materials, the food is exposed to the sun directly. To turn sunlight into heat in these types of systems, a black-painted surface containing the crop to be dried is positioned within. To boost efficiency, these dryers could incorporate glass lid coverings and vents (46). Despite being widely available, solar energy has an intermittent nature and is not constant (47). The solar drying system can be more dependable, thanks to hybridization, which enables it to dry numerous agri-food products well even when there isn't enough sunlight, such as at night or during overcast weather (48).

Regarding the mentioned problems, a hybrid dryer that utilized 80% thermal solar and low enthalpy aerogeothermal energy and 20% electric power was developed in order to optimize the drying process and guarantee a high-quality dry product (49). Additionally, geothermal-assisted solar dryer for tomato paste drying was conducted by Hadibi et al. (48). In their system, the recorded drying temperature was greatly raised by the geothermal water heat exchanger integration and reached 53.41 °C, an increase of 11.5 °C above the conventional drier. Furthermore, after sunset and during the night, the drying temperature within the drying chamber was 30.8 °C higher than the surrounding air temperature. By ensuring a continuous drying process, geothermal water heat exchanger reduced drying time from the standard dryer's 18 hours over the course of three days to just 22 hours. They claimed that, in addition to reducing drying time, the hybrid system with geothermal water heat exchanger ensures a continuous drying operation during the night.

5.4 GEOTHERMAL ENERGY SOURCE IN FOOD DRYING

The process of drying agricultural products holds great significance, particularly in developing countries. It serves to prevent spoilage, guarantee the availability of nutrient-rich food throughout the year, and ensure resilience during natural calamities and periods of drought. The heat generated as a by-product of geothermal power production can be harnessed as a source of thermal energy for agricultural drying purposes, utilizing either steam or hot water from the geothermal well (1, 28).

Table 5.1 presents research papers published between 1992 and 2023 years, and focuses on food products, equipment type and/or technology, system requirements and/or process conditions, besides research highlights. Different kinds of materials were subjected in articles, and most popular group was undoubtedly tomato (plus tomato derivatives), as also visualized in Figure 5.2a. Although tomato is a fruit, it was not shown under fruit category in order to see its prepotency. Tomatoes are a widely consumed foods globally, but due to high moisture contents, they are highly perishable which lead to significant losses in production and consumption (50). Moreover, further

TABLE 5.1
A Comprehensive Summary of Research Articles Focused on Geothermal Drying for Fresh Produce in Last 31 Years

Material	Equipment Type/ Technology	Process Conditions/ System Configurations	Remarkable Results	Reference
Seaweed	Conveyor type dryer	$\dot{m}_{air}$: 110, 124, 135 kg air/min/m^2 $T_{air,in}$: 70 °C	Diverse potential applications are anticipated for utilizing low-temperature geothermal heat, which include cultivating single-celled and large-scale algae, extracting chemicals from a range of seaweed and agricultural products, and concentrating solutions that are sensitive to heat	(61)
Rice	Cabinet type dryer	$T_{air,in}$: 35 °C $T_{w,in}$: 75 °C $T_{w,out}$: 50 °C	Concrete district heating schemes can benefit from using drying units, which offer a straightforward geothermal solution. The annual heating load factor can be improved without the need for costly geothermal water distribution lines. It is highly recommended to utilize these units for optimal results	(32)
Fruit	Multistacked-tray dryer	V_{air}: 215 m^3/min $T_{w,in}$: 120 °C $T_{w,out}$: 100 °C T_{drying}: max 70 °C	The implementation of a cascading system would result in heightened efficacy and decreased expenses in production and utilization of geothermal resource	(62)
Chili and garlic	Cabinet-type dryer	*Chili:* $\dot{m}_{air}$: 1.00 kg/s $\dot{m}_{w}$: 1.00 kg/s $T_{ambient}$: 28.3–30.0 °C T_{air}: 44.9–50.4 °C *Garlic:* $\dot{m}_{air}$: 1.00 kg/s $\dot{m}_{w}$: 0.04–0.09 kg/s $T_{ambient}$: 25.4–27.0 °C T_{air}: 34.8–40.6 °C	According to economic analysis, related dryer presents a compelling option, as the associated expenses are reasonable and the system's functionality is impervious to atmospheric conditions	(63)

(Continued)

TABLE 5.1 *(Continued)*

A Comprehensive Summary of Research Articles Focused on Geothermal Drying for Fresh Produce in Last 31 Years

Material	Equipment Type/ Technology	Process Conditions/ System Configurations	Remarkable Results	Reference
Tomato	Rack tunnel dryer	$T_{w,in}$: 59 °C $T_{w,out}$: 51–53 °C V_w: 25 m³/h V_{air}: 10,000–12,000 m³/h P_{fan}: 7 kW v_{air}: 1.7 m/s	* Geothermal energy that operates at low temperatures can be utilized with great efficiency and dependability in heating of air required for the dehydration process. * Geothermal dehydration helps to preserve the vibrant red color, essential nutrients, and distinct flavors of fresh tomatoes	(17)
Capelin fish	Rack tunnel dryer Conveyor-type dryer	*Primary drying:* T_{air}: 18–25 °C v_{air}: 3 m/s RH: 30–50% *Secondary drying:* T_{air}: 22–26 °C v_{air}: 0.5–1.0 m/s RH: 20–50%	*The process of division in drying results in reduced production costs, as it enables the drying of greater quantities of cod-heads. * The utilization of geothermal energy for the purpose of drying is significantly contingent upon the prevailing rates of crude oil and electricity, as well as the prevailing market prices of dehydrated fish products	(29)
Tea leaves	Tube-bank heat exchanger dryer	$T_{air,in}$: 15 °C $T_{w,in}$: 180 °C T_{vapor}: 140 °C $T_{condensate}$: 140 °C	* Among three models evaluated, Model III stands out as the most viable option, when considering all process parameters, despite the fan only operating at 66% of its maximum capacity. * The current ducting system is still compatible with all three calculation models	(36)
Tomato	Rack tunnel dryer	$T_{air,in}$: 55–58 °C v_{air}: 1.5–2.0 m/s	A way to dry food in a tunnel has been created and tested using a specific product, which can be used to design and improve a system for drying tomatoes continuously using geothermal water at a lower temperature	(38, 64)
Olive leaves	Cabinet-type dryer	$T_{air,\ in}$: 37.9 °C $T_{air,\ out}$: 40.9 °C $T_{w,in}$: 62 °C $T_{w,out}$: 56.5 °C m_{air}: 0.20 kg/s	* The fan experienced the highest level of exergy destruction, with the heat exchanger and dryer following closely behind. * An economic evaluation should be conducted to determine the feasibility of implementing a recirculation system in commercial drying facilities	(33)
Tomato	Drying room/ rack tunnel dryer	$T_{ambient}$: 18 °C T_{room}: 50 °C $T_{w,in}$: 70 °C $T_{w,out}$: 50 °C $\dot{m}_{air}$: 0.813 kg/s $\dot{m}_w$: 2.84 kg/s	The utilization of this approach is technically viable. To dry 1 ton/day tomatoes, 7 kg/s of brine is required, with a T_{in} of 180 °C from the separator and a T_{min} of 85.1 °C from heat exchanger	(17, 65)
Tomato, avocado, green chili	Geothermal cascade system with five alternatives	*For alternative S_1, S_2, S_4:* $T_{resource}$: 120 °C *For alternative S_3, S_5:* $T_{resource}$: 92 °C $\dot{m}$: 200–500 kg dry product/day	* The cascade principle has potential to be applied in a variety of settings, including those where temperatures exceed from 90 °C to 120 °C, such as waste heat and solar energy. This versatility allows for a wide range of applications, both direct and in combination with other processes.	(66)

(Continued)

TABLE 5.1 ***(Continued)***
A Comprehensive Summary of Research Articles Focused on Geothermal Drying for Fresh Produce in Last 31 Years

Material	Equipment Type/ Technology	Process Conditions/ System Configurations	Remarkable Results	Reference
			* Among avocado, tomato, and pepper, the dehydration of tomatoes was found to be the most cost-effective and energy-efficient method. * CO_2 reduction were at significant level for all cases	
Milk powder	Spray drier	$T_{ambient}$: 25 °C $T_{resource}$: 110 °C $\Delta T_{w,in\ and\ out}$: 70 °C $T_{milk,in\ and\ out}$: 76 °C	*Exergetic efficiency: 9–83% *Overall energy efficiency: 85.4% *Overall exergy efficiency: 57.45% * Evaporator and heater exhibit a greater degree of influence in the implementation of enhancement strategies	(12)
Coffee	Tube-bank heat exchanger dryer	$T_{w,in}$: 80 °C T_{system}: 50 °C T_w: above 147.5 °C $\dot{m}$: 1,000 kg dry product/12.5 h	The use of geothermal brine as an energy source for drying apparatus is a viable alternative to diesel oil	(55)
Cereal	Hybrid geothermal–solar dryer	$T_{ambient}$: about 27 °C v_{air}: 1.5 m/s $T_{collector}$: 50 °C $\Delta T_{air,in\ and\ out}$: 10–14 °C	The operational efficiency of the equipment was approximately 60%, achieving a peak $T_{operating}$ of 52 °C and exhibiting system stabilization within the first hour. The prototype's design is highly adaptable and reproducible	(49)
Olive leaves	Cabinet-type dryer	T_{air}: 40–50–60 °C v_{air}: 0.5–1.0–1.5 m/s $T_{w,in}$: 62–65 °C	Elevated temperatures and extended drying periods were found to significantly impact the reduction in total phenolics and antioxidants. T_{air} of 50 °C and v_{air} of 1 m/s provided the optimal drying conditions resulting in minimal loss of bioactives	(67)
NA	Hybrid geothermal–solar dryer	T_w: 70 °C V_w: 3 L/min Time period: 09:00 a.m. (1st day)–09:00 a.m. (2nd day)	*A noteworthy enhancement in the thermal efficiency of the direct solar dryer, particularly during the post-sunset period, was demonstrated. *The inclusion of a heat exchanger within a solar dryer guarantees the uninterrupted progress of the drying procedure, even in conditions of cloudiness or during the nighttime	(68)
Grain (wheat and maize)	Cabinet-type dryer	T_{air}: 40 °C T_w: between 50 °C and 75 °C v_{air}: between 4 m/s and 9 m/s $T_{resource}$: 160 °C	Various types of grains necessitate distinct drying parameters, such as v_{air} and T. Inadequate T_{drying} may impede proper desiccation and foster the growth of rot, whereas excessively high temperatures may lead to both cracking and case hardening, and potentially extinguish the life present within the seed grain	(69)
Coffee	Tube-bank heat exchanger dryer	T_{air}: 50–60–70 °C v_{air}: 0.2–0.4–0.6 m/s V_w: 18 L/s	The implementation of a heat pipe heat exchanger, fueled by geothermal energy, is a poetic solution to lessen the consumption of fuel and electricity in drying processes	(56)

(Continued)

TABLE 5.1 *(Continued)*
A Comprehensive Summary of Research Articles Focused on Geothermal Drying for Fresh Produce in Last 31 Years

Material	Equipment Type/ Technology	Process Conditions/ System Configurations	Remarkable Results	Reference
Wheat	Hybrid geothermal–solar dryer	$T_{w,in}$: 15 °C $T_{air,\ in}$: 60 °C T_{drying}: 50 °C m: 45, 50, 55 kg product/h A: 70 m^2	The efficiency of drying process is significantly influenced by the volume of tank, surface characteristics of collectors, and nature of the soil	(53)
Coffee bean	Tube-bank heat exchanger dryer	V_{pump}: 18 L/m v_{air}: 0.6 m/s T_{drying}: max 53 °C	* The utilization of both solar energy and low-temperature geothermal energy in drying is more expeditious than relying on either solar or geothermal energy. * The drying time of coffee beans can be expedited by about 23%	(57)
Tea leaves	Tube-bank heat exchanger dryer	$T_{w,in}$: 50 °C v_{air}: 0.2–0.4–0.6 m/s V_{pump}: max 18 L/m	Page model is the superior for expressing drying behavior of tea leaves among Henderson and Pabis, modified Page and Lewis models	(70)
Tomato paste	Hybrid geothermal–solar dryer	T_{air}: 35.75–53.41 °C $T_{ambient}$: 18.08–23.05 °C	The implementation of hybrid system, which incorporates a geothermal water heat exchanger, facilitates continuous drying during night hours and significantly reduces drying time. An economic analysis of hybrid dryer demonstrated its high viability, as it exhibits a short period for recouping initial investments	(48)
Tomato paste	Hybrid geothermal–solar dryer	T_{air}: 39–50–70 °C $T_{w,in}$: 67 °C V_w: 0.05 L/s	The innovative dehydration process, which combines solar/geothermal drying, mechanical dewatering, and instant controlled pressure drop texturing/swell drying, had a notable impact on enhancing drying kinetics, considerably diminishing energy consumption, and augmenting visual appeal	(54)
Tomato paste	Hybrid geothermal–solar dryer	T_{drying}: 51.28–54.55 °C $T_{ambient}$: 40.16 °C $T_{w,in}$: 67 °C $\Delta T_{w,in\ and\ out}$: 4 °C V_w: 0.05 L/s	The application of geothermal energy in conjunction with ventilation is suggested as a viable approach for drying of tomato paste, given its advantages in terms of time, exergy, and energy efficiency, economic feasibility, and preservation of quality attributes	(51)
Corn kernel	Cabinet-type dryer	T_{room}: about 38 °C T_w: max 50 °C m: 960–1,280 kg/day	Additional research and refinement ought to be conducted to fully leverage the potential of Internet-of-Things devices, including the management of stirrer blade velocity and water volume in container, to regulate temperature within the oven	(71)

(Continued)

TABLE 5.1 (Continued)
A Comprehensive Summary of Research Articles Focused on Geothermal Drying for Fresh Produce in Last 31 Years

Material	Equipment Type/ Technology	Process Conditions/ System Configurations	Remarkable Results	Reference
Pineapple	Food dehydrator system	T_{drying}: 64.2 °C, 67.1 °C, 70.0 °C, 71.4 °C, 72.0 °C, 78.4 °C, 80.0 °C	The primary nutritional attributes inherent in dehydrated pineapple offer an auspicious prospect to reproduce and expand the technology in other regions that possess an agricultural landscape and a substantial capacity in the utilization of geothermal power	(72)
Tomato	Cabinet-type dryer	T_{air}: 40–50–60 °C v_{air}: 0.5–1.5 m/s	* Energy efficiency: 6.6% * Exergy efficiency: 22.31%	(34)

NA: Not available.

research is necessary to investigate the production of tomato paste, as there is potential to obtain multiple derivative products simultaneously. The process involves mechanically separating the wet tomato from its seeds and skin, followed by dewatering to produce paste. The method employed for mechanical dewatering involves low energy consumption, leading to a higher price for the resulting tomato juice, seeds, and peel (51). In second rank, cereals stand out. Cereal grains are renowned for their abundant energy, carbohydrates, and plant proteins, making them a crucial dietary staple worldwide. However, it may come as a surprise that a mere 41% of grains are actually utilized for human consumption. This presents an opportunity to tap into the underutilized potential of cereals as a source of environmentally sustainable and nutritious plant proteins, paving the way for a more sustainable food system that prioritizes healthy diets (52).

Figure 5.2b demonstrated geothermal dryer types with their frequencies, and hybrid geothermal–solar dryer was the superior, according to the data of Table 5.1. Related type of geothermal dryer was used for dehydrating cereals (49, 53) and tomato pastes (48, 51, 54). Apart from this, geothermal tube-bank heat exchanger dryer, rack tunnel, and cabinet dryers were also favorable. Tube-bank heat exchanger dryer was mostly utilized in decreasing the moisture content of coffee

FIGURE 5.2 Utilization of geothermal energy in food drying. (a) Categories of food products, (b) Popularity of dryers

beans (55–57). Coffee is considered a crop commodity that possesses a considerable economic worth. Due to its intricate chemical makeup, coffee is a highly sought-after beverage worldwide, renowned for its distinct flavor and alluring aroma. Scientific studies have identified over 1,000 volatile compounds in coffee, comprising both aroma-active and non-aroma-active substances (58–60). On the other hand, in geothermal drying, air temperature at inlet and outlet, water temperature at inlet and outlet, drying temperature, mass flow rate of air and water, volumetric flow rate of water, and relative humidity of air are crucial factors and they are very diverse (Table 5.1).

5.5 CONCLUDING REMARKS AND FUTURE RECOMMENDATIONS

The ability of geothermal energy to provide communities from various economic and political backgrounds with uninterrupted sustainable renewable energy has been demonstrated. Many countries with rich geothermal energy reserves are currently using geothermal energy for dehydration of agricultural materials. Different types of dryers used for drying various kinds of materials with geothermal energy were comprehensively reviewed in this chapter. Among all reviewed studies, tomato (plus tomato derivatives) came first, followed by cereals, plants, fruits, vegetables, seafoods and others, respectively. Geothermal tube-bank heat exchanger, cabinet, ruck tunnel, convective, hybrid geothermal–solar, deep-bed, and conveyor types of dryers were used for drying agricultural products. The most preferred one among these has been the hybrid geothermal–solar dryer.

On the other hand, there are some challenges related to the adoption of geothermal drying by developing countries. Its high upfront cost is one of the major obstacles to using geothermal drying technology. Additionally, despite being a recognized renewable energy source, relatively few governments in developing nations have clear regulations and laws governing the use of geothermal energy. Since most geothermal dryers are used in conjunction with geothermal power plants, the lack of geothermal plants in developing nations is another significant challenge. In order to overcome such difficulties, some regulations, including government support and tax break, are needed to attract producers.

Despite all these challenges, using geothermal energy in drying is a promising method. Especially for developing countries, it is possible to take advantage of this energy potential in drying, both to open new business areas and to make more efficient use of existing energy resources. Thus, by reducing fossil fuel consumption, foreign currency savings will be achieved and the share of countries in clean energy production will increase. Geothermal energy use in power plants and hybridized energy systems may also be used in the future to address power inconsistency challenges. In conclusion, it is hoped that future studies and projects would result in a world economy that is only marginally dependent on fossil fuels. Geothermal energy is unquestionably a key part of the solution that will enable the globe to achieve that objective.

NOMENCLATURE

A	surface area
$\dot{\mathbf{E}}$	total energy
h	enthalpy
i	interest rate
g	gravitational acceleration
$\dot{\mathbf{m}}$	mass flow rate
N	system life
P	pressure
Q	heat input rate (net)
s	specific entropy
T	temperature
v	velocity

V	volumetric flow rate
$\dot{\mathbf{W}}$	network output rate
z	elevation

ABBREVIATIONS

CRF	capital recovery factor
Ex	exergy
$\Delta_f G^0$	Gibbs free energy (formation)
LCHP	levelized cost of hydrogen production
LCE	levelized cost of electricity
NA	not available
RH	relative humidity

GREEK SYMBOLS

Φ	annual costs for maintenance and operating
η	system efficiency
υ	process velocity
ν	moles of element in chemical formation's equation
ψ	exergy efficiency
$\hat{Z}$	initial equipment cost

SUBSCRIPTS

ch	chemical
in	inlet
kn	kinetic
out	outlet
ph	physical
pt	potential
w	water
0	reference or ambient conditions

REFERENCES

1. Islam MT, Nabi M, Arefin M, Mostakim K, Rashid F, Hassan N et al. Trends and prospects of geothermal energy as an alternative source of power: A comprehensive review. Heliyon. 2022;8(12):e11836.
2. Moosavian SF, Borzuei D, Zahedi R, Ahmadi A. Evaluation of research and development subsidies and fossil energy tax for sustainable development using computable general equilibrium model. Energy Science & Engineering. 2022;10(9):3267–80.
3. Esmaeilion F, Soltani M, Nathwani J. Assessment of a novel solar-powered polygeneration system highlighting efficiency, exergy, economic and environmental factors. Desalination. 2022;540:116004.
4. Yaghoubirad M, Azizi N, Ahmadi A, Zarei Z, Moosavian SF. Performance assessment of a solar PV module for different climate classifications based on energy, exergy, economic and environmental parameters. Energy Reports. 2022;8:68–84.
5. Sharmin T, Khan NR, Akram MS, Ehsan MM. A state-of-the-art review on geothermal energy extraction, utilization, and improvement strategies: Conventional, hybridized, and enhanced geothermal systems. International Journal of Thermofluids. 2023;18:100323.
6. Borzuei D, Moosavian SF, Ahmadi A, Ahmadi R, Bagherzadeh K. An experimental and analytical study of influential parameters of parabolic trough solar collector. Journal of Renewable Energy and Environment. 2021;8(4):52–66.

7. Azizi N, Esmaeilion F, Moosavian SF, Yaghoubirad M, Ahmadi A, Aliehyaei M et al. Critical review of multigeneration system powered by geothermal energy resource from the energy, exergy, and economic point of views. Energy Science & Engineering. 2022;10(12):4859–89.
8. Ozgener L, Hepbasli A, Dincer I. Energy and exergy analysis of geothermal district heating systems: An application. Building and Environment. 2005;40(10):1309–22.
9. IRENA. Global Geothermal Installed Capacity by International Renewable Energy Agency. Available at: https://www.irena.org/publications.
10. Cao Y, Mihardjo LWW, Farhang B, Ghaebi H, Parikhani T. Development, assessment and comparison of three high-temperature geothermal-driven configurations for power and hydrogen generation: Energy, exergy thermoeconomic and optimization. International Journal of Hydrogen Energy. 2020;45(58):34163–84.
11. Dincer I, Ozturk M. Geothermal energy systems: Elsevier; 2021.
12. Yildirim N, Genc S. Energy and exergy analysis of a milk powder production system. Energy Conversion and Management. 2017;149:698–705.
13. Bayrakci AG, Kocar G Utilization of renewable energies in Turkey's agriculture. Renewable & Sustainable Energy Reviews. 2012;16(1):618–33.
14. Minh VN, Arason S, Gissurarson M, Pálsson PG. Uses of geothermal energy in food and agriculture: Opportunities for developing countries: Food and Agriculture Organization of the United Nations (FAO); 2015.
15. Lund JW, Freeston DH. World-wide direct uses of geothermal energy 2000. Geothermics. 2001;30(1):29–68.
16. Popovska-Vasilevska S. Drying of agricultural products with geothermal energy. International Summer School on Direct Application of Geothermal Energy. 2003.
17. Andritsos N, Dalampakis P, Kolios N. Use of geothermal energy for tomato drying: Oregon Institute of Technology Geo-Heat Center Bulletin; 2003.
18. Ozgener L, Hepbasli A, Dincer I Performance investigation of two geothermal district heating systems for building applications: Energy analysis. Energy and Buildings. 2006;38(4):286–92.
19. Balta MT, Dincer I, Hepbasli A Energy and exergy analyses of a new four-step copper-chlorine cycle for geothermal-based hydrogen production. Energy. 2010;35(8):3263–72.
20. Bina SM, Jaliiinasrabady S, Fujii H. Exergoeconomic analysis and optimization of single and double flash cycles for Sabalan geothermal power plant. Geothermics. 2018;72:74–82.
21. Yao S, Zhang YF, Yu XH. Thermo-economic analysis of a novel power generation system integrating a natural gas expansion plant with a geothermal ORC in Tianjin, China. Energy. 2018;164:602–14.
22. Dincer I, Rosen MA. Exergy analysis of heating, refrigerating and air conditioning: Methods and applications: Academic Press; 2015.
23. Bejan A, Tsatsaronis G, Moran MJ. Thermal design and optimization: John Wiley & Sons, Inc.; 1995.
24. Moosavian SF, Borzuei D, Ahmadi A. Energy, exergy, environmental and economic analysis of the parabolic solar collector with life cycle assessment for different climate conditions. Renewable Energy. 2021;165:301–320.
25. Emadi MA, Mahmoudimehr J. Modeling and thermo-economic optimization of a new multi-generation system with geothermal heat source and LNG heat sink. Energy Conversion and Management. 2019;189:153–66.
26. Hoseinzadeh S, Garcia DA. Techno-economic assessment of hybrid energy flexibility systems for islands' decarbonization: A case study in Italy. Sustainable Energy Technologies and Assessments. 2022;51:101929.
27. Qu H, Masud MH, Islam M, Khan MIH, Ananno AA, Karim A. Sustainable food drying technologies based on renewable energy sources. Critical Reviews in Food Science and Nutrition. 2022;62(25):6872–86.
28. Vasquez NC, Bernardo RO, Cornelio RL. Industrial uses of geothermal energy a framework for application in a developing country. Geothermics. 1992;21(5–6):733–43.
29. Arason S. (Ed.). The drying of fish and utilization of geothermal energy: The Icelandic experience. International Geothermal Conference, Reykjavík, Iceland; 2003: Citeseer.
30. Gupta HK, Roy S. Geothermal Energy: An Alternative Resource for the 21st Century: Elsevier; 2006.
31. Hasan Masud M, Karim A, Ananno AA, Ahmed A. Sustainable drying techniques for developing countries. Sustainable Food Drying Techniques in Developing Countries: Prospects and Challenges. 2020:81–168.
32. Lund JW, Rangel M. (Eds.). Pilot fruit drier for the Los Azufres geothermal field, Mexico. Proceedings of the World Geothermal Congress, Florence, Italy; 1995.

33. Helvacı HU, Akkurt GG. Thermodynamic performance evaluation of a geothermal drying system. Progress in Exergy, Energy, and the Environment: Springer; 2014, pp. 331–41.
34. Helvaci HU, Keles N, Akkurt GG. Energy and exergy analysis of a geothermal energy sourced hot-air drying system. International Journal of Exergy. 2023;41(1):1–19.
35. Abdullah K, Gunadnya I. (Eds.). Use of geothermal energy for drying and cooling purposes. Proceedings World Geothermal Congress; 2010: Citeseer.
36. Suyanto TS, Atmojo JP, Prasetyo BT. (Eds.). Design of a geothermal energy dryer for tea withering and drying in Wayang Windu geothermal field. Proceedings World Geothermal Congress 2010, Bali, Indonesia; 2010.
37. Lund JW. Direct heat utilization of geothermal resources worldwide 2005. ASEG Extended Abstracts. 2006;(1):1–15.
38. Kostoglou M, Chrysafis N, Andritsos N. Modelling tomato dehydration in a tunnel dryer using geothermal energy. Drying Technology. 2013;31(1):5–16.
39. Sreekumar A. Techno-economic analysis of a roof-integrated solar air heating system for drying fruit and vegetables. Energy Conversion and Management. 2010;51(11):2230–38.
40. Sharma A, Chen C, Lan NV. Solar-energy drying systems: A review. Renewable and Sustainable Energy Reviews. 2009;13(6–7):1185–210.
41. Jairaj K, Singh S, Srikant K. A review of solar dryers developed for grape drying. Solar Energy. 2009;83(9):1698–1712.
42. Ekechukwu OV, Norton B. Review of solar-energy drying systems II: An overview of solar drying technology. Energy Conversion and Management. 1999;40(6):615–55.
43. Sodha MS, Chandra R. Solar drying systems and their testing procedures: A review. Energy Conversion and Management. 1994;35(3):219–67.
44. Doymaz I. Air-drying characteristics of tomatoes. Journal of Food Engineering. 2007;78(4):1291–97.
45. Ertekin C, Yaldiz O. Drying of eggplant and selection of a suitable thin layer drying model. Journal of Food Engineering. 2004;63(3):349–59.
46. Chauhan PS, Kumar A, Tekasakul P. Applications of software in solar drying systems: A review. Renewable and Sustainable Energy Reviews. 2015;51:1326–37.
47. Benhamza A, Boubekri A, Atia A, El Ferouali H, Hadibi T, Arıcı M. et al. Multi-objective design optimization of solar air heater for food drying based on energy, exergy and improvement potential. Renewable Energy. 2021;169:1190–1209.
48. Hadibi T, Boubekri A, Mennouche D, Benhamza A, Kumar A. Economic analysis and drying kinetics of a geothermal-assisted solar dryer for tomato paste drying. Journal of the Science of Food and Agriculture. 2021;101(15):6542–51.
49. Delgado-Plaza E, Peralta-Jaramillo J, Quilambaqui M, Gonzalez O, Reinoso-Tigre J, Arevalo A. et al. Thermal evaluation of a hybrid dryer with solar and geothermal energy for agroindustry application. Applied Sciences. 2019;9(19):4079.
50. Cesar LVE, Lilia CMA, Octavio GV, Isaac PF, Rogelio BO. Thermal performance of a passive, mixed-type solar dryer for tomato slices (*Solanum lycopersicum*). Renewable Energy. 2020;147:845–55.
51. Hadibi T, Boubekri A, Mennouche D, Benhamza A, Kumar A, Bensaci C et al. Effect of ventilated solar–geothermal drying on 3E (exergy, energy, and economic analysis), and quality attributes of tomato paste. Energy. 2022;243:122764.
52. Poutanen KS, Karlund AO, Gomez-Gallego C, Johansson DP, Scheers NM, Marklinder IM et al. Grains: A major source of sustainable protein for health. Nutrition Reviews. 2022;80(6):1648–63.
53. Ismaeel HH, Yumrutas R. Investigation of a solar assisted heat pump wheat drying system with underground thermal energy storage tank. Solar Energy. 2020;199:538–51.
54. Hadibi T, Boubekri A, Mennouche D, Benhamza A, Besombes C, Allaf K. Solar–geothermal drying/instant controlled pressure drop-swell drying of mechanically dewatered tomato paste. Journal of Food Process Engineering. 2021;44(10);e13811.
55. Prasetyo RM, Wicaksono A, Biddinika MK, Takahashi F. (Eds.). Study of geothermal direct use for coffee drying at Wayang Windu geothermal field. 2nd International Conference on Chemistry, Chemical Process and Engineering (IC3PE), Yogyakarta, Indonesia; August 14, 2018.
56. Hakim II, Putra N, Margono KT, Gunawan Y. Utilization the heat pipe heat exchanger techniques at low enthalpy geothermal energy to coffee drying process. Journal of Advanced Research in Fluid Mechanics and Thermal Sciences. 2020;74(2):43–53.
57. Gunawan Y, Margono KT, Rizky R, Putra N, Al Faqih R, Hakim II et al. Enhancing the performance of conventional coffee beans drying with low-temperature geothermal energy by applying HPHE: An experimental study. Open Agriculture. 2021;6(1):807–18.

58. Shofinita D, Lestari D, Aliwarga L, Sumampouw GA, Ambarwati SA, Gunawan KC et al. Drying methods of coffee extracts and their effects on physicochemical properties: A review. Food and Bioprocess Technology. 2023;17:47–72.
59. Marhaenanto B, Soedibyo DW, Farid M. Penentuan lama sangrai kopi berdasarkan variasi derajat sangrai menggunakan model warna rgb pada pengolahan citra digital (Digital image processing). Jurnal Agroteknologi. 2015;9(02):102–11.
60. Mahmud MMC, Shellie RA, Keast R. Unravelling the relationship between aroma compounds and consumer acceptance: Coffee as an example. Comprehensive Reviews in Food Science and Food Safety. 2020;19(5):2380–2420.
61. Hallsson SV. Drying of seaweeds by geothermal heat in Iceland. Geothermics. 1992;21(5–6):717–31.
62. Popovski K, Dimitrov K, Andrejevski B, Popovska S. Geothermal rice drying unit in Kotchany, Macedonia. Geothermics. 1992;21(5–6):709–16.
63. Hirunlabh J, Thiebrat S, Khedari J. Chilli and garlic drying by using waste heat recovery from geothermal power plant. RERIC International Energy Journal. 1999;21(2).
64. Kostoglou M, Chrysafis N, Andritsos N. Modeling a tomato dehydration process using geothermal energy. Drying Technology. 2013;31(1):5–16.
65. Afuar W, Sibarani B, Abdurrahman G, Hendrarsakti J. (Eds.). Design of tomato drying system by utilizing brine geothermal. 5th ITB International Geothermal Workshop (IIGW), Bandung, Indonesia; March 27–April 2, 2016.
66. Ambriz-Diaz VM, Rubio-Maya C, Ibarra JJP, Gonzalez SRG, Patino JM. Analysis of a sequential production of electricity, ice and drying of agricultural products by cascading geothermal energy. International Journal of Hydrogen Energy. 2017;42(28):18092–102.
67. Helvaci HU, Menon A, Aydemir LY, Korel F, Akkurt GG. Drying of olive leaves in a geothermal dryer and determination of quality parameters of dried product. Proceedings of the 2nd International Conference on Sustainable Energy and Resource Use in Food Chains Including Workshop on Energy Recovery Conversion and Management; 2019;161:108–14.
68. Sandali M, Boubekri A, Mennouche D, Gherraf N. Improvement of a direct solar dryer performance using a geothermal water heat exchanger as supplementary energetic supply: An experimental investigation and simulation study. Renewable Energy. 2019;135:186–96.
69. Njuguna E. (Ed.). Evaluation of geothermal grain dryers: Case study of Menengai grain dryer. 8th African Rift Geothermal Conference (ARGE-C8), Nairobi; 2020.
70. Gunawan Y, Putra N, Hakim II, Agustina D, Mahlia TMI. Withering of tea leaves using heat pipe heat exchanger by utilizing low-temperature geothermal energy. International Journal of Low-Carbon Technologies. 2021;16(1):146–55.
71. Situmorang V, Pratama Y, Tambunan IH, Hutagalung FW, Sagala A. (Eds.). Internet of Things–based food product drying oven using shallow geothermal heat. 2022 IEEE International Conference of Computer Science and Information Technology (ICOSNIKOM); 2022: IEEE.
72. Perez-Gonzalez E, Severiano-Perez P, Avina-Jimenez HM, Velazquez-Madrazo OD. Geothermal food dehydrator system, operation and sensory analysis, and dehydrated pineapple quality. Food Science & Nutrition. 2023;11(11):6711–27.

6 Energy Storage-Based Hybrid Solar Drying Systems

Abhipriya Patra and V. Arun Prasath
Department of Food Process Engineering, National Institute of Technology, Rourkela, India

Hemanatha Jayasuriya and Pankaj B. Pathare
Department of Soils, Water and Agricultural Engineering, College of Agricultural & Marine Sciences, Sultan Qaboos University, Muscat, Oman

6.1 INTRODUCTION

Drying is an essential step in reducing postharvest food losses and improving shelf stability. It was often achieved by direct exposure to sunlight and was one of the first techniques for preserving and storing food. Solar energy is a prominent renewable source that has become very popular in recent decades. The solar drying devices have been broadly used in the food and agriculture segments to harvest solar energy and dry products like fruits, vegetables, and grains. Traditional solar drying is beyond cost-effective and ecologically friendly, but it predominantly changes on the amount of sunlight and other weather conditions [1, 2]. Hybrid sun drying systems with thermal energy storage (TES) have been established to overcome this restriction and make certain drying even in cloudy or nighttime circumstances. This technique amplifies drying efficiency, improves product quality, and allows operation despite solar energy accessibility [3]. The inclusion of TES, which collects surplus thermal energy throughout periods of utmost sunlight and releases it during periods of gentle solar energy, is a substantial revolution in hybrid solar drying systems [3, 4]. With an emphasis on TES, the idea, elements, and operation of energy storage–based hybrid solar drying systems have been studied in this chapter.

6.2 ENERGY STORAGE–BASED HYBRID SOLAR DRYING SYSTEMS WITH THERMAL ENERGY STORAGE

A hybrid solar drying system links solar energy with a supplementary energy source (such as electricity, biomass, or natural gas) to confirm continuous drying even when sunlight is not present. The chief objectives of these systems are to preserve constant drying temperatures and minimize reliance on inconsistent weather patterns [5]. In order to supply an additional reliable and energy-efficient drying process, TES is necessary while it keeps extra heat through sunny periods and releases it when required [5, 6].

There are two major modes of operation for the hybrid sun drying system:

- *Solar-Only Mode:* During sunny conditions, the system utilizes direct solar energy to dry the products.
- *Hybrid Mode:* To sustain drying conditions, the system uses an extra energy source or stored thermal energy (from the TES unit) during low- or no-sunlight times (cloudy days, nighttime).

In this case, the TES system retains the heat energy gathered by solar collectors and uses it to increase drying time and process efficiency overall.

DOI: 10.1201/9781003344087-6

6.2.1 Components

A hybrid solar drying system with TES typically consists of the following major components:

- *Solar Collector:* The sun radiation is captured by this apparatus and transformed into thermal energy. Flat-plate collectors, evacuated tube collectors, and parabolic trough collectors are common solar collector types used in drying systems. The content type, surface area, and tilt angle all affect the collector's efficiency [6, 7].
- *Drying Chamber:* The drying chamber is where the products to be dried are placed. It can be either a direct or an indirect type:
  - In direct solar drying, the product is exposed to sunlight directly.
  - In indirect solar drying, heat exchangers or air ducts are used to move the heat produced by the solar collector into the drying chamber. Generally speaking, indirect systems provide more control over the drying process and shield the product from outside pollutants [7].

6.2.2 Thermal Energy Storage

Systems for TES, which store excess thermal energy for later use, are essential to energy management. To close the gap between energy supply and demand, TES is widely used in a variety of industries, including building heating and cooling, industrial processes, and renewable energy systems [8, 9]. TES improves energy efficiency, lowers operating costs, and advances environmental sustainability by capturing and storing thermal energy when it is abundant (such as during the day with solar energy) and releasing it during periods of low energy availability (e.g., nighttime or cloudy conditions) [9, 10]. TES stores excess heat energy collected by the solar collector during sunny periods. Various materials, including sensible heat storage (SHS) (e.g., water, rocks, sand) or latent heat storage (LHS) (e.g., phase change materials [PCMs]), can be used for TES [8, 10].

The TES system ensures continuous drying when solar energy is unavailable by releasing stored heat into the drying chamber. In indirect systems, effective heat distribution is achieved without subjecting the product to direct sunlight by means of the heat exchanger, which transmits thermal energy from the solar collector or TES to the drying chamber [5]. When solar energy and stored thermal energy both are insufficient, a backup energy source—such as electricity, biomass, or natural gas—is integrated to provide heat [5]. The system's functioning is governed by a control unit that controls thermal energy flow, alternates between solar-only and hybrid modes, and maintains ideal drying temperatures. To maintain constant drying conditions, sensors may be employed to track the temperature and humidity inside the drying chamber [7].

6.2.3 Types of TES Systems

Based on the heat storage technique, TES systems can be divided into three primary categories: thermochemical energy storage (TCES), LHS, and SHS [11]. Each has unique benefits, and the choice is made in accordance with particular application needs, such as the required temperature range, storage capacity, and system complexity.

- *Sensible Heat Storage:* SHS, the most basic and popular type of TES, stores energy by increasing a material's temperature (such as water, pebbles, or sand). As the material's temperature rises, the heat is retained without going through a phase shift [11, 12]. SHS systems are simple to install and reasonably priced, but because of their poor energy density in comparison to other techniques, they need huge amounts of material to store substantial amounts of energy. Despite this, they are frequently employed in large-scale applications where space constraints are not a serious issue, such as solar thermal plants [13].

- *Latent Heat Storage:* Energy is stored by LHS systems during phase transitions, most frequently between the solid and liquid phases. PCMs are used in these systems because they absorb or release significant amounts of latent heat at a temperature that is almost constant during the phase change [11]. LHS systems require less capacity to store the same amount of energy and are more efficient than SHS systems because PCMs have the ability to store and release thermal energy at certain, stable temperatures [12]. Familiar PCMs involve organic molecules, salt hydrates, and paraffin. LHS is specifically beneficial in purposes such as solar drying systems and developing HVAC systems where preserving a persistent temperature is essential [13].
- *Thermochemical Energy Storage:* It is a more superior type of TES that employs revocable chemical reactions to save energy. Energy is gathered in this manner when a reaction is powered in one direction (endothermic) and released when the reaction is inverted (exothermic) [11]. TCES provides extremely high energy density and long-term energy storage without appreciable losses. These systems are typically employed in specialized applications needing long-term energy storage, but they are also more expensive and harder to build [12].

6.2.4 Applications of TES

TES systems are employed in various fields, providing critical benefits in terms of energy efficiency and sustainability. TES systems are essential in solar power plants, especially concentrated solar power (CSP) plants, to store thermal energy during daylight hours for electricity generation at night or during cloudy periods. This improves the capacity factor of solar power plants and enables more reliable energy supply [8, 11]. TES can store excess heat or cold generated by HVAC systems during off-peak hours and release it during peak hours, reducing energy costs and balancing the load on the power grid. This application is particularly useful in district heating systems and commercial buildings [13]. Many industrial processes require large amounts of thermal energy, often at specific temperatures. TES allows industries to store heat during periods of low demand or when excess heat is available (e.g., from waste heat recovery) and use it during peak periods, improving energy efficiency [11, 12]. TES is combined into hybrid solar drying systems, where surplus solar heat is collected for use during phases of low sunlight. This confirms permanent drying, lowering product decay and increasing energy use productivity [13].

6.2.5 Working of Energy Storage–Based Hybrid Solar Drying Systems

The working of a hybrid solar drying system with TES can be divided into several phases, depending on the availability of solar energy and the drying requirements.

6.2.5.1 Daytime Operation with Solar Energy (Solar-Only Mode)

The solar collector absorbs solar radiation and transforms it into heat during the day when sunlight is present. The drying chamber is heated either directly or indirectly using this thermal energy [10]. Any extra heat is simultaneously stored for later use in the TES unit. In this mode, the system runs entirely on solar power, guaranteeing economical and environment-friendly drying. To encourage moisture evaporation from the product, a fan or blower circulates the air heated by the solar collector through the drying chamber. The hot air engages moisture, which is then ejected from the system via exhaust vents [8, 10].

6.2.5.2 Heat Storage Phase

The solar collector engages solar radiation and transforms it into heat during the day when sunlight is present. The compartment for drying is heated either directly or indirectly managing this thermal energy. Any additional heat is concurrently deposited for afterward use by the TES section. The system lasts completely on solar power in this method, ensuring reasonably and naturally reliable drying [8]. The air heated by the solar collector is issued through the drying chamber by a fan or blower to organize moisture evaporation from the product. The hot air absorbs moisture, which is later banished from the system through eject vents [3, 5].

6.2.5.3 Nighttime or Cloudy Weather Operation (Hybrid Mode)

In the lack of solar energy, the system changes to hybrid mode. Using the thermal energy stored in the TES unit, the drying chamber is heated to accomplish the drying process [5]. Steady drying conditions are retained by the TES unit's constant and controlled energy release. The supplementary energy source is turned on to keep the drying process in the event that the TES unit runs out of energy. This promises nonstop drying despite the weather [10].

6.2.5.4 Regeneration of TES

When solar energy is once again presented, the excess heat stored by the solar collector is kept to "recharge" the TES unit. This charging and discharging cycle facilitates the hybrid drying system to operate continuously over an unlimited period of time, encouraging overall efficiency and confirming more efficient dried products. [3].

6.2.6 Advantages of Energy Storage–Based Hybrid Solar Drying System

The incorporation of TES grants the system to continue drying products even when there is no sunlight, which leads to quicker and more dependable drying cycles. Moreover, by absorbing and storing excess solar energy, hybrid systems decrease dependence on external energy sources, ensuing in lower operating costs and enhanced energy efficiency. Furthermore, hybrid systems afford dependable and controlled drying conditions by avoiding over- or under-drying and reducing the probability of product spoilage or nutrient loss. However, the system lessens carbon emissions and raises sustainability by significantly lowering the requirement for fossil fuels. Additionally, hybrid systems with TES progressively reduce energy costs by amplifying the use of free solar energy and minimizing the need for exterior energy sources.

6.3 THERMAL ENERGY STORAGE WITH PCM IN ENERGY STORAGE–BASED HYBRID SOLAR DRYING SYSTEMS

By collecting thermal energy for subsequent use, TES improves the drying process's efficacy and reliability, making it a critical part of energy organization systems, particularly hybrid solar drying systems [14]. Because PCMs have a great latent heat capacity, their integration into these systems requires an improved solution for effective energy storage [14, 15]. In the circumstances of hybrid solar drying systems, this chapter will assemble on the design, principles, and performance of TES using PCMs, analyzing how they improve drying procedures and increase energy expenditure.

When a material undergoes a phase transition, usually from a solid to a liquid or vice versa, it absorbs and releases thermal energy. Because the sensible heat stored by traditional materials (such as water or sand) is significantly smaller than the latent heat involved in this transition, PCMs are very useful for TES applications [15]. In hybrid solar drying systems, PCMs are perfect because they can store thermal energy at a temperature that is almost constant throughout the phase change. The PCM absorbs or releases a substantial amount of heat during melting or freezing, enabling effective temperature control and guaranteeing a steady drying environment [14, 16].

6.3.1 Types of PCMs

PCMs can be broadly classified into three categories based on their chemical nature:

- *Organic PCMs:* The stability, noncorrosiveness, and congruent melting (without segregation) of organic materials, like fatty acids and paraffin waxes, make them widely used.

Due to their low cost and high LHS capacity, paraffins are especially well-liked. Although fatty acids may cost more, they also have good phase change properties [14, 15].

- *Inorganic PCMs:* The inorganic PCMs most frequently utilized in TES applications are salt hydrates. When it comes to energy storage density and thermal conductivity, they are superior to organic PCMs. However, supercooling—where the substance stays liquid below its melting point—and phase segregation—where the substance splits into distinct components during melting and solidification—are two issues that salt hydrates may encounter [15, 16].
- *Eutectic Mixtures:* In order to ensure that both components undergo a phase change at the same time, eutectic PCMs are mixtures of two or more materials that melt and solidify at a single, precise temperature. Materials that are either organic, inorganic, or a combination of both can form eutectic mixtures. They provide customized melting points and can resolve phase segregation problems in PCMs made of just one component [3].

6.3.2 Working

The ability of PCMs to absorb and release significant amounts of latent heat during phase transitions is the basic mechanism by which they function in TES systems. PCMs absorb extra heat in a hybrid solar drying system during the day when there is a lot of solar radiation [3, 14]. Latent heat is the energy that the PCM material stores as it melts. When there is little to no solar radiation, such as at night or during cloudy situation, the PCM solidifies, emitting the latent heat that has been collected to preserve a constant drying temperature in the chamber [17].

The procedure can be divided into several phases. The PCM absorbs heat during the day and changes from a solid to a liquid when solar energy is abundant. This phase transition, which occurs at a temperature that is nearly constant and does not substantially change the drying conditions, enables efficient heat absorption [14, 18]. Then again, the absorbed energy is stored as latent heat by the PCM, which is still liquid. Long-term availability of the stored energy makes it a dependable thermal reservoir [17]. Furthermore, when solar energy is no longer there, PCM creates to solidify, distributing the collected latent heat. This heat is then used to preserve a constant temperature in the drying chamber, confirming constant drying and avoiding product damage [19]. Additionally, once solar energy is accessible again, the PCM is reheated, repeating the cycle.

6.3.3 Advantages of PCMs in Hybrid Solar Drying Systems

There are numerous noteworthy benefits to using PCMs in TES in hybrid solar drying systems. First, a sizable amount of thermal energy can be stored in a relatively small volume, thanks to PCMs' latent heat capacity. This enables the development of compact, efficient TES units with sufficient heat storage for ongoing drying [1, 20]. Second, in drying systems, when maintaining a precise temperature range is crucial for product quality, the phase change process takes place at a practically constant temperature. By preventing temperature swings that can interfere with the drying process, the PCM makes sure that the drying chamber temperature is constant [16]. Also, in order to minimize operating expenses and increase overall energy efficiency, PCMs store extra heat during periods of high solar radiation and release it when needed. However, by optimizing the utilization of solar energy, PCMs contribute to a decrease in dependency on nonrenewable energy sources. This helps create dryers that are more ecologically friendly and sustainable, especially in the food and agriculture sectors where energy use is a factor [14, 15]. Moreover, large industrial drying applications to small-scale agricultural dryers can benefit from both the scalability and integration of PCM-based TES systems. Because of PCMs' adaptability in terms of melting points and energy storage capacities, customized solutions that satisfy particular drying requirements can be created [16].

6.3.4 Design Considerations for PCM-Based TES in Hybrid Solar Drying Systems

When designing a hybrid solar drying system with PCM-based TES, several factors must be considered to ensure optimal performance. The temperature range needed for the drying process determines which PCM to use. For efficient heat storage and release, the melting point of the PCM should coincide with the intended drying temperature. For many agricultural drying applications, paraffin wax with a melting point of 50–60°C is appropriate; however, salt hydrates can be utilized in systems that need higher temperatures [5, 20]. The thermal conductivity of the PCM and the heat exchange system are essential for effective energy storage and recovery. Heat exchangers or materials with better transmission, such as metal fins or graphite, can be added to PCMs to improve their low thermal conductivity [3, 20]. The hybrid drying system must effortlessly incorporate the TES unit. To continue constant drying situation, this implies ensuring the solar collector, drying chamber, and PCM storage unit all function concurrently. For the PCM to use energy as efficiently as feasible, the control system must control its cycles of charging and discharging [14]. PCMs should be strong sufficient to survive numerous melting and solidification cycles lacking undergoing evident degradation. Materials with durable thermal properties are better for long-term use while they develop system reliability and decrease preservation losses [15, 16].

6.3.5 Applications of PCM-Based TES in Solar Drying Systems

Industries that require reliable drying procedures over long eras of time advantage significantly from PCM-based TES systems. In agricultural drying systems, PCMs are managed to dry products like grains, fruits, vegetables, and herbs. By applying TES, the shelf life of products is extended by keeping them from destroying and ensuring they continue to dry during periods of low sunlight [3, 17]. Maintaining ideal drying specifications is necessary for the food business to retain texture, flavor, and nutritional value. Precise temperature control is provided by PCM-based TES systems, guaranteeing superior dried foods [14]. Low-temperature drying is required for many herbs and medicinal plants in order to protect their active components. In order to preserve these circumstances and ensure that the foods maintain their potency and therapeutic properties, PCMs with low melting points are perfect [16]. PCM-based TES systems assist maintaining consistent drying temperatures and lower energy usage in large-scale industrial drying applications, such as wood, textiles, and ceramics, by employing stored solar heat during slack periods [21].

Despite the many benefits of PCM-based TES systems, there are specific issues that must be solved. To begin with, some high-performance PCMs, especially the inorganic types, can be expensive and difficult to achieve for large-scale uses. Researchers are working to form PCMs that are more cost-effective and environment-friendly [15, 21]. Moreover, PCMs have a lot of heat phases in which to consider their qualities. Over time, the system's efficiency may reduce due to phase separation, supercooling, or corrosion-induced PCM degradation. To increase PCM steadiness, new materials and encapsulating methods are being researched [3]. Additionally, it takes careful planning and management to involve PCM-based TES systems into hybrid solar dryers. For best results, sophisticated control systems with effective charging and discharging cycles are necessary [14, 16].

6.4 RECENT STUDIES ON SOLAR DRYERS WITH TES

Studies examining the incorporation of TES in solar dryers have attracted a lot of attention in recent years. Researchers are looking into how TES systems, especially those that use PCM, might improve the reliability and effectiveness of solar drying processes [3, 8]. This would consent for constant drying during periods when there is insignificant to no solar energy. This chapter will examine current developments and research in the field of solar dryers employing thermoelectric sensing (TES), emphasizing the ways in which different TES materials and systems have been used to improve drying effectiveness, energy economy, and product quality. The design, functionality,

TABLE 6.1
List of Drying Foods in Different Types of Dryers along with Their Performances

S. No.	Solar Dryer Types	Food Dried	Key Findings	Reference
1	Indirect solar dryer	Apples	The TES system increased drying efficiency, allowing for quicker drying times without sacrificing the texture and color preservation of the final product	[32]
2	Direct solar dryer	Mango slices	Notably shorter drying times with better energy efficiency and product quality, particularly when it comes to flavor compound preservation	[33]
3	Hybrid solar dryer	Carrots	When compared to open solar drying, TES reduced drying time and improved overall energy efficiency by extending the drying hours into the night	[34]
4	Solar greenhouse dryer	Pineapples	Stable drying temperatures, lower energy usage, and enhanced final quality in terms of texture and color all were made possible by the use of PCM	[35]
5	Solar tunnel dryer	Okra	PCM incorporation in the dryer increased drying efficiency and considerably reduced microbiological activity in the final product, extending product shelf life	[36]
6	Mixed mode solar dryer	Grapes	Intensified drying speed and equality, along with a constant heat source at night, result in lower postharvest losses and better drying performance overall	[6]
7	Solar cabinet dryer	Figs	Smaller drying times and increased sensory qualities for the dried figs can be attained by using TES to continuously dry intermittent sunlight	[37]
8	Solar–electric hybrid dryer	Peppers	Raised energy effectiveness and drying consistency, which reduce drying times and maintain bioactive compounds like capsaicin	[38]
9	Solar tray dryer	Garlic	Significant energy investments and a faster drying rate allow for constant drying still in cloudy situations	[39]
10	Solar-assisted hybrid dryer	Medicinal leaves	The aromatic and medicinal qualities of the leaf were maintained by preserving constant drying temperatures, which declined the thermal degradation of essential oils	[40]

and use of these technologies all will be covered in the conversation, along with some new findings from simulation and experimentation investigations. However, various studies have been carried out recently to dry several types of foods in different types of dryers and their performances on the properties of dried foods which is enlisted in Table 6.1.

Solar dryers are machines that use solar radiation to extract the moisture out of industrial and agricultural goods. The drying process of traditional solar dryers is sporadic and less efficient under overcast or nocturnal conditions due to its need of sunshine, which is a common limitation. In order to get over these restrictions, TES devices are built into solar dryers so that extra thermal energy is stored at times of intense sunshine and released throughout the day [7, 12]. TES can shorten drying times, keep drying conditions constant, and stop product spoiling. The most popular TES materials used in solar drying systems are PCMs, which during phase transitions store thermal energy as latent heat [11, 15]. Recent studies have shown promising results in using PCM-based TES systems to enhance the drying process. In addition to PCMs, other TES approaches, such as SHS (using materials like water or rocks), have been studied.

6.4.1 Recent Advances in PCM-Based TES for Solar Dryers

Recent studies have focused on augmenting the design and performance of PCM-based TES systems in solar dryers. These systems make use of the extensive latent heat capability of PCMs, which permits them to store and release thermal energy at almost constant temperatures. This stuff is essential for maintaining stable drying conditions. Here are a few of the main conclusions from current research.

6.4.1.1 Types of PCMs Used in Solar Dryers

Numerous kinds of PCMs have been studied in the circumstances of solar drying. The choice of PCM is largely influenced by the desired operating temperature of the dryer and the thermal properties of the product being dried. Some commonly used PCMs include the following:

- *Paraffin-Based PCMs:* Their high latent heat capacity, stability, and lack of toxicity make them widely employed. Paraffin wax, for example, has been extensively researched for agricultural drying applications, such as drying fruits, vegetables, and cereals, with a melting temperature of approximately 55–60°C [22, 23].
- *Fatty Acids and Salt Hydrates:* It has been demonstrated that inorganic PCMs like calcium chloride hexahydrate and fatty acids like stearic and lauric acid have superior heat transfer capabilities and higher thermal conductivity than paraffins [24].
- *Eutectic Mixtures:* Eutectic combinations of inorganic and organic PCMs have also been studied by researchers in an effort to develop specialized TES systems with melting points that correspond to the necessary drying temperatures [14].

Abuelnuor et al. [23] investigated the use of paraffin wax as a PCM in a solar dryer for drying tomatoes. The results showed that the dryer with TES improved the drying efficiency and product quality, particularly during periods of intermittent solar radiation.

6.4.1.2 Performance Improvements with PCM-Based TES

Numerous studies have shown that PCM-based TES devices can greatly enhance solar dryer performance. Sayehvand et al. [25] directed an investigational study evaluating a solar dryer with PCM-based TES to a conservative solar dryer. They determined that improving PCM authorized for extended drying times, a 30% decrease in drying time and a complete enhancement in the system's energy efficacy are achieved. In a similar study, Constantino-Robles et al. [1] studied the benefit of a hybrid solar dryer with PCM and a supplementary heating resource. As per the study, even at night, the system could persist to dry herbs at temperatures that were ideal for doing, thus lacking the requirement for a backup power source. As a result, energy usage was 25% lower than in non-TES systems.

6.4.1.3 Optimization of TES Design and Integration

The degree to which TES systems integrate with solar dryers depends on several design factors, such as the size and location of the TES unit, the heat transfer mechanism, and the type of PCM used. To optimize the performance of PCM-based TES systems in solar dryers, a variety of designs have been studied. A study by Pandey et al. [26] established a solar dryer with a PCM–TES system that used paraffin wax fixed in a thermal storage bed. In order to expand the heat exchange between the PCM and the drying chamber, the system was constructed. The outcome showed that employing a finned heat exchanger lowered drying times for agricultural products by 40% and improved heat transfer

19. Bhardwaj, A. et al., *Experimental investigation and performance evaluation of a novel solar dryer integrated with a combination of SHS and PCM for drying chilli in the Himalayan region*. Thermal Science & Engineering Progress, 2020. **20**: p. 100713.
20. Shalaby, S. et al., *The effect of drying sweet basil in an indirect solar dryer integrated with phase change material on essential oil valuable components*. Energy Reports, 2020. **6**: p. 43–50.
21. Brahma, B., *Design and development of phase change material integrated solar air heater for drying application in agriculture*. 2023: Tezpur University.
22. Tarachand, R.R., *Experimental Studies on Solar Cabinet Dryer Integrated with Thermal Storage System Using Phase Change Material*. 2017: Dr. Panjabrao Deshmukh Krishi Vidyapeeth.
23. Abuelnuor, A.A. et al. *Experimental Study on Tomato Drying Using a Solar Dryer Integrated with Reflectors and Phase Change Material*. In: *2020 International Conference on Computer, Control, Electrical, and Electronics Engineering (ICCCEEE)*. 2021: IEEE.
24. Rathod, M.K., and J. Banerjee, *Thermal stability of phase change materials used in latent heat energy storage systems: A review*. Renewable & Sustainable Energy Reviews, 2013. **18**: p. 246–258.
25. Sayehvand, H.-O., S. Abolfathi, and B. Keshavarzian, *Investigating heat transfer enhancement for PCM melting in a novel multi-tube heat exchanger with external fins*. Journal of Energy Storage, 2023. **72**: p. 108702.
26. Pandey, S., A. Kumar, and A. Sharma, *Sustainable solar drying: Recent advances in materials, innovative designs, mathematical modeling, and energy storage solutions*. Energy, 2024. **308**: p. 132725.
27. Shalaby, S., and M. Bek, *Experimental investigation of a novel indirect solar dryer implementing PCM as energy storage medium*. Energy Conversion & Management, 2014. **83**: p. 1–8.
28. Bareen, A. et al., *Experimental investigation of an indirect solar dryer with PCM-integrated solar collector as a thermal energy storage medium*. Environmental Science & Pollution Research, 2024. **31**(12): p. 18209–18225.
29. Demissie, Y.A. et al., *Advancements in solar greenhouse dryers for crop drying*. Energy Reports, 2024. **11**: p. 5046–5058.
30. Bhardwaj, A. et al., *Energy and exergy analyses of drying medicinal herb in a novel forced convection solar dryer integrated with SHSM and PCM*. Sustainable Energy Technologies & Assessments, 2021. **45**: p. 101119.
31. Razak, A.A. et al., *Recent advances in solar drying system: A review*. International Journal of Engineering Technology & Sciences, 2021. **8**(1): p. 1–13.
32. Singh, D., S. Mishra, and R. Shankar, *Experimental investigation and drying kinetics of mixed type solar dryer with thermal energy storage material for drying of apple slices*. Energy Sources, Part A: Recovery, Utilization, & Environmental Effects, 2022. **44**(2): p. 4763–4782.
33. Kannapiran, K. et al. *An experimental investigation of a direct cabinet solar dryer integrated with a phase change material for drying mangoes*. In: *AIP Conference Proceedings*. 2024. AIP Publishing.
34. Gilago, M.C., V.R. Mugi, and C. Velayudhan Parvathy, *Analysis and comparison of the performance parameters of passive and active indirect solar dryers with heat storage facility while drying carrot*. Environmental Science & Pollution Research, 2023. **30**(19): p. 56246–56258.
35. Gilago, M.C., and V. Chandramohan, *Study of drying parameters of pineapple and performance of indirect solar dryer supported with thermal energy storage: Comparing passive and active modes*. Journal of Energy Storage, 2023. **61**: p. 106810.
36. Yadav, D.K., V.K. Arora, and V. Yadav, *Investigation of thermal performance, drying characteristics and environomical analysis: Direct flow evacuated tube solar drying of okra*. Environmental Science & Pollution Research, 2024. **31**(23): p. 34214–34233.
37. Ekka, J.P. et al., *Performance studies on mixed-mode forced convection solar cabinet dryer under different air mass flow rates for drying of cluster fig*. Solar Energy, 2021. **229**: p. 39–51.
38. Azaizia, Z. et al., *Experimental study of a new mixed mode solar greenhouse drying system with and without thermal energy storage for pepper*. Renewable Energy, 2020. **145**: p. 1972–1984.
39. Malakar, S., V.K. Arora, and P.K. Nema, *Design and performance evaluation of an evacuated tube solar dryer for drying garlic clove*. Renewable Energy, 2021. **168**: p. 568–580.
40. Khallaf, A.E.-M., and A. El-Sebaii, *Review on drying of the medicinal plants (herbs) using solar energy applications*. Heat & Mass Transfer, 2022. **58**(8): p. 1411–1428.
41. Bennamoun, L., and J. Li, Drying process of food: Fundamental aspects and mathematical modeling. In: Natural and Artificial Flavoring Agents and Food Dyes. 2018: Elsevier. p. 29–82.
42. Chen, X.D., and A.S. Mujumdar, *Drying technologies in food processing*. 2009: John Wiley & Sons, Inc.
43. Adeyeye, S.A.O., T.J. Ashaolu, and A.S. Babu, *Food drying: A review*. Agricultural Reviews, 2022. **1**(8). doi: 10.18805/ag.R-2537.
44. Dutta, C. et al., *Drying characteristics and quality analysis of pre-treated turmeric (Curcuma longa) using evacuated tube solar dryer with and without thermal energy storage*. Solar Energy, 2023. **251**: p. 392–403.

7 Refractive Window Drying for Food Materials

Preetinder Kaur and Ruchika Zalpouri
Department of Processing and Food Engineering,
Punjab Agricultural University, Ludhiana, Punjab, India

Manpreet Singh
Department of Renewable Energy Engineering,
Punjab Agricultural University, Ludhiana, Punjab, India

ABBREVIATIONS

CD	convective drying
DD	drum drying
FD	freeze-drying
HAD	hot-air drying
IRD	infrared drying
IT	inlet air temperature
MHAD	microwave-assisted hot-air drying
MSBD	microwave spouted bed drying
OD	oven drying
OT	outlet air temperature
q_{cond}	heat transfer by conduction
q_{conv}	heat transfer by convection
q_{evap}	heat transfer by evaporation
q_{radi}	heat transfer by radiation
RWD	refractance window drying
SBD	spouted bed drying
SD	spray drying
T_{air}	ambient air temperature
TD	tray drying
t_{tpm}	thickness of the transparent plastic material
T_{water}	water temperature
t_{wm}	thickness of the wet material
T_1 **and** T_2	transparent plastic film temperature below and above the surface
T_3	product surface temperature
UICHD	ultrasound- and infrared-assisted conductive hydro-drying

7.1 INTRODUCTION

The evolution of drying technology progresses from basic solar energy utilization to cutting-edge advancements, categorized into four generations. First-generation dryers rely on hot air circulation to extract moisture from solid materials like grains, sliced fruits, or vegetables. Second-generation dryers use heat energy to turn slurries and purees into powders and flakes, while third-generation dryers use nonthermal energy to dry fruits and vegetables. The highest level of innovation can be

DOI: 10.1201/9781003344087-7

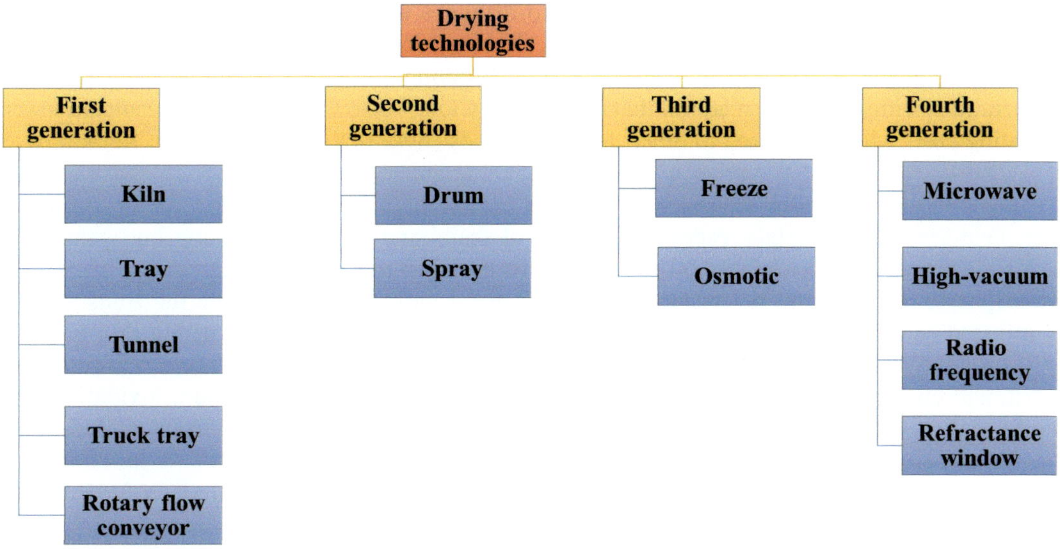

FIGURE 7.1 The development of drying methods. (*Source:* Vega-Mercado et al., 2001.)

found in fourth-generation dryers, which prioritize maintaining product quality for the intended use. This progression, outlined by Vega-Mercado et al. (2001), illustrates the journey from traditional methods to state-of-the-art technology. Baeghbali et al. (2018) discuss the shift toward nonthermal energy in third-generation dryers, emphasizing advancements in preserving food quality. Additionally, Chou and Chua (2001) underscore the significance of fourth-generation dryers, highlighting their role in elevating drying technology standards (Figure 7.1).

Drying is an essential step in the process of turning liquid or semiliquid materials into flakes, powders, or leathers. For this, a variety of drying technologies, each with its own set of drawbacks, such as convective hot air, drum, spray dryer, and freeze dryers are frequently used. Direct exposure to hot air during convective hot-air drying can alter the physicochemical quality (Maisnam et al., 2017). Drum drying is not recommended for sticky or high-sugar foods because it usually involves temperatures between 120°C and 170°C, which can cause significant quality degradation (Karthik et al., 2017). However, it has been discovered that drum drying is a better option than spray drying for some products, such as potato flakes. In order to quickly remove moisture, spray drying, a popular industrial technique, converts liquid feed into an atomized mist that is then exposed to hot air. However, rotary disk atomizers may experience internal corrosion, and spray dryer nozzles are prone to clogging (Sosnik & Seremeta, 2015). Particles of powder sticking to the walls of the chamber also raise cleaning costs.

Freeze-drying stands out as a superior technique for producing high-quality dried products with desirable crystal structure, color, flavor, and excellent rehydration capacity. However, it is considerably more expensive than convective hot air drying, typically costing three to seven times more (Waghmare, 2021). Furthermore, freeze-dried materials tend to be porous and highly hygroscopic, especially in humid environments (Oyinloye & Yoon, 2020). Despite their benefits, these methods entail laborious processes, require costly equipment installation, and demand high levels of maintenance, making them financially burdensome. Additionally, they often lead to the loss of flavor, color, and bioactive properties in the final product (Srivastav et al., 2020; Verma et al., 2019).

Researchers have focused on refractance window drying (RWD), a novel drying technique that falls under the categories of contact, indirect, or film drying technique, in an effort to overcome these difficulties. Also known as conductive hydro-drying, RWD represents a variation of cast-tape drying (da Silva Simão et al., 2021). This novel technology was first created for commercial use by MCD Technologies Inc. in Tacoma and Mt. Capra, Washington (Kumar et al., 2021). It works at temperatures within the food material that are relatively low—below 70°C—which results in lower operating costs when compared to other drying methods (Aykın-Dinçer et al., 2021; Karadbhajne et al., 2019).

The RWD uses a quick-drying, low-temperature method that works well for drying delicate, heat-sensitive materials, especially purees of fruits and vegetables. Using hot water as a heat transfer medium below its boiling point is essential to this method (Hashem et al., 2019). Atmospheric pressure and hot water below 99°C are usually used. To enable the transmission of infrared energy at the speed of light, a transparent film is evenly coated with a thin layer of food products. By drying at atmospheric pressure, RWD preserves the color, flavor, aroma, and nutritional quality of the products. Additionally, dried products are less prone to oxidation and the formation of free radicals, thereby enabling longer storage periods. In terms of food quality, RWD outperforms drum drying, spray drying, and freeze-drying due to its lower drying temperatures, shorter drying times, and lower energy consumption (Kumar et al., 2021; Raghavi et al., 2018). Moreover, water from the dryer is reclaimed and reheated in the reservoir rather than being wasted, enhancing the thermal efficiency of the RWD process. According to Zotarelli et al. (2015), RWD technology uses 50% less energy and requires 50–70% less capital investment than freeze-drying technology.

This chapter aims to delve into the current trends in RWD of food products. It provides a comprehensive examination of the design and fundamental principles behind RWD. In addition, it provides a thorough examination of energy consumption and product quality throughout the RWD process, contrasting them with different drying methods used on different goods. Through this chapter, readers gain insight into the intricacies of RWD, its efficiency in preserving product quality, and how it stands against other drying methods in terms of energy consumption and overall performance across different types of products.

7.2 DESIGN OF A RW DRYER

Among the cutting-edge technologies for turning liquids and semi-liquids into flakes or powders is the RWD technique. The main parts of a normal RWD are shown in Figure 7.2, and they include a water tank, heating unit, hot-water pump, hopper, transparent plastic conveyor belt, and stainless

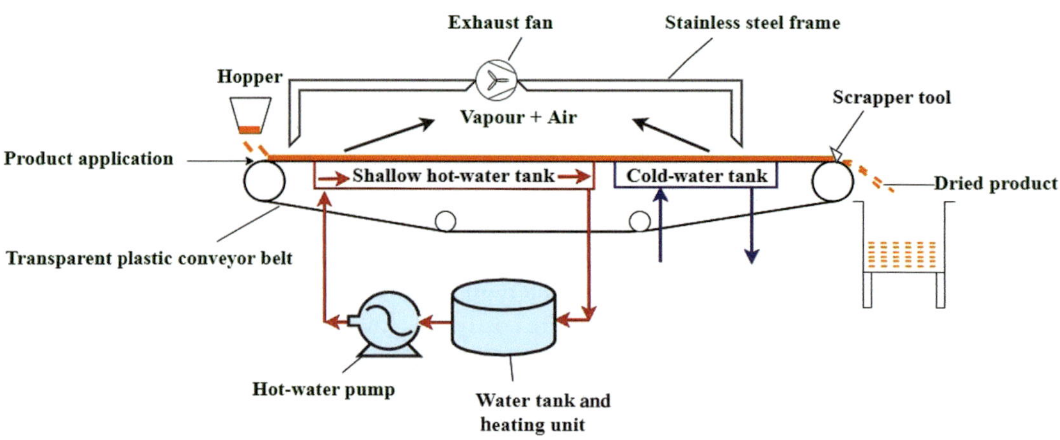

FIGURE 7.2 Schematic diagram of a RWD technique. (*Source:* Zalpouri et al., 2022.)

steel frame. The conveyor belt is positioned above a shallow hot-water tank. RW dryers utilize circulating hot water (temperature: 95–97°C) at atmospheric pressure to provide thermal energy to the wet material undergoing drying. The wet material is spread thinly over the conveyor belt, facilitating the transfer of heat energy from the hot-water surface and expediting the drying process of the sample. The belt moves across a cold-water tank as the material dries, bringing the product's temperature down ideally below the glass transition point, and preventing it from adhering to the belt. A scraper tool is provided at the end of the conveyor belt to facilitate the easy removal of the dried product (Raghavi et al., 2018; Waghmare, 2021; Yoha et al., 2019).

7.3 WORKING PRINCIPLE OF THE RWD

During the process of RWD, thermal energy transfers from hot water to the product through three modes: conduction, convection, and radiation. Illustrated in Figure 7.3, this heat transfer begins with the conveyor belt, which rapidly adopts the temperature of the hot water beneath it due to its thinness. The main mechanisms facilitating the drying of the sample involve convection (from water to conveyor belt), conduction (from conveyor belt to wet material on the conveyor belt), and radiation (from conveyor belt to wet material) (see Figure 7.3) (Aragón-Rojas et al., 2019; Hernández-Carrión et al., 2021). The use of hot water just below the boiling temperature and a thin conveyor belt with infrared transmission within the wavelength range matching water's absorption spectrum synergistically enable rapid drying (Amer, 2011; Smith, 1994). Water exhibits high infrared absorption within wavelengths ranging from 3.0 μm to 15.3 μm (van 't Land & Raes, 2019). Minimizing refraction at the plastic–puree interface permits radiant thermal energy to pass through the plastic and into the product when a wet material is placed on the conveyor belt. The absorptivity of infrared thermal energy is dependent on several factors, including the temperature of the heating medium, the plastic material's thickness, and the moisture content of the material being dried (Ratti & Mujumdar, 1995). Due to convective heat transfer to the ambient air above the material and evaporative cooling, the product temperature in RWD is limited to 60–70°C. This effect may also be attributed to the creation of an air gap. Saturated vapor in RWD limits the interaction between the sample and oxygen, producing high-quality dried products (Waghmare, 2021).

Thermal radiation, particularly within relevant wavelengths, effectively permeates the plastic material, making it an efficient transmitter of thermal radiation. An ideal situation for RWD is to create a "window" that allows for the transfer of mass and heat. When wet material is spread over the conveyor, this "window" makes it easier for infrared energy from hot water to pass through it. The majority of the radiant energy incident at the plastic–water interface, however, is reflected back into the hot water when there is no wet material above the conveyor due to the mismatch in refractive index between the hot water and air. As a result, the conveyor transmits very little radiant energy, essentially closing the "window" (Zotarelli et al., 2015).

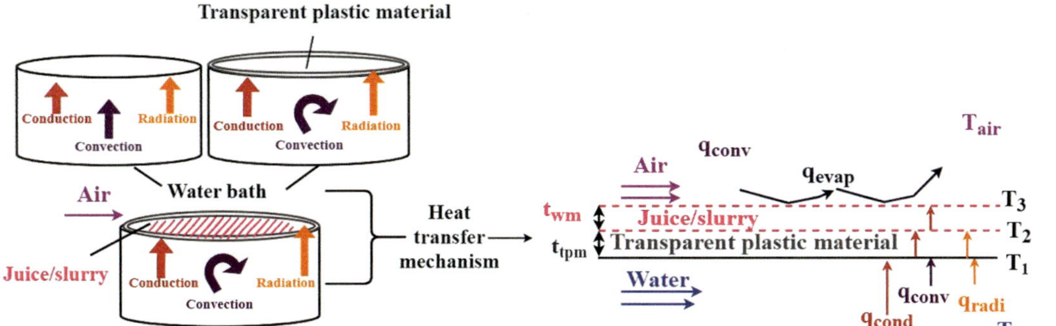

FIGURE 7.3 Heat transfer mechanism in RW dryer. (*Source:* Zalpouri et al., 2022.)

As the wet material is distributed over the conveyor during RWD, the refractive index mismatch between the wet material and water is greatly decreased at the water–plastic–wet material interface. The "windows" are open when the refractive indices are closer because there is less reflection and more radiant heat transmission into the food. But as the product dries out and its moisture content drops, the refractive index mismatch rises and more light is reflected back into the water, closing the "window" (Nindo & Tang, 2007; Ortiz-Jerez et al., 2015).

Until the product hits critical moisture content, conduction effects continue to support moisture evaporation throughout the drying process. The related evaporative cooling procedure reduces the rate of heat transfer during the last drying stages, protecting the product from overheating. Inhibiting the majority of vegetative bacterial pathogens, cutting drying costs, and preserving relatively low temperatures are the main advantages of this drying method (Ortiz-Jerez et al., 2015).

7.4 HISTORY OF THE DEVELOPMENT OF RWD

In the field of RWD, numerous developments and refinements have been detailed in scholarly literature. Magoon (1986) pioneered a specialized apparatus tailored for RW drying, specifically targeting pulp and fruit juice to yield leathers. Expanding upon this innovation, Bolland (2000) enhanced the RWD by integrating a cold-water tank, facilitating the cooling of the dried product before extraction. Notably, Nindo et al. (2003a) employed a commercial full-size RW dryer in their experiments, a significant departure from the conventional pilot-scale dryers, being five times larger in scale. A batch RWD system with a polyester film plastic membrane and an exhaust fan for effective water vapor collection was introduced by Baeghbali et al. (2010). A pilot-scale RW dryer with a glass plate cover and an insulated water basin was created by Ghanem (2010). Baeghbali et al. (2019) showcased further innovation by creating a continuous pilot-scale RW dryer that integrated infrared and ultrasound technologies. Zalpouri et al. (2020) presented a pilot-scale RWD with water heaters and a flexible metalized polyester belt that was designed for processing potato puree. A lab-scale batch RW dryer with a static Mylar belt and hot-water trough was built by Ssenyimba et al. (2021). A PV-T system-assisted RW dryer that uses solar power for electric heaters was presented by Seyfi et al. (2021). Hernández-Carrión et al. (2021) optimized an RW dryer with various components, including a water pump, heating unit, and polyethylene terephthalate plastic belt, achieving a drying surface area of 0.43 m^2. Recent contributions include Mandale et al. (2023), who developed a solar-assisted continuous RW dryer, incorporating components such as a hot-water bath, dryer stand, and an evacuated tube solar water heater. Additionally, Zalpouri (2023) devised a batch-type pilot-scale solar-assisted RW dryer employing an evacuated tube water heating system alongside an auxiliary heater of 4 kW, with modifications to the water trough transforming it into a drying cabinet.

7.5 ENERGY EFFICIENCY

Many studies have compared the energy consumption and efficiency of different drying technologies to that of RWD. The main energy consumption during RWD operation is due to water evaporation from the air–product interface. Equation 7.1 states that the energy needed to heat the product and evaporate any water from it can be divided by the energy input from the water heater to determine the thermal efficiency of a RW dryer (Abonyi et al., 1999). RWD transfers thermal energy to the material being dried by means of recirculating hot water. Among other drying techniques, freeze-drying, drum drying, spray drying, and hot-air drying are less energy-efficient than RWD.

This increased effectiveness results from RWD's recycling and reheating of the processing water, which reduces losses. Nindo and Tang (2007) reported that the thermal efficiency of RWD ranges from 52% to 77%, equivalent to that of drum drying (35–78%), while rotary dryer efficiency (25–50%) matches that of a spray dryer (20–51%). Baeghbali et al. (2016a) assessed how RWD, spray drying, and freeze-drying affected the overall energy efficiency of pomegranate juice drying. The overall energy efficiencies for pomegranate juice after freeze-drying, spray drying, and RWD

TABLE 7.1
The Energy Efficiency of Different Dryers for Different Commodities

Crop	Type of Dryer	Overall Energy Efficiency (%)	Reference
Pomegranate juice	Freeze	1.12	Baeghbali & Niakousari, 2018; Baeghbali et al., 2016a
	Spray	12.92	
	RWD	31.56	
Apple slices	Freeze	2.02	Baeghbali et al., 2019
	Cabinet	27.62	
	RWD	37.66	
Aloe vera gel	Hot-air	7.00	Seyfi et al., 2021
	RWD	20.00	

were 1.12%, 12.92%, and 31.56%, respectively (Table 7.1). The RW dryer had a higher overall energy efficiency than the freeze dryer and spray dryer. In particular, the RWD of pomegranate juice samples used about 1/3 and 1/30 of the energy used for spray drying and freeze-drying, respectively. It is important to remember that freeze-drying is an extremely expensive drying technique.

$$\text{TE}\,(\%) = \frac{E_{ev}}{\text{water heater energy input}} \times 100 \tag{7.1}$$

$$\text{Eev: } Q \times \text{hfg} \tag{7.2}$$

where TE is the overall thermal efficiency of dryer, Q is the moisture removed per unit time, and hfg is the latent heat of vaporization of water.

7.6 QUALITY OF RW-DRIED PRODUCTS

7.6.1 Color

High-quality food products must have a good color because it has a direct impact on consumer acceptance. Color is measured using a variety of color coordinate systems, including RGB (red, green, blue) and L* a * b * from the Commission Internationale de l'Eclairage (CIE) and Hunter Lab. In these systems, reddish colors have positive values denoted by the parameter a*, greenish colors have negative values denoted by the parameter b*, and yellowish colors have positive values denoted by the parameter b*. L*, which ranges from black to white in brightness, is frequently used as a rough indicator of luminosity (Costa et al., 2011; Pathare et al., 2013).

During thermal processing, various reactions can cause changes in color. The kind of drying process used and the surrounding environment can have a big impact on how a product's color changes. Furthermore, the color of food products is directly related to the concentration of pigment (Giuliani et al., 2016). Gentle drying, which reduces the creation of unwanted pigments and variations in color parameters, is a characteristic of RWD (Zalpouri et al., 2020). Abonyi et al. (2001) compared the color characteristics of carrot purees dried using various methods, including spray drying, drum drying, freeze-drying, and RW drying. They found that freeze-dried carrot purees exhibited the brightest color (highest L*), while drum-dried purees had a darker product, indicated by higher b*/a* and hue angle values. Higher L*, a*, b*, and chroma values in RW-dried carrot purees indicated more intense and saturated red and yellow colors.

When compared to other drying methods, Nindo et al. (2003b) found that powdered asparagus puree dried using RW drying produced a vibrant green color. In an identical manner, Caparino et al. (2012)

investigated the color properties of mango powder made using various drying techniques and discovered that while spray-dried powder had a lighter color than drum-dried powder, there was no noticeable distinction between RW and freeze-dried powder. RW-dried samples displayed noticeably brighter colors than drum-dried samples, according to similar results reported for dried paprika and potato samples. Research comparing the color of dried products from RW and freeze-dried goldenberry pulp revealed little variations, suggesting that RW drying is an efficient method of preserving color that is comparable to freeze-drying (Nayak et al., 2011; Puente-Díaz et al., 2020; Tontul et al., 2018).

Additionally, it has been demonstrated that RW drying produces cherry powder with superior color compared to convective drying techniques, effectively preserving the orange color of carrot slices longer than traditional air drying. These results imply that RW drying can yield high-quality products at a lower cost that are at par with those produced by freeze-drying. Furthermore, when compared to convective tray-dried potato flakes and mushroom slices, RW-dried flakes and slices showed higher L* and b* values and lower a* values. This was explained by the intense heat treatment used during convective tray drying, which resulted in nonenzymatic reactions and the caramelization of sugars (Hernández-Santos et al., 2016; Tontul et al., 2018; Zalpouri, 2018; Zalpouri et al., 2021; Kumar et al., 2025).

7.6.2 Textural Properties

A vital component of food is texture, which is the way that its structure manifests to the senses and reacts to forces applied to it through the senses of sight, touch, and sound (Szczesniak, 1990). The texture of a meal plays a significant role during mastication, influencing the touchable and physical sensations experienced, making it a key quality indicator (Martynenko & Janaszek, 2014). Food items' textures are greatly influenced by the temperature and rate of drying. When it comes to drying liquid or semiliquid foods, the RWD method works better than cubes or slices of fruit or vegetable. Less shrinkage happens during RWD of sliced fruits and vegetables, improving the textural quality and pore count (Waghmare, 2021).

Research comparing oven-dried at 80–100°C and RW-dried kiwifruit samples showed that the RW-dried kiwifruit samples had less textural hardness than the oven-dried ones (Jafari et al., 2016). Similar findings were observed for RW-dried mango leather (Shende et al., 2020). Rising RWD temperature and pectin concentration were found to increase the hardness of sapota bars by Jalgaonkar et al. (2018), however an interaction effect led to a decrease in hardness. The development of a gel-like structure as a result of the interaction between pectin and pulp ingredients and case hardening at higher water temperatures were the reasons given for this hardening. Furthermore, sapota bars became more cohesive at higher pectin concentrations and water temperatures.

In a study by Tontul and Topuz (2017), it was found that pomegranate leather that had been air-dried had a better texture than leather that had been hot-air-dried with microwave assistance. Hardness was highly influenced by drying temperature; when the drying temperature in the microwave-assisted hot-air dryer was raised from 50°C to 60°C, a sharp decline in hardness and chewiness was seen. This phenomenon was ascribed to the quick removal of moisture, which caused gaps and channels to form as the material dried. Remarkably, drying methods and temperature had no discernible effects on springiness.

7.6.3 Ascorbic Acid

The ascorbic acid content of fruits is influenced by a number of factors, such as light, temperature, oxygen, moisture content, metal ion catalysis, and processing time. Because ascorbic acid is highly unstable and heat-sensitive, it has a major impact on the nutrient quality index of foods. According to several studies (Kaya et al., 2010; Rajoriya et al., 2020; Tontul et al., 2018), ascorbic acid is more retained in RW-dried samples than in samples dried using convective drying techniques. This is mainly because ascorbic acid degrades thermally and oxidatively over longer drying times.

Baeghbali et al. (2020) used three different methods to dry okra: a freeze-dryer (temperature = −50°C), a hot-air oven (temperature = 82.0°C), and an ultrasonic and infrared-assisted conductive hydro-dryer (water temperature = 82.5°C). The freeze-dried sample showed the highest ascorbic acid retention, while the hot-air oven-dried sample showed the lowest. As compared to freeze-dried samples, ultrasound- and infrared-assisted conductive hydro-dried samples produced results that were comparatively closer. The decrease in ascorbic acid was ascribed to overheating brought on by infrared radiation.

Likewise, Rajoriya et al. (2021) discovered that banana samples dried at 90°C in a RW dryer had a higher ascorbic acid content than samples dried at 80°C and 70°C, respectively. This is explained by the fact that ascorbic acid degrades less oxidatively at higher drying temperatures. RWD-dried samples showed higher ascorbic acid retention (about 98%) compared to other drying methods, according to a study by Nindo et al. (2003b) on the retention of ascorbic acid in dried asparagus spears using five different drying methods (freeze, microwave spouted bed, spouted bed, and tray drying). In contrast to these results, Nemzer et al. (2018) found that freeze-dried strawberries, tart cherries, and cranberries retained more ascorbic acid than RW-dried samples. RW-dried samples, as opposed to freeze-dried samples, demonstrated superior ascorbic acid retention for blueberries. These observations highlight the variability in ascorbic acid retention depending on the specific fruit or vegetable and the drying method employed.

7.6.4 Rehydration Ratio

When it comes to determining the sensory qualities and preparation time of dehydrated products, the volume and rate of water absorption during reconstitution are critical factors (Maskan, 2001). Rehydration properties are often employed as a quality measure, revealing physical and chemical changes induced by drying conditions, sample pretreatment, and composition. The rehydration ratio, defined as the ratio of the drained weight of the rehydrated sample to the weight of the dried sample, is commonly used for quality evaluation after drying. Nonetheless, according to Abul-Fadl and Ghanem (2011), it is a complex process that reflects the physical and chemical alterations brought about by drying techniques.

Extended drying times and high temperatures cause the dried product to form an impermeable layer that prevents it from fully rehydrating, which has a substantial effect on the rehydration ratio. The rehydration ratio of tomato powder dried with RW dryers at 60–90°C and a convective air dryer at 60°C was examined by Abul-Fadl and Ghanem (2011). Samples that were allowed to air-dry had greater rehydration ratios (3.14:3.24) than tomato powder that had been dried by convection. This is because convection drying creates an impermeable layer.

Zalpouri et al. (2021) found that when it came to potato flakes, RW-dried flakes showed a higher rehydration ratio than those that were dried in a convective tray dryer. This was explained by the formation of an impermeable layer as a result of longer drying times and overheating during convective drying. According to Azizi et al. (2016), as drying temperature and slice thickness increased, there was a significant decrease in the rehydration ratio of dried kiwifruit slices in RW dryers. The phenomenon was ascribed to the caramelization of sugar at elevated temperatures, resulting in the obstruction of surface pores and decreased absorption of water. These studies underscore the critical role of drying conditions in influencing the rehydration properties of dehydrated products, with RWD often resulting in superior rehydration ratios compared to convective drying methods.

7.7 MICROSTRUCTURE ANALYSIS

Exposing materials to extreme temperatures can induce changes in their microstructure, thereby affecting physical characteristics like texture. According to Athmaselvi et al. (2014), the scanning

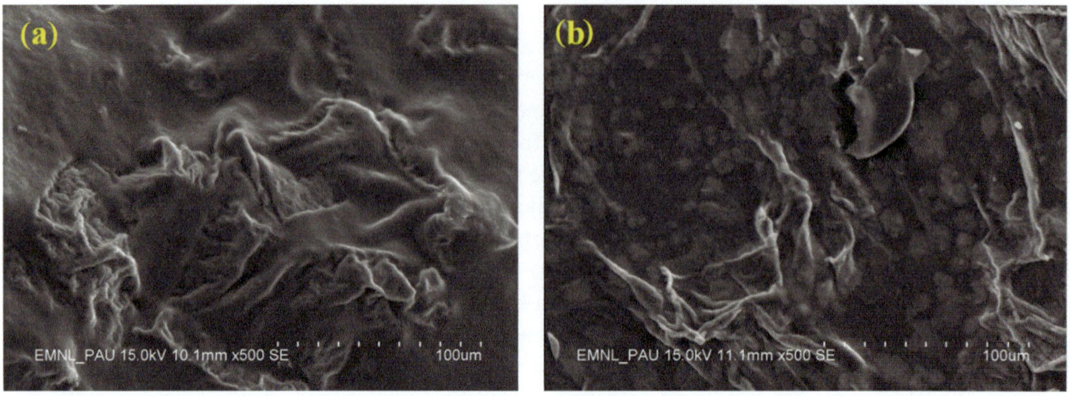

FIGURE 7.4 Micrographs of (a) convective-dried and (b.) RW-dried potato flakes using SEM at bar length of 100 μm. (*Source:* Zalpouri et al., 2021.)

electron microscope (SEM) has shown usefulness in detecting and tracking caking phenomena on powder surfaces. The consistent drying of RW-dried powder/flakes is ensured by controlled puree feeding through a spreader tool at the RW dryer inlet, resulting in a smooth and uniform thickness. Furthermore, the uniformly thick sheet at the end of drying is produced by the thinly distributed material on the conveyor belt remaining undisturbed (Shende & Datta, 2019).

Mango purees dried by RWD and freeze-dried by Caparino et al. (2013) showed comparable structures. However, when RW-dried mango flakes were ground into powder, the resulting irregular particles had smooth edges and a constant thickness, suggesting a lower risk of oxidation because of the smaller surface area. Mango powder that had been freeze-dried, on the other hand, showed a skeletal structure that indicated places that had been occupied by ice prior to the process of freeze-drying, preventing collapse and shrinkage and producing a porous dry material (Caparino et al., 2012). Comparing RW-dried mango powder to tray-dried mango powder, which is characterized by unevenly formed, rounded, or irregular edges and inconsistent thickness due to unequal shrinkage, bubble formation, and structural collapse during convective drying, Shende and Datta (2020) reported similar findings.

Additionally, the analysis shown in Figure 7.4 demonstrated that RW-dried potato flakes had smooth surfaces with small pockets, indicating potential protein retention, while convective tray-dried potato flakes showed uneven shrinkage (Zalpouri et al., 2021). RWD did not negatively affect starch granules in a study on unripe green banana flour, but hot-air drying had an effect on starch structure (Table 7.2) (Padhi and Dwivedi, 2022).

7.8 LIMITATIONS OF RWD TECHNIQUE

The main barrier to the RWD technique's scaling up is the sizable area needed for heat exchange and drying when a large amount of product is being dehydrated. Even though the RWD process is fast, the process capacity is limited because the product layer's thickness must be kept low (Shende & Datta, 2019). An additional challenge is that the air that passes over the surface and after the finished products are removed from the belt can contaminate the RW-dried product (Waghmare, 2021). In addition, the belt needs to be cleaned both before and after each drying cycle. Since high-sugar purees tend to stick to the conveyor belt, the RWD is not the best option for drying them (Karadbhajne et al., 2019). Furthermore, not much study has been done on the chemical behavior of commodities with strong structural integrity, like tubers and nuts.

TABLE 7.2
Effect of Different Drying in Comparison with RWD on Quality Parameters (Physical Properties, Bioactive Compounds, and Rehydration Characteristics) of Dried Product

Sample	Drying Method	Process Conditions	Observations	Reference
Fruits				
Apple	Hot air Refractance window Freeze	60°C, 70°C, 80°C, 90°C 60°C, 70°C, 80°C, 90°C Temperature: *Heating plate:* 20°C; *condenser:* −80°C	Maximum bioactive concentration was preserved in RWD samples as compared to HAD samples 80% retention of bioactives in RWD samples at 90°C comparable with 83% in FD samples	Rajoriya et al., 2019
Blueberries, tart cherries, Strawberries, cranberries	Refractance Window Hot air Freeze	88°C 70°C Temperature *Heating plate*: 20°C; *condenser*: −64°C Sample thickness: 50–80 μm	Higher retention of total phenolic compounds in FD samples followed by RWD and HD samples	Nemzer et al., 2018
Carrot	Refractance window Spray Drum Freeze-drying	95°C IT = 190, OT = 95°C 138°C Temperature *Heating plate:* 20°C; *condenser:* −64°C	Better color retention in RWD carrot puree	Abonyi et al., 2001
	Refractance window Convective	74°C and 94°C 74°C and 94°C *Slice thickness*: 0.2 and 0.4 cm	Higher retention of total phenolic content in RWD than CD Fresh and RWD samples had equivalent total phenolic compounds	Hernández-Santos et al., 2016
Cherry	Refractance window Convective	90°C, 95°C, and 98°C 50°C, 60°C, and 70°C	Enhanced retention of color with higher L and chroma values as well as ascorbic acid content in RWD samples as compared to CD ones	Tontul et al., 2018
Goldenberry	Refractance window Convective infrared	70°C 70°C 70°C *Pulp thickness:* 4 mm	Better color retention in RWD than in CD and IRD samples No thermal hydrolysis or enzymatic degradation during RWD retained total phenolic content in the samples	Puente et al., 2020
Kiwifruits	Refractance window Oven	80°C, 90°C, and 100°C 80°C, 90°C, and 100°C *Slice thickness:* 0.8, 1.6, and 2.4 mm	Less color changes and shrinkage in RWD product than OD method	Jafari et al., 2016
	Refractance window	80–100°C *Slice thickness:* 0.8, 1.6, or 2.4 mm	Drying temperature and slice thickness significantly affected the rehydration ratio of product	Azizi et al., 2016

(Continued)

TABLE 7.2 (Continued)

Effect of Different Drying in Comparison with RWD on Quality Parameters (Physical Properties, Bioactive Compounds, and Rehydration Characteristics) of Dried Product

Sample	Drying Method	Process Conditions	Observations	Reference
Mango	Refractance window Spray Drum Freeze	95–97°C IT: 190°C, OT: 90°C 152°C Temperature *Heating plate:* 20°C; *condenser:* −60°C	Lighter color of RWD, FD and SD powder in comparison to DD powder. RWD (18.0 ± 0.36%) and FD (18.0 ± 0.19%) mango powder had similar hygroscopicity DD powder had highest hygroscopicity (20.1 ± 0.88%) and SD powder had least (16.5 ± 0.06%)	Caparino et al., 2012
Pomegranate	Refractance window Hot-air Microwave-assisted hot air	90°C, 95°C, and 98°C 50°C, 60°C, or 70°C 50°C, 60°C, or 70°C (90 W or 180 W)	RWD and HAD pomegranate leather has a better texture than MHAD samples RWD samples were lighter in color with higher L value followed by HAD and MHAD samples	Tontul & Topuz, 2017
	Refractance window Spray Freeze	91°C IT = 140°C, OT = 75°C Temperature: *Condenser*: −40°C	Total phenolic content of RWD, SD, and FD samples was observed as 534.2, 511.9, and 571.0 mg GAE, respectively.	Baeghbali et al., 2016b
Unripe green banana	Refractance window Hot-air	85°C 60°C	Higher total phenolic content in RWD sample (0.85 ± 0.40 mg/g dw) as compared to HAD sample (0.62 ± 0.3 mg/g dw) Higher hygroscopicity of HAD sample (6.94 ± 0.45 g/gdw) wrt RWD sample (5.37 ± 0.50 g/gdw)	Padhi & Dwivedi, 2022
Vegetables				
Asparagus	Refractance window Freeze Microwave spouted bed Spouted bed Tray	95–97°C Temperature: *Heating plate*: 20°C; *condenser*: −64°C Magnetron of 1.5 kW operating at 2,450 MHz, 60°C 60°C 60°C	RWD sample was closest in greenness to the FD product samples Highest ascorbic acid was found for RWD followed by FD, MSBD, SBD, and least in TD	Nindo et al., 2003b
Okra	Ultrasound- and infrared-assisted conductive hydro Freeze Oven	82.5°C Temperature: *Condenser*: −50°C 82.0°C	Lower total color difference observed in UICHD samples than other methods Maximum retention of ascorbic acid in FD and least in OD samples UICHD samples had an ascorbic acid value closer to FD samples	Baeghbali et al., 2020

(Continued)

TABLE 7.2 *(Continued)*

Effect of Different Drying in Comparison with RWD on Quality Parameters (Physical Properties, Bioactive Compounds, and Rehydration Characteristics) of Dried Product

Sample	Drying Method	Process Conditions	Observations	Reference
Potato	Refractance window	70°C, 80°C, and 90°C	RWD samples had least browning index and highest L value	Zalpouri et al., 2021
	Convective	50°C, 60°C, and 70°C	RWD potato flakes had a higher rehydration ratio than CD potato flakes	
Tomato	Refractance window	60°C, 75°C, and 90°C	Rehydration ratio varied from 3.14 to 3.24 for RWD samples and 2.95 for CD samples	Abul-Fadl & Ghanem, 2011
	Convective	60°C	RWD samples (0.122–0.149g/g dw) were highly hygroscopic as compared to CD samples (0.113 g/g dw)	

7.9 CONCLUSION

It appears that the RWD is a promising technology that has a great deal of potential for maintaining food products' quality and nutritional attributes. According to research, RWD functions at combinations of temperature and time that preserve the product's quality. It is appropriate for crops that are sensitive to heat because of its indirect drying mode and lower product temperature when compared to other dryers like drum and spray dryers. The temperature and sample thickness have a major impact on the physicochemical properties and quality of the final dried product. Furthermore, RWD is unique as an energy-saving drying method that yields fine powders or flakes. With comparatively simple and inexpensive equipment, it can produce dried materials of a quality similar to freeze-dried materials.

In addition to maintaining more sensory and nutritional value than other drying methods, RWD has benefits over alternative drying methods in terms of energy efficiency, safety, processing expenses, and environmental impact. The integration of RWD with sophisticated processing methods like osmotic dehydration, infrared, microwaves, or ultrasounds as pretreatments is the subject of ongoing research. Computational fluid dynamics analysis is necessary to fully comprehend the RWD system in order to evaluate the relative contributions of radiation, convection, and conduction to the drying process.

On the other hand, the RWD systems that are currently on the market are large and need a large amount of space for installation in commercial and pilot plant settings. Therefore, in order to improve RWD systems' applicability in the food industry, more research and design modifications are required. In the food industry, RW drying has the potential to become one of the most well-liked and widely accepted methods for creating high-quality, high-value food powders and flakes, as consumers seek out more nutritious foods at competitive prices.

REFERENCES

Abonyi, B. I., Feng, H., Tang, J., Edwards, C. G., Chew, B. P., Mattinson, D. S., & Fellman, J. K. (2001). Quality retention in strawberry and carrot purees dried with refractance window TM system. *Journal of Food Science*, *67*, 1051–1056.

Abonyi, B. I., Tang, J., & Edwards, C. G. (1999). Evaluation of energy efficiency and quality retention for the Refractance Window TM drying system, Research report, Pp 1-38, Department of Biological Systems Engineering, Washington State University, Pullman, WA.

Abul-Fadl, M. M., & Ghanem, T. H. (2011). Effect of refractance-window (RW) drying method on quality criteria of produced tomato powder as compared to the convection drying method. *World Applied Sciences Journal*, *15*(7), 953–965.

Amer, B. M. A. (2011). Mathematical modeling of temperature and heat profiles in pilot refractance window drying system. *Misr Journal of Agricultural Engineering*, *28*, 999–1020.

Aragón-Rojas, S., Quintanilla-Carvajal, M. X., Hernández-Sánchez, H., Hernández-Álvarez, A. J., & Moreno, F. L. (2019). Encapsulation of *Lactobacillus* fermentum K73 by refractance window drying. *Scientific Reports*, *9*(1), 1–16. https://doi.org/10.1038/s41598-019-42016-0

Athmaselvi, K. A., Kumar, C., Balasubramanian, M., & Roy, I. (2014). Thermal, structural, and physical properties of freeze dried tropical fruit powder. *Journal of Food Processing*, *2014*, 1–10. https://doi.org/10.1155/2014/524705

Aykın-Dinçer, E., Özdemir, M., & Topuz, A. (2021). Quality characteristics of bone broth powder obtained through Refractance Window™ drying. *LWT – Food Science and Technology*, *147*, 111526. https://doi.org/10.1016/j.lwt.2021.111526

Azizi, D., Jafari, S. M., Mirzaei, H., & Dehnad, D. (2016). The influence of refractance window drying on qualitative properties of kiwifruit slices. *International Journal of Food Engineering*. https://doi.org/10.1515/ijfe-2016-0201

Baeghbali, V., Ngadi, M., & Niakousari, M. (2020). Effects of ultrasound and infrared assisted conductive hydro-drying, freeze-drying and oven drying on physicochemical properties of okra slices. *Innovative Food Science and Emerging Technologies*, *63*, 102313. https://doi.org/10.1016/j.ifset.2020.102313

Baeghbali, V., & Niakousari, M. (2018). A review on mechanism, quality preservation and energy efficiency in refractance window drying: A conductive hydro-drying technique. *Journal of Nutrition, Food Research and Technology*, *1*(2), 50–54. https://doi.org/10.30881/jnfrt.00011

Baeghbali, V., Niakousari, M., & Farahnaky, A. (2016a). Refractance window drying of pomegranate juice: Quality retention and energy efficiency. *LWT – Food Science and Technology*, *66*, 34–40. https://doi.org/10.1016/j.lwt.2015.10.017

Baeghbali, V., Niakousari, M., & Kiani, M. (2010). Design, manufacture and investigating functionality of a new batch refractance window system. 5th International Conference on Innovations in Food and Bioprocess Technology, pp. 1–7. https://doi.org/10.13140/2.1.2882.1448

Baeghbali, V., Niakousari, M., & Ngadi, M. (2018). Introduction of a new ultrasound and infrared assisted conductive hydro-dryer. Paper No. CSBE18-216, CSBE/SCGAB 2018 Annual Conference University of Guelph, Guelph.

Baeghbali, V., Niakousari, M., Ngadi, M. O., & HadiEskandari, M. (2019). Combined ultrasound and infrared assisted conductive hydro-drying of apple slices. *Drying Technology*, *37*(14), 1793–1805. https://doi.org/10.1080/07373937.2018.1539745

Bolland, K. M. (2000). A new low-temperature/short-time driving process. *Cereal Foods World*, *45*(7), 293–296.

Caparino, O. A., Sablani, S. S., Tang, J., Syamaladevi, R. M., & Nindo, C. I. (2013). Water sorption, glass transition, and microstructures of refractance window- and freeze-dried mango (Philippine "Carabao" var.) powder. *Drying Technology*, *31*, 1969–1978. https://doi.org/10.1080/07373937.2013.805143

Caparino, O. A., Tang, J., Nindo, C. I., Sablani, S. S., Powers, J. R., & Fellman, J. K. (2012). Effect of drying methods on the physical properties and microstructures of mango (Philippine "Carabao" var.) powder. *Journal of Food Engineering*, *111*(1), 135–148. https://doi.org/10.1016/j.jfoodeng.2012.01.010

Chou, S. K., & Chua, K. J. (2001). New hybrid drying technologies for heat sensitive foodstuffs. *Food Science and Technology*, *12*, 359–369. https://doi.org/10.1016/S0924-2244(01)00102-9

Costa, C., Antonucci, F., Pallottino, F., Aguzzi, J., Sun, D. W., & Menesatti, P. (2011). Shape analysis of agricultural products: A review of recent research advances and potential application to computer vision. *Food and Bioprocess Technology*, *4*(5), 673–692. https://doi.org/10.1007/s11947-011-0556-0

da Silva Simão, R., de Moraes, J. O., Monteiro, R. L., Schaidt, A. L., Carciofi, B. A. M., & Laurindo, J. B. (2021). Conductive drying methods for producing high-quality restructured pineapple-starch snacks. *Innovative Food Science and Emerging Technologies*, *70*, 102701. https://doi.org/10.1016/j.ifset.2021.102701

Ghanem, T. H. (2010). Modeling of refractance window film dryer for liquids. *Misr Journal of Agricultural Engineering*, *27*(2), 676–687.

Giuliani, A., Cerretani, L., & Cichelli, A. (2016). Colors: Properties and determination of natural pigments. In *Encyclopedia of Food and Health* (pp. 273–283). Elsevier. https://doi.org/10.1016/B978-0-12-384947-2.00189-6

Hashem, H. A., Tayel, S. A., Younes, O. S., & Mageed, A. O. A. (2019). Effect of refractance-window drying method on quality characteristics of balady menthol. *Journal of Food and Dairy Sciences*, *10*(8), 265–269. https://doi.org/10.21608/jfds.2019.58143

Hernández-Carrión, M., Moyano, M., & Quintanilla-Carvajal, M. X. (2021). Design of high-oleic palm oil nanoemulsions suitable for drying in Refractance Window^{TM}. *Journal of Food Processing and Preservation*, *45*(1), 1–9. https://doi.org/10.1111/jfpp.15076

Hernández-Santos, B., Martínez-Sánchez, C. E., Torruco-Uco, J. G., Rodríguez-Miranda, J., Ruiz-López, I. I., Vajando-Anaya, E. S., Carmona-García, R., & Herman-Lara, E. (2016). Evaluation of physical and chemical properties of carrots dried by refractance window drying. *Drying Technology*, *34*(12), 1414–1422. https://doi.org/10.1080/07373937.2015.1118705

Jafari, S. M., Azizi, D., Mirzaei, H., & Dehnad, D. (2016). Comparing quality characteristics of oven-dried and refractance window-dried kiwifruits. *Journal of Food Processing and Preservation*, *40*(3), 362–372. https://doi.org/10.1111/jfpp.12613

Jalgaonkar, K., Mahawar, M. K., Vishwakarma, R. K., Shivhare, U. S., & Nambi, V. E. (2018). Optimization of process condition for preparation of sapota bar using refractance window drying method. *Drying Technology*, *38*(3), 269–278. https://doi.org/10.1080/07373937.2018.1482314

Karadbhajne, S., Thakare, V. M., Kardile, N. B., & Thakre, S. M. (2019). Refractance window drying: An innovative drying technique for heat sensitive product. *International Journal of Recent Technology and Engineering*, *8*(4), 1765–1771. https://doi.org/10.35940/ijrte.c6091.118419

Karthik, P., Chhanwal, N., & Anandharamakrishnan, C. (2017). Drum drying. In Handbook of Drying for Dairy Products (pp. 43–56). Wiley-Blackwell. https://doi.org/10.1002/9781118930526.ch4

Kaya, A., Aydin, O., & Kolayli, S. (2010). Effect of different drying conditions on the vitamin c (ascorbic acid) content of hayward kiwifruits (*Actinidia deliciosa* Planch). *Food and Bioproducts Processing*, *88*(2–3), 165–173. https://doi.org/10.1016/j.fbp.2008.12.001

Kumar, C., Singh, M., Zalpouri, R., Kaur, P. and Singh, S. (2025), Optimization of Refractance Window Drying for Nutrient-Rich Mushroom Slices: A Comparative Study With Convective and Solar Drying. *Heat Transfer*. https://doi.org/10.1002/htj.23340

Kumar, N., Kumar, S., Sudhakar, A., Verma, D. K., Shankar, S., Thakur, M., Singh, S., Tripathy, S., Gupta, A. K., & Srivastav, P. P. (2021). Refractance Window^{TM}-drying vs. other drying methods and effect of different process parameters on quality of foods: A comprehensive review of trends and technological developments. *Future Foods*, *3*, 100024. https://doi.org/10.1016/j.fufo.2021.100024

Magoon, R. E. (1986). Method and Apparatus for Drying Fruit Pulp and the Like. US Patent 4,631,837.

Maisnam, D., Rasane, P., Dey, A., Kaur, S., & Sarma, C. (2017). Recent advances in conventional drying of foods. *Journal of Food Technology and Preservation*, *1*(1), 25–34.

Mandale, N. M., Attkan, A. K., Kumar, S., & Kumar, N. (2023). Drying kinetics and quality assessment of refractance window dried beetroot. *Journal of Food Process Engineering*, *46*(7), e14332. https://doi.org/10.1111/jfpe.14332

Martynenko, A., & Janaszek, M. A. (2014). Texture changes during drying of apple slices. *Drying Technology*, *32*(5), 567–577. https://doi.org/10.1080/07373937.2013.845573

Maskan, M. (2001). Drying, shrinkage and rehydration characteristics of kiwifruits during hot air and microwave drying. *Journal of Food Engineering*, *48*, 177–182. https://doi.org/10.1016/S0260-8774(00)00155-2

Nayak, B., Berrios, J., de, J., Powers, J. R., Tang, J., & Ji, Y. (2011). Colored potatoes (*Solanum tuberosum* L.) dried for antioxidant-rich value-added foods. *Journal of Food Processing and Preservation*, *35*(5), 571–580. https://doi.org/10.1111/j.1745-4549.2010.00502.x

Nemzer, B., Vargas, L., Xia, X., Sintara, M., & Feng, H. (2018). Phytochemical and physical properties of blueberries, tart cherries, strawberries, and cranberries as affected by different drying methods. *Food Chemistry*, *262*, 242–250. https://doi.org/10.1016/j.foodchem.2018.04.047

Nindo, C. I., Feng, H., Shen, G. Q., Tang, J., & Kang, D. H. (2003a). Energy utilization and microbial reduction in a new film drying system. *Journal of Food Processing and Preservation*, *27*, 117–136. https://doi.org/10.1111/j.1745-4549.2003.tb00506.x

Nindo, C. I., Sun, T., Wang, S. W., Tang, J., & Powers, J. R. (2003b). Evaluation of drying technologies for retention of physical quality and antioxidants in *Asparagus* (*Asparagus officinalis* L.). *LWT – Food Science and Technology*, *36*(5), 507–516. https://doi.org/10.1016/S0023-6438(03)00046-X

Nindo, C. I., & Tang, J. (2007). Refractance window dehydration technology: A novel contact drying method. *Drying Technology*, *25*(1), 37–48. https://doi.org/10.1080/07373930601152673

Ortiz-Jerez, M. J., Gulati, T., Datta, A. K., & Ochoa-Martínez, C. I. (2015). Quantitative understanding of Refractance Window™ drying. *Food and Bioproducts Processing*, *95*, 237–253. https://doi.org/10.1016/j.fbp.2015.05.010

Oyinloye, T. M., & Yoon, W. B. (2020). Effect of freeze-drying on quality and grinding process of food produce : A review. *Processes*, *8*, 354. https://doi.org/10.3390/pr8030354

Padhi, S., & Dwivedi, M. (2022). Physico-chemical, structural, functional and powder flow properties of unripe green banana flour after the application of refractance window drying. *Future Foods*, *5*, 100101. https://doi.org/10.1016/J.FUFO.2021.100101

Pathare, P. B., Opara, U. L., & Al-Said, F. A. J. (2013). Colour measurement and analysis in fresh and processed foods: A review. *Food and Bioprocess Technology*, *6*, 36–60. https://doi.org/10.1007/s11947-012-0867-9

Puente, L., Vega-Gálvez, A., Ah-Hen, K. S., Rodríguez, A., Pasten, A., Poblete, J., Pardo-Orellana, C., & Muñoz, M. (2020). Refractance window drying of goldenberry (*Physalis peruviana* L.) pulp: A comparison of quality characteristics with respect to other drying techniques. *LWT – Food Science and Technology*, *131*, 109772. https://doi.org/10.1016/j.lwt.2020.109772

Puente-Díaz, L., Spolmann, O., Nocetti, D., Zura-Bravo, L., & Lemus-Mondaca, R. (2020). Effects of infrared-assisted Refractance Window™ drying on the drying kinetics, microstructure, and color of physalis fruit purée. *Foods*, *9*(3), 343. https://doi.org/10.3390/foods9030343

Raghavi, L. M., Moses, J. A., & Anandharamakrishnan, C. (2018). Refractance window drying of foods: A review. *Journal of Food Engineering*, *222*, 267–275. https://doi.org/10.1016/j.jfoodeng.2017.11.032

Rajoriya, D., Bhavya, M. L., & Hebbar, H. U. (2021). Impact of process parameters on drying behaviour, mass transfer and quality profile of refractance window dried banana puree. *LWT – Food Science and Technology*, *145*, 111330. https://doi.org/10.1016/j.lwt.2021.111330

Rajoriya, D., Shewale, S. R., Bhavya, M. L., & Hebbar, H. U. (2020). Far infrared assisted refractance window drying of apple slices: Comparative study on flavour, nutrient retention and drying characteristics. *Innovative Food Science and Emerging Technologies*, *66*, 102530. https://doi.org/10.1016/j.ifset.2020.102530

Rajoriya, D., Shewale, S. R., & Hebbar, H. U. (2019). Refractance window drying of apple slices: Mass transfer phenomena and quality parameters. *Food and Bioprocess Technology*, *12*, 1646–1658. https://doi.org/10.1007/s11947-019-02334-7

Ratti, C., & Mujumdar, A. S. (1995). Handbook of Industrial Drying (pp. 567–588). Marcel Dekker, Inc, New York.

Seyfi, A., Rezaei Asl, A., & Motevali, A. (2021). Comparison of the energy and pollution parameters in solar refractance window (photovoltaic-thermal), conventional refractance window, and hot air dryer. *Solar Energy*, *229*, 162–173. https://doi.org/10.1016/j.solener.2021.05.094

Shende, D., & Datta, A. K. (2019). Refractance window drying of fruits and vegetables: A review. *Journal of the Science of Food and Agriculture*, *99*(4), 1449–1456. https://doi.org/10.1002/jsfa.9356

Shende, D., & Datta, A. K. (2020). Optimization study for refractance window drying process of Langra variety mango. *Journal of Food Science and Technology*, *57*, 683–692. https://doi.org/10.1007/s13197-019-04101-0

Shende, D., Kour, M., & Datta, A. K. (2020). Evaluation of sensory and physico-chemical properties of Langra variety mango leather. *Journal of Food Measurement and Characterization*, *14*(6), 3227–3237. https://doi.org/10.1007/s11694-020-00558-2

Smith, T. M. (1994). Heat transfer dynamics. *Tappi Journal*, *77*(8), 239–245.

Sosnik, A., & Seremeta, K. P. (2015). Advantages and challenges of the spray-drying technology for the production of pure drug particles and drug-loaded polymeric carriers. *Advances in Colloid and Interface Science*, *223*, 40–54. https://doi.org/10.1016/j.cis.2015.05.003

Srivastav, P. P., Verma, D. K., Patel, A. R., & Al-Hilphy, A. R. (2020). Emerging Thermal and Nonthermal Technologies in Food Processing. CRC Press.

Ssenyimba, S., Kigozi, J., Tumutegyereize, P., Muyonga, J. H., & Mutumba, R. (2021). Design and evaluation of a refractance window lab-scale dryer. *Journal of Engineering, Design and Technology*, *20*(6). https://doi.org/10.1108/JEDT-12-2020-0523

Szczesniak, A. S. (1990). Texture: Is it still an overlooked food attribute? *Food Technology (USA)*, *44*(9), 86–95.

Tontul, I., Eroğlu, E., & Topuz, A. (2018). Convective and refractance window drying of cornelian cherry pulp: Effect on physicochemical properties. *Journal of Food Process Engineering*, *41*(8), 1–8. https://doi.org/10.1111/jfpe.12917

Tontul, I., & Topuz, A. (2017). Effects of different drying methods on the physicochemical properties of pomegranate leather (pestil). *LWT – Food Science and Technology*, *80*, 294–303. https://doi.org/10.1016/j.lwt.2017.02.035

van't Land, M., & Raes, K. (2019). Refractance window drying of fish silage: An initial investigation into the effects of physicochemical properties on drying efficiency and nutritional quality. *LWT – Food Science and Technology*, *102*, 71–74. https://doi.org/10.1016/j.lwt.2018.12.001

Vega-Mercado, H., Góngora-Nieto, M., & Barbosa-Cánovas, G. V. (2001). Advances in dehydration of foods. *Journal of Food Engineering*, *49*, 271–289. https://doi.org/10.1016/S0260-8774(00)00224-7

Verma, D. K., Billoria, S., Mahato, D. K., Swarnakar, A. K., & Srivastav, P. P. (2019). Effects of thermal processing on nutritional composition of green leafy vegetables: A review. Engineering Interventions in Foods and Plants (pp. 157–208). Apple Academic Press. https://www.researchgate.net/publication/326668875

Waghmare, R. (2021). Refractance window drying: A cohort review on quality characteristics. *Trends in Food Science & Technology*, *110*(50), 652–662. https://doi.org/10.1016/j.tifs.2021.02.030

Yoha, K. S., Priyadarshini, S. R., Moses, J. A., & Anandharamakrishnan, C. (2019). Refractance window drying and its applications in food processing. In Technologies for Value Addition in Food Products and Processes (1st ed., pp. 61–72). Apple Academic Press Inc. https://doi.org/10.1201/9780429242847-4

Zalpouri, R. (2018). Development and evaluation of refraction based system for dehydration of potato. Punjab Agricultural University, Ludhiana, India.

Zalpouri, R. (2023). Development and Evaluation of Solar Assisted Refractance Window Dryer for Vegetable Purees. Punjab Agricultural University, Ludhiana, India.

Zalpouri, R., Kaur, P., & Kaur, A. (2020). Influence of developed refractance based drying method on physical parameter of potato flakes. *International Journal of Chemical Studies*, *8*, 2833–2838. https://doi.org/10.22271/chemi.2020.v8.i3ao.9645

Zalpouri, R., Kaur, P., Kaur, A., & Sidhu, G. K. (2021). Comparative analysis of optimized physiochemical parameters of dried potato flakes obtained by refractive and convective drying techniques. *Journal of Food Processing and Preservation*, *45*, e15077. https://doi.org/10.1111/jfpp.15077

Zalpouri, R., Kaur, P., & Sain, M. (2020). Refractive window drying: A better approach to preserve the visual appearance of dried products. *Pantnagar Journal of Research*, *18*(1), 90–94.

Zalpouri, R., Singh, M., Kaur, P., & Singh, S. (2022). Refractance window drying: A revisit on energy consumption and quality of dried bio-origin products. *Food Engineering Reviews*, *14*, 257–270. https://doi.org/10.1007/s12393-022-09313-3

Zotarelli, M. F., Carciofi, B. A. M., & Laurindo, J. B. (2015). Effect of process variables on the drying rate of mango pulp by refractance window. *Food Research International*, *69*, 410–417. https://doi.org/10.1016/j.foodres.2015.01.013

8 Sustainable Drying Technologies for Marine Products

Darshana Deka, Prashant M. Singh, and Shivanand S. Shirkole
Institute of Chemical Technology Mumbai, ICT-IOC Odisha Campus, Bhubaneswar, Odisha, India

8.1 INTRODUCTION

Aquatic products have long been recognized as an affordable source of high-quality nutrients. It is a rich source of bioavailable macro- and micronutrients – proteins, healthy fats like omega-3 fatty acids, minerals, and vitamins and so forth. Hence, it is a prerequisite for a well-balanced diet and a wholesome lifestyle. But fish is a highly perishable food product with moisture content (*mc*) about 70–80% (wet basis weight) and water activity (a_w) around 0.9 that contributes to post-harvest loss (PHL). This not only affects the potential income of fishers and traders but also leads to the wastage of highly nutritious food sources for consumers, thereby contributing to ever-increasing environmental concerns. Various factors are responsible for PHL, but the most important factor is temperature, which contributes to the accelerated spoilage of foods. Annually, about 10–12 million tonnes of marine products are lost due to spoilage [1]. Cold chain is the widely adopted solution for highly perishable marine products. However, the high energy consumption and cost associated with this technology are not economically or environmentally feasible in general.

The method of drying marine products is an age-old technique of preservation adopted by various communities, especially isolated rural populations lacking access to highly equipped infrastructures. The principle behind the process involves the removal of moisture from the product to the environment due to the vapour pressure gradient between the product and the surrounding. As the amount of free moisture in the food is reduced to a safer level (mostly water activity below 0.6), it slows down the proliferation of microbes like moulds, yeast, or bacteria and inhibits any enzymatic reaction, thereby enhancing the shelf life of the product. Drying not only helps to enhance the shelf life of perishable foods but is also a cost-effective preservation technique as compared to cold storage.

Dehydrated fish and fish products are a major source of affordable nutritious food for poor and remote populations. Also, they play a vital role in the fish processing industry, generating countless employment opportunities for people residing in coastal districts and women folks. The dehydrated fish products are also beneficial during the fish ban season ensuring a continuous flow of income even during fish scarcity. They also contribute significantly to the revenue of a nation. As per a report by Marine Products Export Development Authority (MPEDA), the Indian dry fish sector had a contribution of about $156.94 million to the national economy in the year 2020–2021 [2].

The traditional methods of drying aquaculture produce mostly involve harnessing available renewable energy sources, such as sunlight, wind, or smoke for the removal of moisture. Although these methods are cost-effective, they are often time-consuming and the end products lack quality. As the growing global population drives the global food demand, the implementation of better drying technologies became a necessity. The most common industrially adopted thermal dryers incorporate fossil-based, convective hot-air drying, which is deemed highly energy-intensive.

DOI: 10.1201/9781003344087-8

They contribute about 20% of the total energy consumed during the production process and 90% of the aggregated production cost [3]. It is also liable to generate up to 45% of waste heat in the form of hot exhaust gases which can add to GHG emissions like CH_4, CO_2, and NO_x. In this era of a substantial energy crisis, low energy-efficient technologies could be a central issue contributing to climate consequences. A paradigm shift towards more sustainable drying methods using renewable sources of energy could add significantly towards reduced energy consumption as well as minimized environmental effects.

8.2 CRITERIA FOR ASSESSING THE SUSTAINABILITY OF DRYING TECHNOLOGIES

Sustainability means fulfilling the requirements of the present generation without comprising the needs of the future generation. The term sustainable drying technology implies the application of dryers designed to comply with higher energy efficiency, reduced cost of production, and superior quality products with limited contribution towards carbon footprint. Since the food industry is considered to be a huge consumer of energy and up to 30% of this energy is mostly expended on drying operations, it is imperative to regulate their contribution to total CO_2 emission [4]. Hence, it is important to incorporate sustainable drying technologies in various food processing units. Assessing the sustainability of a process or technology does not have a well-laid standardized procedure so far and it can vary based on the goals or scope of assessment. But generally, the following domains are considered to analyse the sustainability of drying technologies.

8.2.1 Thermodynamic Efficiency

A technology design should consume minimum energy and provide higher economic output with minimized energy loss (waste heat), thereby reducing the overall environmental impact. This would involve optimizing the design and operation of the technology to attain reduced energy losses and increased energy recovery. There are various ways of calculating efficiency η (Equation 8.1), but generally it is expressed as a ratio between energy utilized by the technology (E_1) for moisture evaporation at time t to the total energy (E_0) supplied at that time:

$$\eta = E_1/E_0 \tag{8.1}$$

Although investigating the energy efficiency of a technology gives us the idea about the overall energy utilized, aspects like quality or availability of the useful energy to perform the work is not explained. Therefore, another parameter exergy (E_x) is used to measure the sustainability of a dryer. Exergy is defined as the maximum potential of energy to perform useful work. It provides a more comprehensive and accurate measure of the quality of energy than conventional energy measures such as energy consumption or η. It is an efficient tool to determine the sustainability of energy systems by highlighting and reducing energy wastage, operational cost, and the level of pollutants generated. The E_x (Equation 8.2) of technology can be calculated using the second law of thermodynamic equation:

$$E_x = mc_p\left[(T - T_c) - T_c \ln\left(\frac{T}{T_c}\right)\right] \tag{8.2}$$

where m is the mass flow rate of the product, c_p is the specific heat of the product (kJ/kg·K), T_c is the reference temperature (usually set to 298 K or 25°C), and T is the drying temperature. The c_p can be calculated using Equation 8.3:

$$c_p = 4.187mc + 1.424c + 1.549p + 1.675f + 0.837a \tag{8.3}$$

where *mc*, *c*, *p*, *f*, and *a* stand for moisture content, carbohydrate content, protein content, lipid content, and ash content, respectively.

Further inflow $E_{x_{in}}$ (Equation 8.4), outflow $E_{x_{ou}}$ (Equation 8.5), and E_x loss (Equation 8.6), respectively, can be calculated as follows:

$$E_{x_{in}} = mc_p\left[(T_i - T_c) - T_c \ln\left(\frac{T_i}{T_c}\right)\right] \quad (8.4)$$

$$E_{x_{ou}} = mc_p\left[(T_o - T_c) - T_c \ln\left(\frac{T_o}{T_c}\right)\right] \quad (8.5)$$

$$E_{x_{loss}} = E_{x_{in}} - E_{x_{ou}} \quad (8.6)$$

Other parameters are also evaluated like exergy efficiency (η_{Ex}) (Equation 8.7), which is the ratio of convertible energy available for work to the amount of total energy supplied, exergy sustainability index (ESI) (Equation 8.8), which explains the sustainability of a dryer by taking into account the amount of exergy supplied for work and loss in exergy, and exergy improvement potential (EIP) (Equation 8.9), which describes the potential to increase exergy efficiency in order to reduce the environmental impact of technology. Mathematically, these parameters are expressed as follows:

$$\eta_{Ex} = \frac{E_{x_{ou}}}{E_{x_{in}}} \quad (8.7)$$

$$ESI = \frac{1}{1 - \eta_{Ex}} \quad (8.8)$$

$$EIP = E_{x_{loss}} * (1 - \eta_{Ex}) \quad (8.9)$$

8.2.2 Environmental Impact

The design should not add to the pollution and reduce any carbon emissions. There are various environmental assessment methods that can be used to analyse the sustainability of a process or technology.

8.2.2.1 Environmental Impact Assessment (EIA)

This evaluates the overall impact of technology on the surroundings. It is mostly inclusive of effects on the local environmental aspects like water, air, land, biodiversity, etc.

8.2.2.2 Life Cycle Assessment (LCA)

It is a comprehensive approach to investigating the environmental impact of a process or technology by assessing its entire life cycle from its creation to its ultimate disposal. This methodology identifies, quantifies, and assesses all the resources used and wastes generated during the entire life span of the technology. It is also the most common method applied to investigate the sustainability of a dryer. The LCA framework commonly uses three different components to analyse the environmental effect of technology such as energy payback time (EPT), global warming potential (GWP), and embodied energy (EE).

The EPT is defined as the total time required to recuperate the energy utilized for the manufacturing of the technology. It can be calculated as follows:

$$EPT = \frac{EE\ (\text{kWh})}{AE(\text{kWh/yr})} \quad (8.10)$$

Here, *EE* is the embodied energy and *AE* is the annual thermal energy savings/generation, which can be calculated using the following equation:

$$AEO = DEO\,(\text{kWh}) * N_d \tag{8.11}$$

Here, N_d is the hours of operation in a year and DEO is the daily thermal energy output. DEO can further be calculated as follows:

$$DEO = \frac{\text{Moisture evaporated (kg)} * \text{Latent heat of vaporization}\left(\frac{\text{J}}{\text{kg}}\right)}{3.6 * 10^6} \tag{8.12}$$

The GWP measures the impact of greenhouse gas emissions on global warming and climate change. It assesses the GHGs like CO_2 emitted by the technology over its entire life cycle. It can be assessed by evaluating the carbon dioxide emission, annual carbon dioxide mitigation, and carbon credit earned by the dryer.

The CO_2 emitted per year is calculated as follows:

$$CO_2 \text{ emitted per year} = \frac{EE\,(\text{kWh}) * e_f\,(\text{kg})}{\text{Life span of technology}} \tag{8.13}$$

Here e_f is the emission factor which is the amount of CO_2 produced per unit energy consumed. The emission factor can vary depending on the type of fuel or energy source used, as well as other factors such as combustion efficiency and emission control technology.

Annual CO_2 mitigation is basically the cumulative amount of CO_2 obstructed from being emitted out into the atmosphere by the technology. It can be calculated as follows:

$$\text{Annual } CO_2 \text{ mitigation} = \big((AE * \text{life span of technology}) - EE\big) * e_f \tag{8.14}$$

A carbon credit is earned by technology to mitigate 1,000 kg of emission and is calculated as follows:

$$\text{Carbon credit gained} = \text{Total } CO_2 \text{ mitigated per year} * \text{cost per tonne} \tag{8.15}$$

8.3 TRADITIONAL DRYING METHODS OF MARINE PRODUCTS AND THEIR LIMITATIONS

The most commonly followed traditional drying methods for fish and fish products are sun drying, smoke drying, and salt drying.

8.3.1 Direct Sun Drying

Traditional methods for preserving marine items like fish, prawns, and seaweed include sun drying. It is one of the oldest methods of drying marine products. It entails exposing the goods to the sun directly for a long time until they are adequately dried out. The typical range of moisture content by direct sun drying is between 10% and 20% but the moisture content differs depending upon the season, region, surface area of fish, temperature, humidity, and airflow [5–7]. Sun-dried fish has significantly influenced many people's diets in Asia's coastal regions since it is readily available and has distinct flavour and texture. Apart from this, the cost of sun drying is almost negligible when compared to other methods as it involves drying the marine products directly under the sun either on a bamboo tray or simply on the net used during fishing or hanging big fish between poles through ropes [8, 9]. The main principle behind sun drying involves the absorption of short-wavelength incident sun rays by the surface of the food materials and reflecting the longer wavelengths simultaneously. The shorter wavelength radiations are further converted to thermal energy leading to evaporative moisture loss inside the food. However, the longer wavelength radiation which could have been beneficial in accelerating the drying process is lost significantly due to convection to the surroundings. So, even though this method offers various advantages, it also pertains to certain limitations in terms of efficiency, and other uncontrollable drawbacks such as weather dependency, hygiene and contamination risks, inconsistent drying, nutrient loss, handling, and storage which cannot be controlled [9–11].

8.3.2 Smoke Drying

Like sun drying, smoking is also one of the traditional methods of drying and preserving fish and marine products. In this method, fish are salted and processed with organic ingredients produced from smoke are preserved through the technique of smoking. Different woods are used as smoke is generated by the incomplete burning of biomass. Not all woods are used, as the odour and taste of different woods differ. Different types of woods, like coconut shells and husks, mango wood, sandalwood leaves and dust, paddy husks, cedar, rose, apple, etc., are used in different geographical areas [12]. Also, different ovens, like cylindrical metal ovens, rectangular mud ovens, rectangular/square metal ovens, and chorkor ovens, etc, are used in the African continent. The moisture content of the fish dried by this method depends on the fish species, temperature, and duration, the fish has been subjected to during the drying process, and it is typically between 10–20% [13–15]. For example, the Tilapia fish subjected to smoke drying under different temperatures, i.e. 50°C, 60°C, and 70°C, and at different time intervals of 5, 10, and 15 hours showed a decrease in moisture content from 75.11% to 15.30% and 17.95% after smoke drying at 60°C for 15 hours and at 70°C for 10 hours, respectively [14].

Similarly, in Western countries, different industrial ovens, kilns, smokehouses, and smokers based on halo heat technology are used [12]. Smoke has antibacterial properties and contains volatile aromatic chemicals that give fish flesh distinctive qualities (attractive look, colour, flavour, and aroma). Fish that has been smoked is a flavourful and healthy food that may be consumed with or without additional heating or processing. Smoke contains hazardous chemicals with cancer-causing characteristics in addition to beneficial ones. The type and size of the fish, its diet, its state and preparation before smoking (fresh or frozen), as well as the techniques used for salting, heating, and smoking all affect the quality of the smoked fish [16–19].

Smoking can occur in cold, warm, or hot environments depending on the temperature of the smoking chamber, and different smoking techniques are used based on the setting, including smoking with natural smoke, smoking without smoke, and mixed smoking [19]. Fish is processed in a smoke–air mixture created by the direct incomplete combustion (pyrolysis) of wood when smoking with natural smoke. Smoking without smoke involves the use of preparations made using smoke or its by-products. Because dangerous compounds are removed from smoke preparations before usage, fish that has been smoked with them doesn't contain them. More people are becoming interested in liquid smoke (smoke aqueous solution), which has been adequately examined, is readily available, and has a low risk for toxicity [16, 18].

8.3.3 Salt Drying

Salt drying of fish is the oldest drying method where salts are used as an ingredient for drying. In this method, fish slices are kept in the dry commercial salt (NaCl) of about 30% by weight of fish kept for the curing process. The salts act as a desiccant, lowering the fish's moisture content. The extracted water is removed, and the fish slices are again kept in fresh dry salt. This repeated action minimizes the fish's water content and activity, preventing microbes from growing [20].

The addition of salt is repeated till the desired dryness is achieved. The dried fish should then be stored in an airtight container. Salt-dried fish can be stored without refrigeration. Depending upon the method used for salt drying, i.e. dry salt drying or wet salt drying, the moisture content differs. Typically, in dry salt drying, the process of drying is continued till the moisture content is 25% or below; whereas in wet salt drying, the moisture content is around 50–55% with 25% salt concentration [19]. The flavour, colour, and texture enhance compared with the fresh fish. Such dried fish can be stored for about three months at ambient temperature [19].

Though this method is widely used in rural areas, it has certain limitations. Firstly, the addition of salt at regular intervals might not seem to be cost-effective where large-scale drying needs to be done. Secondly, the process of salt drying can take several days, depending on the fish size. Thirdly, improper storage or stored in contact with moisture may increase the moisture content, which can lead to spoilage of fish. Due to the presence of large amounts of salt in the fish, the non-halophilic microbes will not grow, but it can support the growth of halophilic bacteria, which are highly salt resistant.

All the above traditional methods are easy to perform and cost-effective due to the availability of resources to be used for drying purposes. But these methods take a lot of time as well as some parameters are not under the control of humans, i.e. weather which will affect all the above methods. Also, the shelf life of the products is less and cannot be stored for longer duration. The hygiene might be compromised, which can lead to contamination and spoilage. Therefore, in order to overcome these issues, new and improved sustainable methods are required, which can help in reducing the time, reducing or minimizing the contamination, retaining the nutrient value, and increasing the shelf life of the marine products.

8.4 VARIOUS SUSTAINABLE DRYING METHODS FOR MARINE PRODUCTS

Drying of marine products is a useful unit operation because of the various benefits it provides, for example enhancing the shelf life of a product, reducing the product's volume for convenience during storage and transportation, but most traditional drying technologies are highly time-consuming and thereby demands more consumption of energy. Also, delayed drying process of marine products has drastic effects on its nutritional properties. Therefore, application of sustainable methods of drying marine products is gaining momentum. Some of the sustainable drying methods for marine products are as follows.

8.4.1 Sustainable Methods in Solar Drying

Solar drying technology has been in use since decades for drying of various agricultural and marine products to mitigate post-harvest losses and improve the quality of dehydrated foods. This can be a potential sustainable technology which basically captures the sun's thermal energy to evaporate the moisture from the food products. The principle behind food dehydration using a solar dryer is similar to that of open-sun drying, but the main distinguishing factor about solar drying is the presence of a drying chamber in the dryer which comprises a transparent cover, usually made of glass or polycarbonate sheet that traps the longer wavelength radiations reflected by the food surface, thereby significantly reducing convective heat loss. This technology has proved to be more efficient and advantageous over open (direct)-sun drying in terms of producing high-quality dehydrated foods as well as enhanced drying time mitigating the issue of contamination during processing [21]. Solar drying technology can be considered an efficient sustainable technology since it can help mitigate around 27–80% of unclean energy consumption, thereby reducing overall CO_2 emission [22]. Solar drying is of various types and can be classified based on the mode of air circulation – (a) natural convection (passive) and (b) force convection (active) – and type of drying: (a) direct drying, (b) indirect drying, and (c) mixed-mode drying. The classification is represented in Figure 8.1.

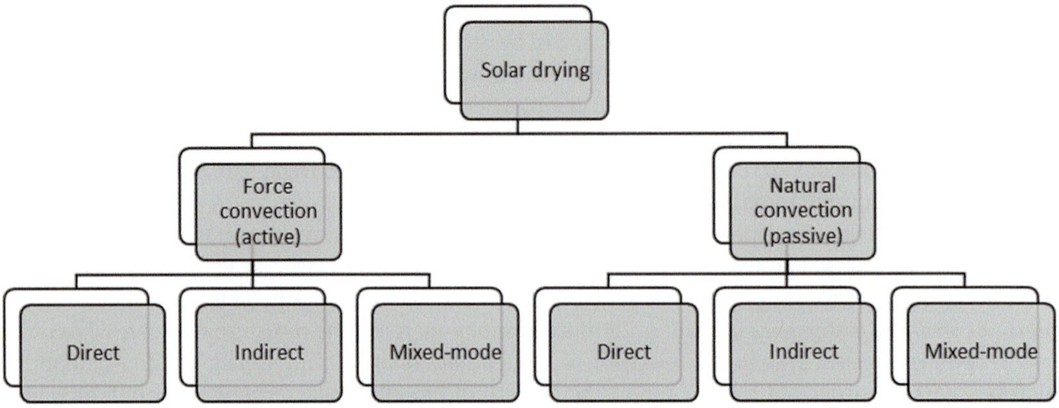

FIGURE 8.1 Flowchart for classification of solar dryers [23].

8.4.1.1 Natural Convection (Passive) Solar Dryer

The direct-type passive solar dryer is the oldest type of solar dryer which mostly applies natural convection as the mode of heat transfer. This type of dryers comprises an insulated hot box as the drying chamber generally made up of inexpensive materials such as plywood, brick, cement, galvanized iron, etc. and the inner part is coated with black paint to increase heat absorptivity. The chamber is provided with a cold-air inlet and a hot-air outlet for natural air circulation and is further covered at the top with single- or double-glazed transparent glass or polycarbonate sheets. It comprises black-painted aluminium trays for good thermal conductivity and heat absorptivity during drying. The incident solar radiation transmits through the transparent cover which is absorbed by the food products placed inside the chamber initiating moisture evaporation (greenhouse effect). The incident solar radiation also heats up the air inside the chamber which enters through the inlet vent located below the dryer creating a thermosyphon effect, i.e. the less dense hot air accumulated with evaporated moisture from the surface of the product rises up and exhausts out from the top outlet vent or chimney, whereas the cold air enters from the bottom and continues the convection cycle due to pressure gradience (Figure 8.2). This type of dryer does not use any external source of energy and the cost of manufacturing is very cheap. The average efficiency of most passive dryers ranges from 20% to 40% which can vary on the basis of drying location, airflow rates, and also type of products dried [24]. Also, due to the presence of a closed glass chamber in this type of dryer, the products heat up faster unlike in open-sun drying, thereby greatly reducing the drying time. But they possess some disadvantages such as they have limited control over drying conditions which might lead to overheating of products and are weather dependent. Moreover, due to direct exposure to sun's radiation, discolouration of food products might take place. Some examples of direct-type passive solar dryers are cabinet dryers and greenhouse dryers.

Indirect-type passive solar dryers utilize solar collectors to capture the solar radiation in order to heat the air inside the dryer and is further circulated through natural convection (Figure 8.3). These dryers comprise mainly three components – a drying chamber, solar collector, and air vents – for air circulation inside the chamber. The drying chamber is separated from the solar collector. Hot air passing through a low-pressure-drop solar collector is transferred to the drying chamber over the drying trays via the air ducts. The moisture collected by the air is further released through the vents or chimneys situated above the dryer. In these types of dryers, the food materials are not exposed to direct solar radiation, thereby helping to maintain the nutrition and colour of the dehydrated products. However, the efficiency of these dryers is 13–25%, which is much lower compared to direct passive dryers [24].

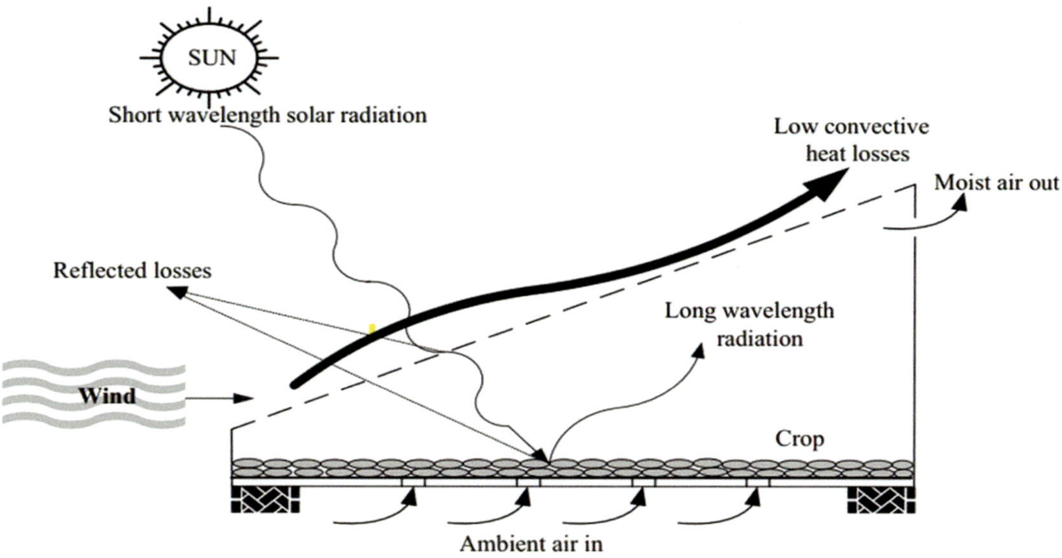

FIGURE 8.2 Schematic diagram of a direct-type passive solar dryer. (Adapted from Ref. [25].)

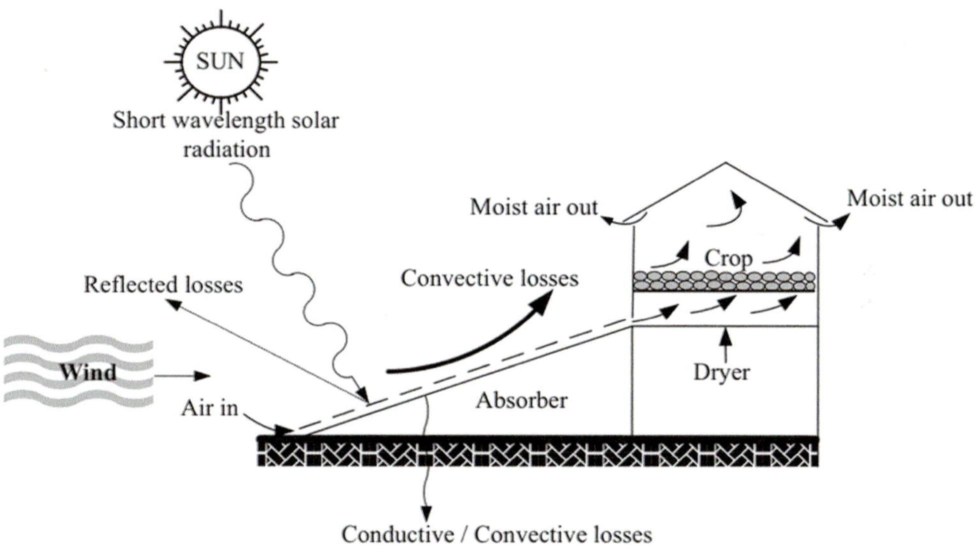

FIGURE 8.3 Schematic diagram of an indirect passive solar dryer. (Adapted from Ref. [25].)

8.4.1.2 Forced Convection (Active) Solar Dyer

The direct-type active solar dryer is similar to direct-type passive dryers (Figure 8.4a and b) except it mostly involves the usage of external sources of energy like blowers and fans that are operated using electricity generated by photovoltaic modules for forced convection circulation of the ambient air inside the chamber. It comprises a drying chamber, a fan or a blower unit, and air ducts. Indirect-type active solar dryers comprises a drying chamber, fans, air ducts along with a solar collector for air heating (Figure 8.5a and b). These dryers are mostly preferable for food products with higher

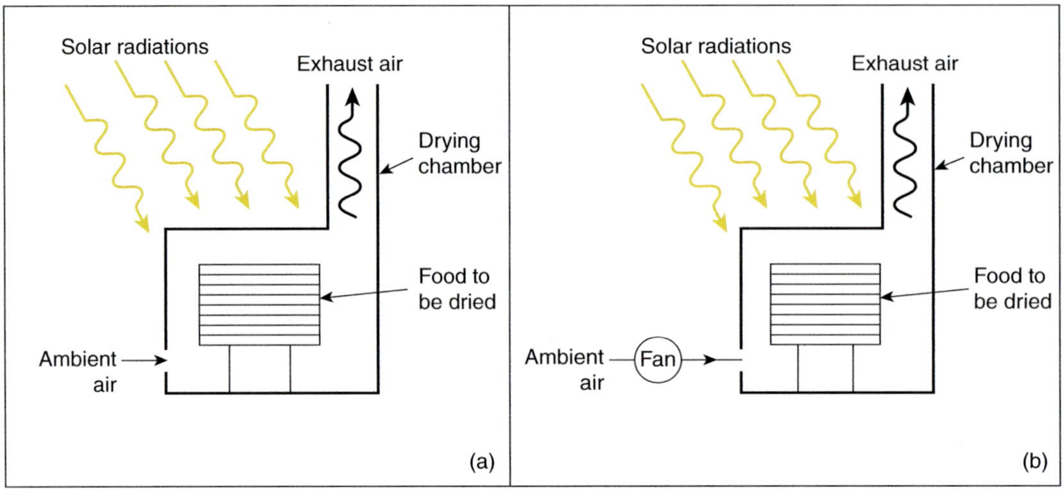

FIGURE 8.4 Schematic diagram showing the difference between (a) direct passive solar dryer and (b) direct active solar dryer [26].

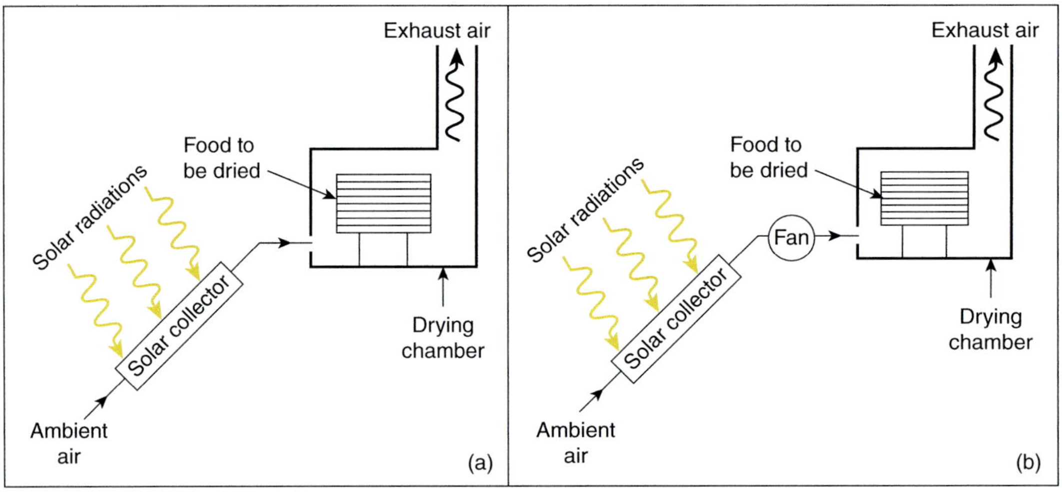

FIGURE 8.5 Schematic diagram showing the difference between (a) indirect passive solar dryer and (b) indirect active solar dryer [26].

moisture content and are utilized for large-scale commercial production. The forced convection solar dryer is advantageous over natural convection dryers in terms of faster drying time, enhanced control over the drying process, independent of weather conditions due to their ability to use electricity to supplement solar power. However, the cost of operation and maintenance is comparatively higher than natural convection dryers.

8.4.1.3 Mixed-Mode Solar Dryers

These types of dryers are an amalgamation of both direct and indirect solar dryers. Drying is carried out either in single mode or in mixed mode. The mode of heating for these dryers can be either by using direct solar radiation exposure or by using forced circulation of solar collected hot air with the help of fans or blower units. Hybrid solar dyers are a type of mixed-mode solar dryers where the source of energy utilized for drying food products generally depends on direct sunlight or on auxiliary sources of energy such as thermal storage energy systems (TES), biomass, etc. to continue drying even during off shine hours or bad weather conditions. These dryers have been reported to be efficient in reducing microbial contamination in dried foods and can be used for drying various perishable products [27, 28].

8.4.2 Microwave Drying

Microwave drying technology utilizes electromagnetic waves with frequencies ranging from 900 MHz to 2,400 MHz to remove moisture from the products [29]. The main principle behind microwave drying is the electric dipole moment of water molecules. The polar water molecules in the food products are attracted to the electromagnetic field which leads to rapid oscillation and reorientation of the water molecules. This generates heat due to intermolecular friction resulting in the evaporation of water from the food products. Microwave drying is different from the conventional drying method as heat transfer due to the high-frequency rotation of the water molecules takes place from the interior to the exterior [30]. It has been also reported that the drying efficiency of microwave drying is far better than conventional hot-air drying by 5–15% [31]. This efficient technology can lead to a substantial reduction of the drying period at the industrial scale [32]. It is an important drying technology as dehydration takes place through targeted heating which reduces the need for excessive heating and thus low energy consumption. This can lead to reduced GHGs and carbon footprint. It helps in the retention of the nutrition of dehydrated products and is also beneficial for drying thermolabile products. This drying process generates vapour in the interior of the dehydrated products which is forced out due to differences in vapour pressure resulting in the puffing effect of the product and reduction of shrinkage during drying [33]. It also enhances the rate of drying during the falling rate period. However, some major disadvantages of microwave drying are as follows: It is not a cost-effective technology due to lack of energy efficiency. Also, non-uniform heating by the microwave field can result in the burnt colour of dehydrated product [34]. To circumvent these issues and improve drying efficiency and its sustainability, microwave is integrated in combination with other drying technologies like vacuum drying, convective drying, fluidized bed, osmotic drying, and so forth. Moreover, microwave drying in combination with reduced air pressure and temperature enhances the retention of temperature and oxidation-sensitive macro- and micronutrients, flavours, and bioactives as compared to other heat-induced drying technologies [35].

8.4.3 Infrared (IR) Drying

IR drying is a promising drying technology which is mainly based on electromagnetic radiation and is mostly used for drying high-moisture food products. IR is a form of radiation energy which lies between light and microwave spectrum. These IR radiations are composed of photons which is absorbed by the materials subjected to drying. They do not require any medium for its transmission. The photons on coming in contact with the molecules inside the food transfers their energy which results in vibration of the molecules at frequency 60,000–150,000 Hz and hence heat generation. This heat further helps in moisture removal. IR drying can be categorized into three types and its wavelengths vary from 0.75 μm to 1,000 μm – near-infrared (0.75–1.4 μm), medium-infrared (1.4–3 μm), and far-infrared (3–1,000 μm) drying. The far IR has been of relevance in drying of food products [36]. However, wavelengths ranging from 1.5 μm to 5 μm are considered to be absorbed easily by most food materials since wavelengths of this range efficiently penetrate a few millimetre deep inside the food [37]. Table 8.1 shows the various IR bands required for heating different food components. This technology has several advantages over other conventional drying techniques such as higher drying efficiency due to the ability of IR to penetrate inside the surface and heat the product uniformly. This also reduces the drying time significantly as compared to

TABLE 8.1
Absorption Bands of Infrared Relevant for Various Groups of Food Components [38]

Food Component	Wavelengths (μm)	Chemical Group
Sugars, water	2.7–3.3	Hydroxyl group (O–H)
Lipids, sugars, proteins	3.25–3.7	Aliphatic carbon–hydrogen bond
Lipids	5.7–5.76	Carbonyl group (C=O) (ester)
Proteins	5.92	Carbonyl group (C=O) (amide)
Proteins	2.83–3.33	Nitrogen–hydrogen group (–NH–)
Unsaturated lipids	4.44–4.76	Carbon–carbon double bond (C=O)

conventional methods. IR drying is also energy-efficient as compared to other traditional drying methods as heat is transferred directly into the subjected material which reduces energy wastage to the surrounding environment, thereby improving efficiency and reducing carbon footprint. This form of targeted heating also reduces energy consumption and operating costs. Also, it can be considered environmentally benign since it produces neither NO_x nor CO, making it a more sustainable drying option. Due to controlled heating, this technology can also result in a good quality final product. The schematic diagram of IR heating of food products is graphically depicted in Figure 8.6.

8.4.4 Refractance Window Drying

This is an innovative non-thermal drying technology that works on the principle of heat transfer through a thin layer of liquid usually circulating at a temperature of 94–98°C. This liquid is placed between two transparent Pyrex glass plate like conveyor belts. The samples to be dried are placed on the upper plate so that the heat can be transferred through the liquid. The water is recirculated and reheated to maintain the uniform drying of the product. At the end of drying, cold water cools down the product (Figure 8.7). It is considered a rapid drying technique (mostly takes 5–10 minutes of drying for most of the food commodities) and can be beneficial for temperature-sensitive marine product drying because of its low-temperature short-time behaviour, these dryers limit oxidation and free

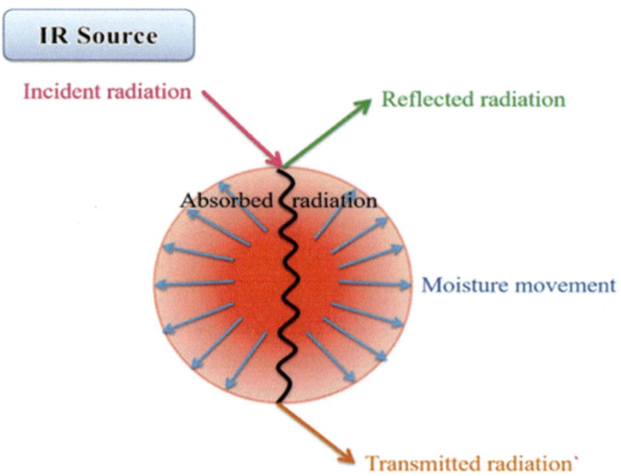

FIGURE 8.6 Schematic diagram of IR heating of food products [39].

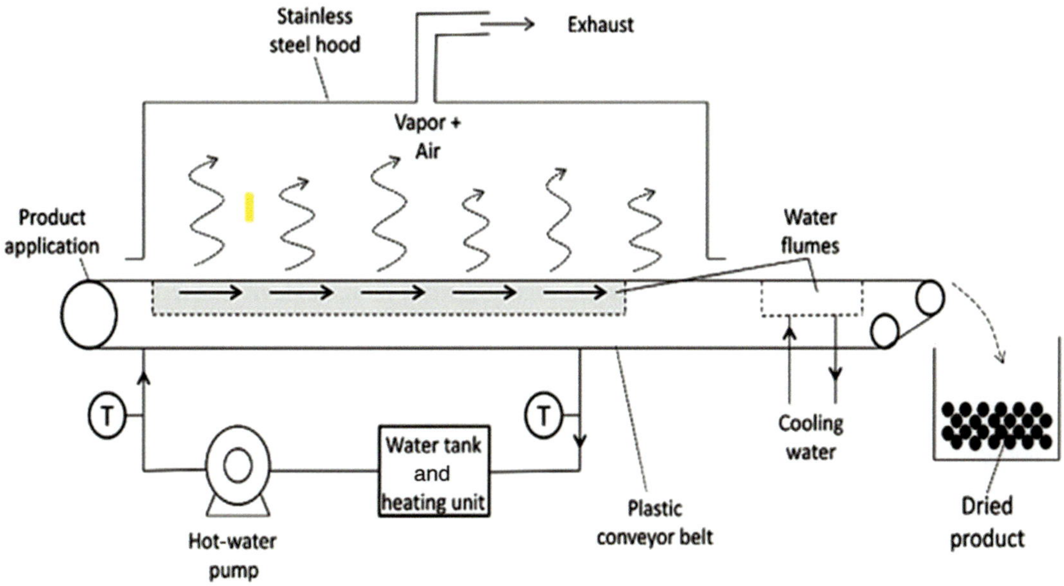

FIGURE 8.7 Schematic diagram of refractance window drying [41].

radical formation [40]. The energy consumption rate of this dryer is very low and thereby has better energy efficiency (28–38%) and the cost of production and utilization is also low, making it a suitable sustainable dryer and a much better technology compared to the conventional drying method.

8.4.5 Hybrid Drying

Hybrid drying technology is an amalgamation of different drying techniques in order to achieve improved drying kinetics, produce better quality products, and sustainability. In hybrid drying technology, the different drying techniques are combined in such a manner that the advantages of each method are enhanced, whereas the shortcomings are reduced significantly. This novel technology of combining two or more techniques could improve the overall efficiency of a system while drastically reducing the cost as well as various environmental effects related to the individual systems. There are various approaches to hybrid drying technologies which combine non-conventional methods with other novel technologies like vacuum drying with microwave drying, microwave-assisted fluidized bed drying, and fluidized bed drying with heat pump drying. The overall energy consumption of a thermal system can also be reduced by combining it with renewable energy systems, mostly systems utilizing solar energy, making the overall drying process clean, efficient, and sustainable. As per reports suggested in literature, the energy efficiency of hybrid drying technology varies within the range of 5–72% [32]. Hybrid drying technology also involves the process of intermittent drying where the drying method relies on varying temperature, pressure, humidity, or even the mode of heat transfer (convection, conduction, radiation, etc.) with respect to time. It does not involve continuous exposure of the products to the drying conditions but rather follows a periodic drying and resting approach. Due to this, the products processed using intermittent approach are uniformly dried, which inhibits the probability of case hardening, usually perceived as a limitation for the convective drying method [42]. Moreover, this method has also been useful in mitigating the browning effect, hydro-thermal mechanical stress, and retention of important bioactive relevant for the production of nutrition-rich dehydrated products [43] and also has proved to be energy-efficient, thereby reducing the environmental impact.

8.5 FACTORS AFFECTING THE KINETICS AND QUALITY OF DRIED MARINE PRODUCTS

Dehydration of marine products is greatly influenced by its drying behaviour (drying kinetics). Drying kinetics of food provides a knowledge on rate of moisture removal during the drying process and hence indicates about the effectiveness of a certain drying technique. It is also referred for optimization of a drying process. There are several parameters associated with the drying process of fish and fish products which play an important role in determining the kinetics and quality of the final product: characteristics of fish being used (i.e. the surface area, thickness, initial moisture content, a_w), drying parameters (temperature, humidity, airflow rate), mode of heat transfer (conduction, convection, or radiation). Fishes are composed of around 80% of moisture and reducing the water activity (a_w) to a level unfavourable for microbial proliferation would help enhance the shelf life of fish. However, this might lead to various microstructure and textural changes in the product, disrupting its palatability, appearance, and overall experience of fish indulgence. Therefore, it is important to optimize the parameters responsible for influencing the drying behaviour and quality of dried marine products.

8.5.1 Physical Parameters Influencing Drying Behaviour of Fish

The following are some crucial factors important to understand the drying behaviour of fish and fish products. These factors also determine the final outcome of the dehydrated product.

8.5.1.1 Structural Dimensions

Surface area and thickness of marine products to be dried are important parameters which ensure uniform drying of the product as well as determine the rate at which the moisture is removed. The surface area of the fish to be dried is positively correlated, whereas thickness is negatively correlated with the drying rate. The more the surface area, the greater the area of the product exposed to the heating medium, facilitating better heat and mass transfer for higher moisture loss in the sample. This also results in the overall reduction of drying time. The surface area of the fish is also a factor responsible for faster moisture removal during the initial period of drying (constant rate period). The increase in thickness of the fish restricts the rapid diffusion of moisture due to the greater distance between the interior of the fish to the surface. Hence, slowing down the rate of drying during the falling rate period. Therefore, scoring/splitting of fish before drying would enhance its drying rate and hence reduce the drying time.

8.5.1.2 Moisture Content

Drying rate is significantly influenced by the amount of moisture present in the sample. Food samples are generally composed of free moisture (available in the interstitial spaces of the molecules) and bound moisture (found mostly bonded to myofibrils of the fish) which determine the phases of drying fish. The drying of fish takes place in two different phases and eventually stops when a change in moisture reaches equilibrium. The initial phase of drying, also known as the 'constant rate period', is the stage where the rate of drying will be faster as the surface is mostly saturated with free moisture. The surrounding air humidity, temperature, and air velocity also play a significant role in moisture evaporation at this stage [44]. During the constant rate period, the freely available moisture easily diffuses into the surrounding air and the rate of drying at this stage is mostly influenced by heat and mass transfer between the fish surface and the environment. The constant rate period ends when the sample reaches its critical moisture content. This leads to the initiation of a falling rate period where the bound moisture is removed. At this stage, the rate of moisture migration reduces since bound moisture has lower mobility as compared to free moisture. Also, the rate of drying is mostly dependent on the diffusion rate of moisture (effective moisture diffusivity) within the fish which is mainly caused due to concentration gradient within the fish.

The effective moisture diffusivity of the fish is greatly influenced by the nature of the fish (fishes with higher lipid content obstruct moisture migration), thickness, temperature of the sample, and moisture content [45].

8.5.1.3 Temperature and Air Velocity

Temperature and air velocity are crucial factors that can influence the drying behaviour of marine products. Marine products are sensitive to these drying conditions and the quality of the final dehydrated products largely depends on these two parameters along with relative humidity (RH). Temperature provides energy for a transition of a liquid to vapour state, thereby leading to an increase in vapour pressure gradient and thus leading to moisture removal. The temperature of the air is also found to be positively correlated with the drying constant and negatively correlated with an equilibrium moisture content of most dehydrated products [46]. It is also an important factor influencing irreversible biochemical reactions in the fishes that regulate the overall nutritional quality and sensorial properties of dehydrated fish products. Similarly, air velocity is also positively correlated with the drying rate of fish. As air velocity increases, the convective heating also rises, resulting in an enhanced drying rate. During drying, a very thin layer of air (boundary layer) forms over the surface of the samples dried. This layer obstructs moisture migration from the surface to the surroundings. However, with increased air velocity, shear stress above the boundary layer increases. This promotes thinning of the boundary layer, thereby providing rapid evaporation of moisture. Increasing the drying rate will also lead to lesser drying time.

8.5.1.4 Relative Humidity

RH is another influential factor which affects the drying rate of the fish. It is mostly the amount of vapour present in the air relative to the amount of vapour the air can hold at a particular temperature. It also determines the preservation of the desirable quality of the final product. Air with lower RH promotes faster evaporation and removal of moisture from the drying surface of the fish as the air is not saturated with moisture and this increases the water vapour holding capacity of air. At lower RH condition, in order to maintain a state of equilibrium, moisture tends to move from a state of higher level (fish surface) to lower level (surrounding environment) due to concentration gradient. This is further enhanced by optimum temperature and drying air velocity. RH is also related to equilibrium moisture content. The lower the RH, the lower the equilibrium moisture content of the sample, which would further boost the moisture content reduction. This is also an important factor dictating the shelf life of dehydrated fish during storage. The RH during drying can be reduced in three different ways: (a) increasing the drying temperature, (b) reducing the moisture content in the air, and (c) combining both [47].

8.5.2 Effect of Drying on Quality of Dehydrated Marine Products

Quality is an important factor for dehydrated marine products because temperature and moisture loss during drying have detrimental effects, which dictates the sensory attributes, nutritional value, and overall acceptability of the final dehydrated product. Various irreversible physical and biochemical changes take place during the drying process of fishes [48], which mitigates the possibility of microbial growth or any enzyme-related spoilage leading to their shelf life enhancement. However, these changes might instigate a negative impact on other quality attributes like colour, texture, aroma, and bioactive composition. Therefore, this section discusses some important quality parameters that must be considered while choosing the appropriate sustainable dryers for the dehydration of marine products.

8.5.2.1 Water Activity (a_w)

The primary goal of dehydrating high moisture-containing fish and fish products is to significantly reduce the a_w of the product. The a_w is the amount of unbound moisture available in a sample which can be utilized for biochemical reactions. It is basically expressed as a ratio of vapour pressure of water present in the sample to the vapour pressure of pure water at same temperature. It can be considered as a critical control point for dehydrated marine products, as it is an important factor which regulates microbial growth within the sample. Microbes require unbound moisture for their growth and proliferation [49]. Marine products with high amount of free moisture create an ideal condition for microorganisms like bacteria, yeasts, and moulds to grow. Most bacteria require a_w of at least 0.9–0.91 for their growth, whereas moulds and yeasts can sustain on even lower a_w such as 0.7–0.8 and 0.87–0.94, respectively. Therefore, in order to mitigate proliferation of spoilage microbes in marine products, it is advisable to reduce the water activity to 0.6 or below [50]. The a_w of dehydrated fish can be reduced using salting of dehydrated marine products [51]. As per literature, a_w also had a correlation with equilibrium moisture content but not with drying temperature and time [52]. However, reducing the a_w further below (around 0.4–0.2) can also induce various unfavourable, irreversible reactions in fishes like lipid oxidation, browning, or might even lead to loss of bioactives which affect the acceptability and overall value of the final product [53, 54].

8.5.2.2 Colour

The colour of dehydrated fish is a valuable parameter that greatly determines its consumer acceptability. During dehydration, alteration in the surface properties of fishes adds to the change in reflectivity and colour in the final product [55]. The colour change has mostly been attributed to Millard browning or lipid/photo-oxidation in fishes [56]. These reactions are often instigated by the drying conditions and the mode of drying. Temperature, time of drying, and power level have been found to be the prime contributors of colour change in fishes [57–59]. Higher temperature or increased power level would accelerate the rate of Millard browning, which results in a more pronounced browning effect on dehydrated fishes. Moreover, it would also lead to the disintegration of thermolabile pigments like carotenoids in marine products which impart the natural colour in fish. thereby resulting in colour change of the final product. Although reducing the temperature of drying could be favourable, this would increase the total drying time of the process which will potentially result in colour change of fishes due to prolonged exposure to heat and moisture. So, it is important to find a balance between these parameters while choosing a suitable sustainable dryer for fish dehydration in order to achieve optimal drying and minimal colour changes.

8.5.2.3 Lipid Oxidation

Fish mostly comprises omega-3 polyunsaturated fatty acids like eicosapentaenoic acid (EPA) and docosahexaenoic acid (DHA) which are highly unstable in presence of oxygen, temperature, and light. Degradation of these fatty acids results in the production of off-flavour and rancidity of dehydrated fish, which would make the final product unfavourable for consumption. Therefore, lipid oxidation is a crucial factor for drying of marine products since oxidative degradation of lipid can be easily triggered during the drying process, especially for fishes with high fat content if the process undergoes prolong drying time or the samples are exposed to higher drying temperature. On studying the effect of drying time and temperature on lipid oxidation of Atlantic salmon fillets, Ortiz et al. observed that extended drying time had more significant effect on oxidation of fats as compared to higher drying temperature [60]. Moisture also plays an important role in inducing lipid oxidation. Improper drying leading to availability of higher unbound moisture would enhance direct contact of oxygen with lipids in the fish promoting lipid oxidation. It would also speed up the rate of triglyceride hydrolysis or transition metal (Fe or Cu) migration from their bound form within the fish tissue to come into contact with the lipids, promoting oxidation reactions [61] and thereby reducing the product's quality, especially during storage.

8.6 CASE STUDIES OF SUSTAINABLE DRYING TECHNOLOGIES FOR AQUATIC PRODUCTS

8.6.1 Fish

Fish is one of the most consumed food throughout the world. As the demand for fish is growing, more efforts are being made to make it available in the market. Fish contains a lot of water apart from other vital nutrients, which start getting spoiled few hours after catching it. Thus, drying fish will help in curbing spoilage as well as it can be stored for a longer duration of time in order to meet consumer demands.

A relatively recent technology that has a number of benefits over conventional drying techniques is the use of IR radiation to dry fish. Using electromagnetic radiation with an IR wavelength to convey heat directly to the fish will result in an effective and quick drying process. When used in vacuum, this method increases the drying efficiency by reducing the time and temperature. Similarly, microwave-assisted vacuum drying helps in reducing time. Results obtained by Viji et al. [62] who carried out microwave vacuum drying (MVD) of mackerel to study its physiochemical characteristics showed that in 1.2 hours and 12 hours, respectively, the moisture content of the MVD and hot-air drying (HAD) samples was reduced to 30–32%. The capacity of MVD samples to rehydrate and their water absorption index both were noticeably higher than those of HAD samples. In comparison to MVD samples, mackerel dried by HAD had considerably larger salt- and water-soluble protein nitrogen percentages. HAD samples were found to have significantly greater hardness and chewiness values. Compared to HAD samples, the uncooked MVD samples had a better colour and look. The microwave vacuum-dried mackerel outperformed the hot-air oven-dried mackerel in terms of rehydration and textural qualities.

Various sustainable drying methods are used, but keeping aside the nutrient availability after drying, the colour and texture of the product need to be accepted by the consumer. Old traditional methods sometimes reduce the colour and texture which the consumer may not accept. Here, the sustainable methods help us better control their appearance as well as nutrient availability. Biochemical quality parameters and organoleptic assessment of commercially dried fish and solar dried fish are given in Table 8.2.

TABLE 8.2
Biochemical Quality Parameters and Organoleptic Assessment of Commercially Dried Fish and Solar Dried Fish [63]

Sample	Biochemical Quality Parameters	Percent	Organoleptic Parameters	Percent
Commercially dried fish	• Moisture (%)	• 53.17 ± 0.38	• Appearance	• 7
	• Protein (%)	• 14.72 ± 0.16	• Colour	• 8
	• Lipid (%)	• 1.04 ± 1.74	• Taste	• 7
	• Carbohydrate (%)	• 0.81 ± 0.26	• Odour	• 8
	• Ash (%)	• 0.72 ± 1.74	• Texture	• 8
			• Overall acceptability	• 7
Solar-dried fish	• Moisture (%)	• 34.39 ± 0.84	• Appearance	• 9
	• Protein (%)	• 17.93 ± 0.29	• Colour	• 9
	• Lipid (%)	• 2.74 ± 0.05	• Taste	• 9
	• Carbohydrate (%)	• 1.06 ± 0.75	• Odour	• 9
	• Ash (%)	• 0.65 ± 0.83	• Texture	• 9
			• Overall acceptability	• 9

TABLE 8.3
Quality Assessment of Shrimps Dried by Different Drying Technologies

Shrimp	Drying Technique	Moisture (%)	Fat (%)	Protein (%)	Ash (%)	Reference
Penaeus notialis	Sun-dried	13.70 ± 0.05	1.98 ± 0.01	58.4 ± 0.51	6.77 ± 0.05	[64]
	Oven-dried at 50°C	10.91 ± 0.04	2.02 ± 0.01	56.5 ± 0.05	8.93 ± 0.05	
	Oven-dried at 60°C	9.70 ± 0.06	2.24 ± 0.01	55.8 ± 0.04	10.56 ± 0.38	
	Oven-dried at 70°C	6.90 ± 0.05	2.36 ± 0.05	53.2 ± 0.35	11.42 ± 0.09	
P. notialis	Fresh	–	6.54 ± 0.20	86.21 ± 0.08	6.93 ± 0.11	[65]
	Air-oven dried	–	5.98 ± 0.15	85.64 ± 0.26	6.82 ± 0.10	
	Solar-dried	–	5.74 ± 0.11	84.89 ± 0.51	6.77 ± 0.28	

8.6.2 Shrimp

Shrimp is one of the world's most significant commercial seafood species and is a member of a vast group of crustaceans with expanded abdomens. Shrimps are susceptible to quick deterioration due to their high protein and moisture content. They quickly go bad after catch unless they are placed in cold storage, which dramatically increases their shelf life. Using cold storage for a long may not be economical, hence drying the shrimps will be helpful in terms of storage and cost-effectiveness.

The treatments applied throughout the drying process significantly impact the fish's moisture content. Numerous studies have been conducted to examine various drying procedures and find ways to reduce moisture content more successfully and effectively. Quality Assessment of Shrimps dried by different drying technologies has been given in Table 8.3.

Alfiya et al. [66] investigated the quality of prawns dried using solar and microwave methods. When compared to solar drying with higher drying rates, microwave drying reduced drying times to 58.3%. For solar drying, the drying and collector efficiencies were 26.3–33.4% and 32.5–41.2%, respectively, but the efficiency for microwave drying was 35.7%. Shrimp that had been solar-dried had better textural qualities than microwave-dried samples. The results of microbiological tests on dried prawns prepared using solar drying and microwave method were 1.65×10^4 CFU and 2.6×10^4 CFU/g, respectively, compared to the 4.3×10^6 CFU/g in raw prawns, which were found to be within acceptable ranges of IS 14950: 2001.

8.6.3 Squid

Squids are cephalopods, one of the most suitable foods for human consumption. Many squid species are used as food in traditional cuisines around the world. Squid is a rich dietary source for zinc and manganese and is high in the required daily intake of copper, selenium, and vitamins. However, due to its high moisture content (more than 80%), fresh squid has a short shelf life [67]. Therefore, drying of squids is of utmost importance in order to preserve it from getting spoiled. One of the research projects carried out by Chen et al. [68] on the effects of intermediate IR-assisted convection drying, low-frequency microwave drying, and hot-air drying on the quality of squid cubes showed quite promising results. The drying rates of both low-frequency microwave drying and intermediate IR-assisted convection drying were faster than hot-air drying under the same power density conditions and the drying rates of 60°C and 70°C. When the moisture level was high, intermediate IR-assisted convection dried materials more quickly than the low-frequency microwave drying near 50°C. While these values of intermediate IR-assisted convection drying at 50°C and 60°C were lower than those of low-frequency microwave drying at 50°C, the drying shrinkage percentages of intermediate IR-assisted convection drying were smaller than those of hot-air drying. A ballooning event was seen in some dried samples employing microwave drying. In terms of rehydration ratio, microwave drying, intermediate IR-assisted convection drying, and hot-air drying were listed from high to low. In terms of colour quality, intermediate IR-assisted convection drying and microwave drying do not clearly outperform hot-air drying. The hardness of the squid cubes dried by microwave drying was also higher than that of the hot-air drying, although there was no discernible difference in hardness between intermediate IR-assisted convective drying and hot-air drying. In terms of the overall product quality, intermediate IR-assisted convection drying at 60°C is a better method for drying squid cubes than microwave drying at around 50°C and hot-air drying methods [68].

Similarly, researchers also used far-infrared radiation-assisted heat pump drying to understand the dried squid fillet's chemical composition and physical properties [67]. The result showed an increase in the crude protein, fat, ash, and moisture contents of raw squids, which were 13.38 ± 1.85, 0.34 ± 0.01, 1.05 ± 0.02, and 83.73 ± 0.39%, based on fresh weight, respectively. These percentages fell within the typical ranges for squid protein content (13.1–19.2%), fat (0.29–2.0%), and moisture (74.0–84.18%) along with all dried squid fillets that had lower lightness (*L*) values in case of dried squid fillets than raw ones but greater redness and yellowness values.

8.7 CONCLUDING REMARKS

Sustainable drying technology, therefore, holds immense importance in the preservation and value enhancement of marine products. Throughout this discussion, we have examined the significance of dehydrated fish and fish products in the market, highlighting the need for efficient drying methods. The background and significance of sustainable drying technology were explored, emphasizing the need for environment-friendly and energy-efficient approaches. The limitations of traditional marine drying practices were discussed and the advantages of novel sustainable technology over these conventional drying practices were highlighted. Also, various factors affecting drying process and quality of the dehydrated marine products were also discussed, emphasizing to understand these important parameters and further optimize them to obtain high-quality dehydrated marine products. In conclusion, sustainable drying technology would be a promising solution for the drying of marine products, addressing the limitations of traditional methods while promoting energy efficiency, product quality, and environmental sustainability.

REFERENCES

1. Ahmed, A. A. Post-Harvest Losses of Fish in Developing Countries. *Nutr. Health*, **2008**, *19* (4), 273–287. https://doi.org/10.1177/026010600801900403
2. MPEDA. Marine Products Exports. https://mpeda.gov.in/?page_id=438 (accessed June 16, 2021)
3. Kluczek, A.; Olszewski, P. Energy Audits in Industrial Processes. *J. Clean. Prod.*, **2017**, *142*, 3437–3453. https://doi.org/10.1016/J.JCLEPRO.2016.10.123
4. Sivakumar, R.; Saravanan, R.; Elaya Perumal, A.; Iniyan, S. Fluidized Bed Drying of Some Agro Products: A Review. *Renew. Sustain. Energy Rev.*, **2016**, *61*, 280–301. https://doi.org/10.1016/j.rser.2016.04.014
5. Ullah, N.; Hazarika, P.; Handique, P. J. Biochemical Quality Assessment of Ten Selected Dried Fish Species of North East India. *IARJSET*, **2016**, *3* (1), 30–33. https://doi.org/10.17148/iarjset.2016.3107
6. Pradhan, N.; Shrestha, M. K.; Rai, S.; Jha, D. K.; Sah, S. K. Evaluation of Different Drying Methods on Shelf-Life Quality of Mrigal *Cirrhinus mrigala* (Hamilton, 1822). *Asian Fish. Sci*, **2020**, *33* (1), 31–41. https://doi.org/10.33997/j.afs.2020.33.1.004
7. Tenyang, N.; Ponka, R.; Tiencheu, B.; Djikeng, F. T.; Womeni, H. M. Effect of Traditional Drying Methods on Proximate Composition, Fatty Acid Profile, and Oil Oxidation of Fish Species Consumed in the Far-North of Cameroon. *Glob. Challenges*, **2020**, *4* (8), 2000007. https://doi.org/10.1002/gch2.202000007
8. Sultana Marine, S.; Sayeed, A.; Barman, P. P.; Begum, R.; Hossain, M.; Alam, T. Traditional Methods of Fish Drying: An Explorative Study in Sylhet, Bangladesh. *Int. J. Fish. Sci. Aquac.*, **2015**, *2* (1), 28–35.
9. Paul, P. C.; Reza, M. S.; Islam, M. N.; Kamal, M. A Review on Dried Fish Processing and Marketing in the Coastal Region of Bangladesh. *Res. Agric. Livest. Fish*, **2018**, *5* (3), 381–390. https://doi.org/10.3329/ralf.v5i3.39587

10. Paul, P. C.; Reza, M. S.; Islam, M. N.; Kamal, M. Quality Assessment of Traditionally Dried Marine Fish of Bangladesh. *Asian Food Sci. J.*, **2018**, *5* (1), 1–11. https://doi.org/10.9734/afsj/2018/44406
11. Wang, Y.; Zhang, M.; Mujumdar, A. S. Trends in Processing Technologies for Dried Aquatic Products. *Dry. Technol.*, **2011**, *29* (4), 382–394. https://doi.org/10.1080/07373937.2011.551624
12. Adeyeye, S. A. O. Smoking of Fish: A Critical Review. *J. Culin. Sci. Technol*, **2019**, *17* (6), 559–575. https://doi.org/10.1080/15428052.2018.1495590
13. Sikoki, F. D.; Aminigo, E. R. Bacteriological and Sensory Properties of Smoke-Dried Fish Stored at Ambient Temperature. *Glob. J. Agric. Sci*, **2002**, *1* (1). https://doi.org/10.4314/gjass.v1i1.2198
14. Aba, P. Effects of Smoke-Drying Temperatures and Time on Physical and Nutritional Quality Parameters of Tilapia (*Oreochromis niloticus*). *Int. J. Fish. Aquac.*, **2013**, *5* (3), 29–34.
15. Iheanacho, S. C.; Nworu, S. A.; Ogueji, E. O.; Nnatuanya, I.; Mbah, C. E.; Anosike, F.; Okoye, C.; Ibrahim, U. B.; Kogi, E.; Haruna, M. Comparative Assessment of Proximate Content and Organoleptic Quality of African Catfish (*Clarias gariepinus*) Processed by Smoking and Solar Drying Methods. *African J. Agric. Res.*, **2017**, *12* (38). https://doi.org/10.5897/ajar2017.12599
16. Belichovska, K.; Belichovska, D.; Pejkovski, Z. Smoke and Smoked Fish Production. *Meat Technol.*, **2019**, *60* (1), 37–43. https://doi.org/10.18485/meattech.2019.60.1.6
17. Salindeho, N. Contamination Concentration in Smoked Golden Sandfish (*Holothuria scabra*) Using Nutmeg Shell Coconut Shell Smoking Materials. *Int. J. ChemTech Res.*, **2018**, *11* (7), 308–313. https://doi.org/10.20902/ijctr.2018.110736
18. Kumar, S. Smoke-Drying Technology in Fish Preservation. ICAR-Central Central Institute of Fisheries Technology, **2010**.
19. Kumar, K. S. Salt Curing, Smoking and Drying of Fishery Products. ICAR-Central Central Institute of Fisheries Technology, **2021**.
20. Chakraborty, S. C. Effect of Ripening Period of Dry Salting Process in Three Freshwater Fish Species of Bangladesh. *Int. J. Adv. Sci. Res.*, **2017**, *1* (5), 16–21.
21. Udomkun, P.; Romuli, S.; Schock, S.; Mahayothee, B.; Sartas, M.; Wossen, T.; Njukwe, E.; Vanlauwe, B.; Müller, J. Review of Solar Dryers for Agricultural Products in Asia and Africa: An Innovation Landscape Approach. *J. Environ. Manage*, **2020**, *268*, 110730. https://doi.org/10.1016/J.JENVMAN.2020.110730
22. Prakash, O.; Kumar, A.; Laguri, V. Performance of Modified Greenhouse Dryer with Thermal Energy Storage. *Energy Rep.*, **2016**, *2*, 155–162. https://doi.org/10.1016/J.EGYR.2016.06.003
23. Sharma, A.; Chen, C. R.; Vu Lan, N. Solar-Energy Drying Systems: A Review. *Renew. Sustain. Energy Rev.*, **2009**, *13* (6–7), 1185–1210. https://doi.org/10.1016/J.RSER.2008.08.015
24. Kumar, M.; Sansaniwal, S. K.; Khatak, P. Progress in Solar Dryers for Drying Various Commodities. *Renew. Sustain. Energy Rev.*, **2016**, *55*, 346–360. https://doi.org/10.1016/J.RSER.2015.10.158
25. Tomar, V.; Tiwari, G. N.; Norton, B. Solar Dryers for Tropical Food Preservation: Thermophysics of Crops, Systems and Components. *Sol. Energy*, **2017**, *154*, 2–13. https://doi.org/10.1016/J.SOLENER.2017.05.066
26. El Hage, H.; Herez, A.; Ramadan, M.; Bazzi, H.; Khaled, M. An Investigation on Solar Drying: A Review with Economic and Environmental Assessment. *Energy*, **2018**, *157*, 815–829. https://doi.org/10.1016/J.ENERGY.2018.05.197
27. Udeinya, B. C.; Ojike, O.; Okonkwo, W. I.; Abada, U. C. Performance Evaluation of Mixed Mode Passive Solar Stock Fish Dryer. *Niger. J. Technol.*, **2022**, *40* (6), 1104–1109. https://doi.org/10.4314/njt.v40i6.13
28. Afzal, A.; Iqbal, T.; Ikram, K.; Anjum, M. N.; Umair, M.; Azam, M.; Akram, S.; Hussain, F.; Ameen ul Zaman, M.; Ali, A. et al. Development of a Hybrid Mixed-Mode Solar Dryer for Product Drying. *Heliyon*, **2023**, *9* (3), e14144. https://doi.org/10.1016/J.HELIYON.2023.E14144
29. Cao, X.; Zhang, M.; Fang, Z.; Mujumdar, A. S.; Jiang, H.; Qian, H.; Ai, H. Drying Kinetics and Product Quality of Green Soybean under Different Microwave Drying Methods. *Dry. Technol.*, **2017**, *35* (2), 240–248. https://doi.org/10.1080/07373937.2016.1170698
30. Jiang, H.; Zhang, M.; Mujumdar, A. S.; Lim, R. X. Changes of Microwave Structure/Dielectric Properties during Microwave Freeze-Drying Process Banana Chips. *Int. J. Food Sci. Technol.*, **2014**, *49* (4), 1142–1148. https://doi.org/10.1111/IJFS.12410
31. Wang, Y.; Li, X.; Chen, X.; Li, B.; Mao, X.; Miao, J.; Zhao, C.; Huang, L.; Gao, W. Effects of Hot Air and Microwave-Assisted Drying on Drying Kinetics, Physicochemical Properties, and Energy Consumption of Chrysanthemum. *Chem. Eng. Process.: Process Intensif.*, **2018**, *129*, 84–94. https://doi.org/10.1016/J.CEP.2018.03.020
32. Acar, C.; Dincer, I.; Mujumdar, A. A Comprehensive Review of Recent Advances in Renewable-Based Drying Technologies for a Sustainable Future. *Dry. Technol.*, **2022**, *40* (6), 1029–1050. https://doi.org/10.1080/07373937.2020.1848858

33. Punathil, L.; Basak, T. Microwave Processing of Frozen and Packaged Food Materials: Experimental. Reference Module in Food Science, Elsevier, **2016**. https://doi.org/10.1016/B978-0-08-100596-5.21009-3
34. Krishnamoorthy, P.; Birla, S. L.; Jones, D.; Subbiah, J. Assessment of Heating Rate and Non-Uniform Heating in Domestic Microwave Ovens. *J. Microw. Power Electromagn. Energy*, **2012**, *46* (4), 229–240. https://doi.org/10.1080/08327823.2012.11689839
35. Durance, T.; Yaghmaee, P. Microwave Dehydration of Food and Food Ingredients. Comprehensive Biotechnology (Second Edition) (vol. 4, pp. 617–628), Academic Press, **2011**. https://doi.org/10.1016/B978-0-08-088504-9.00306-8
36. Zartha Sossa, J. W.; Orozco, G. L.; García Murillo, L. M.; Peña Osorio, M.; Sánchez Suarez, N. Infrared Drying Trends Applied to Fruit. *Front. Sustain. Food Syst.*, **2021**, *5*, 115. https://doi.org/10.3389/FSUFS.2021.650690/BIBTEX
37. Aboud, S. A.; Altemimi, A. B.; Al-HiIphy, A. R. S.; Yi-Chen, L.; Cacciola, F. A Comprehensive Review on Infrared Heating Applications in Food Processing. *Molecules*, **2019**, *24* (22), 4125. https://doi.org/10.3390/MOLECULES24224125
38. Rosenthal, I. *Electromagnetic Radiations in Food Science* (vol. 19), Advanced Series in Agricultural Sciences, Springer: Berlin, Heidelberg, **1992**. https://doi.org/10.1007/978-3-642-77106-4
39. Delfiya, D. S. A.; Prashob, K.; Murali, S.; Alfiya, P. V.; Samuel, M. P.; Pandiselvam, R. Drying Kinetics of Food Materials in Infrared Radiation Drying: A Review. *J. Food Process. Eng.*, **2022**, *45* (6), e13810. https://doi.org/10.1111/JFPE.13810
40. Raghavi, L. M.; Moses, J. A.; Anandharamakrishnan, C. Refractance Window Drying of Foods: A Review. *J. Food Eng.*, **2018**, *222*, 267–275. https://doi.org/10.1016/J.JFOODENG.2017.11.032
41. Nindo, C. I.; Feng, H.; Shen, G. Q.; Tang, J.; Kang, D. H. Energy Utilization and Microbial Reduction in a New Film Drying System. *J. Food Process. Preserv.*, **2003**, *27* (2), 117–136. https://doi.org/10.1111/J.1745-4549.2003.TB00506.X
42. Calín-Sánchez, Á; Lipan, L.; Cano-Lamadrid, M.; Kharaghani, A.; Masztalerz, K.; Carbonell-Barrachina, ÁA.; Figiel, A. Comparison of Traditional and Novel Drying Techniques and Its Effect on Quality of Fruits, Vegetables and Aromatic Herbs. *Foods*, **2020**, *9* (9), 1261. https://doi.org/10.3390/FOODS9091261
43. Duc Pham, N.; Khan, M. I. H.; Joardder, M. U. H.; Rahman, M. M.; Mahiuddin, M.; Abesinghe, A. M. N.; Karim, M. A. Quality of Plant-Based Food Materials and Its Prediction during Intermittent Drying. *Crit. Rev. Food Sci. Nutr.*, **2017**, *59* (8), 1197–1211. https://doi.org/10.1080/10408398.2017.1399103
44. Naderinezhad, S.; Etesami, N.; Najafabady, A. P.; Falavarjani, M. G. Mathematical Modeling of Drying of Potato Slices in a Forced Convective Dryer Based on Important Parameters. *Food Sci. Nutr.*, **2016**, *4* (1), 110. https://doi.org/10.1002/FSN3.258
45. Aghbashlo, M.; Kianmehr, M. H.; Samimi-Akhijahani, H. Influence of Drying Conditions on the Effective Moisture Diffusivity, Energy of Activation and Energy Consumption during the Thin-Layer Drying of Berberis Fruit (Berberidaceae). *Energy Convers. Manag.*, **2008**, *49* (10), 2865–2871. https://doi.org/10.1016/J.ENCONMAN.2008.03.009
46. Krokida, M. K.; Karathanos, V. T.; Maroulis, Z. B.; Marinos-Kouris, D. Drying Kinetics of Some Vegetables. *J. Food Eng.*, **2003**, *59* (4), 391–403. https://doi.org/10.1016/S0260-8774(02)00498-3
47. Djaeni, M.; Van Asselt, C. J.; Bartels, P. V.; Sanders, J. P. M.; Van Straten, G.; Van Boxtel, A. J. Low Temperature Drying with Air Dehumidified by Zeolite for Food Products: Energy Efficiency Aspect Analysis. *Int. J. Food Eng.*, **2011**, *7* (6). https://doi.org/10.2202/1556-3758.1930/MACHINEREADABLECITATION/RIS
48. Traffano-Schiffo, M. V.; Castro-Giráldez, M.; Fito, P. J.; Balaguer, N. Thermodynamic Model of Meat Drying by Infrared Thermography. *J. Food Eng.*, **2014**, *128*, 103–110. https://doi.org/10.1016/J.JFOODENG.2013.12.024
49. Rezaei, F.; van der Gheynst, J. S. Critical Moisture Content for Microbial Growth in Dried Food-Processing Residues. *J. Sci. Food Agric.*, **2010**, *90* (12), 2000–2005. https://doi.org/10.1002/JSFA.4044
50. Brown, M. Processing and Food and Beverage Shelf Life. *Food and Beverage Stability and Shelf Life* (pp. 184–243), Woodhead Publishing, **2011**. https://doi.org/10.1533/9780857092540.1.184
51. Nuwanthi, S. G. L. I.; Madage, S. S. K.; Hewajulige, I. G. N.; Wijesekera, R. G. S. Comparative Study on Organoleptic, Microbiological and Chemical Qualities of Dried Fish, Goldstripe Sardinella (*Sardinella gibbosa*) with Low Salt Levels and Spices. *Procedia Food Sci.*, **2016**, *6*, 356–361. https://doi.org/10.1016/J.PROFOO.2016.02.072

52. Antal, T.; Tarek-Tilistyák, J.; Cziáky, Z.; Sinka, L. Comparison of Drying and Quality Characteristics of Pear (*Pyrus communis* L.) Using Mid-Infrared-Freeze Drying and Single Stage of Freeze Drying. *Int. J. Food Eng.*, **2017**, *13* (4). https://doi.org/10.1515/IJFE-2016-0294/MACHINEREADABLECITATION/RIS
53. Labuza, T. P.; Heidelbaugh, N. D.; Silver, M.; Karel, M. Oxidation at Intermediate Moisture Contents. *J. Am. Oil Chem. Soc.*, **1971**, *48* (2), 86–90. https://doi.org/10.1007/BF02635692/METRICS
54. Ramos, F. M.; Silveira Júnior, V.; Prata, A. S. Impact of Vacuum Spray Drying on Encapsulation of Fish Oil: Oxidative Stability and Encapsulation Efficiency. *Food Res. Int.*, **2021**, *143*, 110283. https://doi.org/10.1016/J.FOODRES.2021.110283
55. Fellows, P. *Food Processing Technology, Principles and Practice* (2nd edition), Woodhead Publishing Limited: Cambridge, **2000**.
56. Qiu, X.; Chen, S.; Lin, H. Oxidative Stability of Dried Seafood Products during Processing and Storage: A Review. *J. Aquat. Food Prod. Technol.*, **2019**, *28* (3), 329–340. https://doi.org/10.1080/10498850.2019.1581317
57. Franco, S.; Jaques, A.; Pinto, M.; Fardella, M.; Valencia, P.; Núñez, H.; Ramírez, C.; Simpson, R. Dehydration of Salmon (*Atlantic salmon*), Beef, and Apple (*Granny Smith*) Using Refractance Window™: Effect on Diffusion Behavior, Texture, and Color Changes. *Innov. Food Sci. Emerg. Technol.*, **2019**, *52*, 8–16. https://doi.org/10.1016/J.IFSET.2018.12.001
58. Deka, D.; Annapure, U. S.; Shirkole, S. S.; Thorat, B. N. Techno-Economics of Solar Assisted Drying of Small Freshwater Fish to Ensure Global Nutritional Security. *Dry. Technol.*, **2023**, *41*, 1214–1228. https://doi.org/10.1080/07373937.2022.2134416
59. Kipcak, A. S.; İsmail, O. Microwave Drying of Fish, Chicken and Beef Samples. *J. Food Sci. Technol.*, **2021**, *58* (1), 281–291. https://doi.org/10.1007/S13197-020-04540-0/FIGURES/5
60. Ortiz, J.; Lemus-Mondaca, R.; Vega-Gálvez, A.; Ah-Hen, K.; Puente-Diaz, L.; Zura-Bravo, L.; Aubourg, S. Influence of Air-Drying Temperature on Drying Kinetics, Colour, Firmness and Biochemical Characteristics of *Atlantic salmon* (*Salmo salar* L.) Fillets. *Food Chem.*, **2013**, *139* (1–4), 162–169. https://doi.org/10.1016/J.FOODCHEM.2013.01.037
61. Kim, J. Y.; Yi, B. R.; Lee, C.; Gim, S. Y.; Kim, M. J.; Lee, J. H. Effects of PH on the Rates of Lipid Oxidation in Oil–Water System. *Appl. Biol. Chem.*, **2016**, *59* (2), 157–161. https://doi.org/10.1007/S13765-015-0146-3/TABLES/1
62. Viji, P.; Shanmuka Sai, K. S.; Debbarma, J.; Dhiju Das, P. H.; Madhusudana Rao, B.; Ravishankar, C. N. Evaluation of Physicochemical Characteristics of Microwave Vacuum Dried Mackerel and Inhibition of Oxidation by Essential Oils. *J. Food Sci. Technol.*, **2019**, *56* (4), 1890–1898. https://doi.org/10.1007/s13197-019-03651-7
63. Patterson, J.; Kailasam, S.; Giftson, H.; Immaculate, J. K. Effect of Drying Technologies on the Biochemical Properties of *Stolephorus commersonnii. Food Qual. Saf.*, **2018**, *2* (3), 153–158. https://doi.org/10.1093/fqsafe/fyy010
64. Oluwagbemiga Osunsanmi, F.; Elijah Zharare, G.; Kotze Basson, A.; Oluyinka Mary, A. Effects of Drying Methods on Quality Attributes of Shrimps. *J. Food Process. Technol.*, **2019**, *10* (1), 1–5.
65. Akonor, P. T.; Ofori, H.; Dziedzoave, N. T.; Kortei, N. K. Drying Characteristics and Physical and Nutritional Properties of Shrimp Meat as Affected by Different Traditional Drying Techniques. *Int. J. Food Sci.*, **2016**, *2016*, 7879097. https://doi.org/10.1155/2016/7879097
66. Alfiya, P. V.; Rajesh, G. K.; Murali, S.; Aniesrani Delfiya, D. S.; Samuel, M. P.; Prince, M. V. Quality Evaluation of Solar and Microwave Dried Shrimps: A Comparative Study on Renewable and Dielectric Heating Methods. *Sol. Energy*, **2022**, *246*, 234–244. https://doi.org/10.1016/j.solener.2022.09.048
67. Deng, Y.; Liu, Y.; Qian, B.; Su, S.; Wu, J.; Song, X.; Yang, H. Impact of Far-Infrared Radiation-Assisted Heat Pump Drying on Chemical Compositions and Physical Properties of Squid (*Illex illecebrosus*) Fillets. *Eur. Food Res. Technol.*, **2011**, 232 (5), 761–768. https://doi.org/10.1007/s00217-011-1441-9
68. Chen, H.; Zhang, M.; Fang, Z.; Wang, Y. Effects of Different Drying Methods on the Quality of Squid Cubes. *Dry. Technol.*, **2013**, *31* (16), 1911–1918. https://doi.org/10.1080/07373937.2013.783592

9 Solar Drying Technology for Meat Products

Piverjeet Kaur Dhillon
Krishi Vigyan Kendra, Guru Angad Dev Veterinary and Animal Sciences University, Ludhiana, Punjab, India

Priya Singla
Desh Bhagat University, District Fategarh Sahib, Punjab, India

9.1 INTRODUCTION

Aiming at reduced livestock produce wastage, meat products with extended shelf life is a dire need in the contemporary times. Thus, in order to preserve meat, drying is the most appropriate method since ages. Nevertheless, conventional direct sun drying does not ensure quality of the dried product. Besides, another reason for practicing this technology is achievement of food and nutritional security by reducing postharvest spoilage of meat (Green & Schwarz, 2001). To exercise drying process, massive energy/fuel is required (around 2.4 MJ or 100 l of oil to evaporate 1 l of H_2O), and in developing countries, it is not viable at commercial scale because of the scanty energy supply. Moreover, global warming/climate change is also a major concern over the globe. On the contrary, solar energy can be a best alternative to conventional direct sun drying with low investment along with good food quality (Dasin et al., 2019). Therefore, solar drying of meat is essential to curtail the drawbacks of open-sun drying such as temperature control, food contamination, manpower, etc. (Dasin et al., 2015). In this process, meat is dried in a closed chamber through conversion of solar energy to heat when sunrays strike onto the dark surface of this chamber/cabinet. This way, the warm air is produced inside the cabinet through adjustable vents and meat is dried. Drying is an inevitable processing step in sausages and hams (fermented meat products) preparation to extend their shelf life as these products get dehydrated during fermentation in a chamber with adjustable humidity and temperature (Dasin et al., 2019).

In tropical countries, direct sun drying is a preferred method of drying meat at small scale since higher investment in the form of space, labor, and equipment is required for controlled drying (Singh et al., 2019). This practice leads to poor quality in terms of low aesthetic value and high contamination of meat products with no satisfactory income to the producers. There are some recommended meat drying methods which perform the drying process in more efficacious manner since these maintain hygienic conditions even at faster pace of drying (Tomar et al., 2017). These involve freeze microwave and hot-air drying through trays; however, huge investment, specific skillset, and consistent electricity supply are required to operate the equipment (Varalakshmi, 2016). Therefore, this method is not viable for small-scale meat processing ventures. Conversely, solar drying technologies provide wide applications in meat processing sector, with lesser manpower and no electricity requirement (Fuller, 2010). These solar systems are introduced as (1) direct solar, (2) indirect solar, (3) and mixed or hybrid solar systems, which are further categorized into natural convection/passive solar dryer (PSD) and forced convection/active solar dryer (ASD) (Eltawil et al., 2018; Iskandar et al., 2017).

For off-season consumption as well as to minimize wastage, meat products, being perishable, need to be preserved. Aiming at meat preservation, drying is the most suitable technology; and sun drying of meat is exercised since times immemorial. To remove moisture under this process, continuous rubbing of meat with salt and condiments is necessary in the sunlight which is pretty tedious task (Varalakshmi, 2016). Ground and flavored meat is also dried in oven at the temperature of 70–80°C

DOI: 10.1201/9781003344087-9

(Chen & Mujumdar, 2008). Nevertheless, as compared to traditional open-air sun drying method, solar drying for meat products is more easier and effective technology (Karim & Hawlader, 2004). In solar dryers, sun radiations are tapped and concentrated, which results in increased temperature and decreased humidity. Solar dryers are viable for rural settings with erratic electricity supply with minimum capital investment. Moreover, solar energy is eco-friendly energy source with lesser CO_2 emission than electric drying (Varalakshmi, 2016). Solar drying prevents the microbial growth, enzymatic browning, and lipid oxidation in meat, thereby ensuring extended shelf life of meat products with sustained organoleptic characteristics (Barnwal & Tiwari, 2008). In contrast, traditional drying leads to contamination of meat (Varalakhshmi, 2016). Hence, an attempt has been made to discuss and compare the different solar technologies for drying of meat in light of existing research.

9.2 MECHANISM

Drying is a process to remove moisture from the product. When heat is applied to the product which augments the vapor pressure of the product's moisture than the surrounding air, it is called thermal drying. In this process, both vapor pressure and thermal gradient make the moisture move towards the product's surface where the moisture evaporates. Hence, water vapor is shifted to air and product continues to dry since the moist air is superseded by dry air (Fuller, 2010).

Two types of moisture (free and bound) are present in the product. Free/unbound moisture can move in unobstructed way in the product while bound moisture is trapped within the cell structure. The latter type cannot shift towards the product's surface. The drying process depends on free moisture content. Critical moisture content refers to inadequate free moisture. The phase, at which moisture movement rate toward the product's surface consistently declines, is called "falling rate period." When the product's moisture content declines and is comparable to wet bulb temperature of drying air, this stage is called "constant rate drying period." Drying rate during falling and constant rate drying period can be computed by the following equations (Fuller, 2010):

$$\text{Drying rate} = \frac{\begin{array}{l}\text{Convective heat teansfer coeficient } (h_T) * \text{Surface area of product } (A_p) \\ \{\text{dry bulb temperature of drying air } (T_a) - \text{Wet bulb temperature of air} \\ \text{at the surface of the meterial } (T_{wb})\}\end{array}}{\text{Heat of vaprization of } (L)} \tag{9.1}$$

$$e^{-kt}(\text{drying constant} * \text{time}) = \frac{\text{Moisture content } (M) - \text{Equilibrium moisture } (M_e)}{\text{Initial moisture content } (M_0) - \text{Equilibrium moisture } (M_e)} \tag{9.2}$$

During constant drying rate period, decisive key factors for the drying rate are product's surface area, velocity, and temperature of air. Currently, the drying rate is dependent chiefly of product's internal parameters, called drying constant or diffusion coefficient.

Moisture content is expressed in percentage (percent), either on dry or wet basis. The first basis is generally practiced by researchers, while the latter is used on commercial scale. Dry basis moisture content is also known as the ratio of product's water amount to dry matter of the product (Fuller, 2010) and calculated as follows (Equation 9.3):

$$\text{Moisture content } (M) = \frac{100 * \text{Mass of moisture } (W_m)}{\text{Mass of bone dry material } (W_{dm})} \tag{9.3}$$

To determine moisture content on wet basis, the ratio of product's water amount at any time to the product's gross weight at same time, is computed as follows (Equation 9.4) (Fuller, 2010):

$$\text{Moisture content } (M) = \frac{100 * \text{Mass of moisture } (W_m)}{\text{Mass of moisture } (W_m) + \text{Mass of bone dry material } (W_{dm})} \tag{9.4}$$

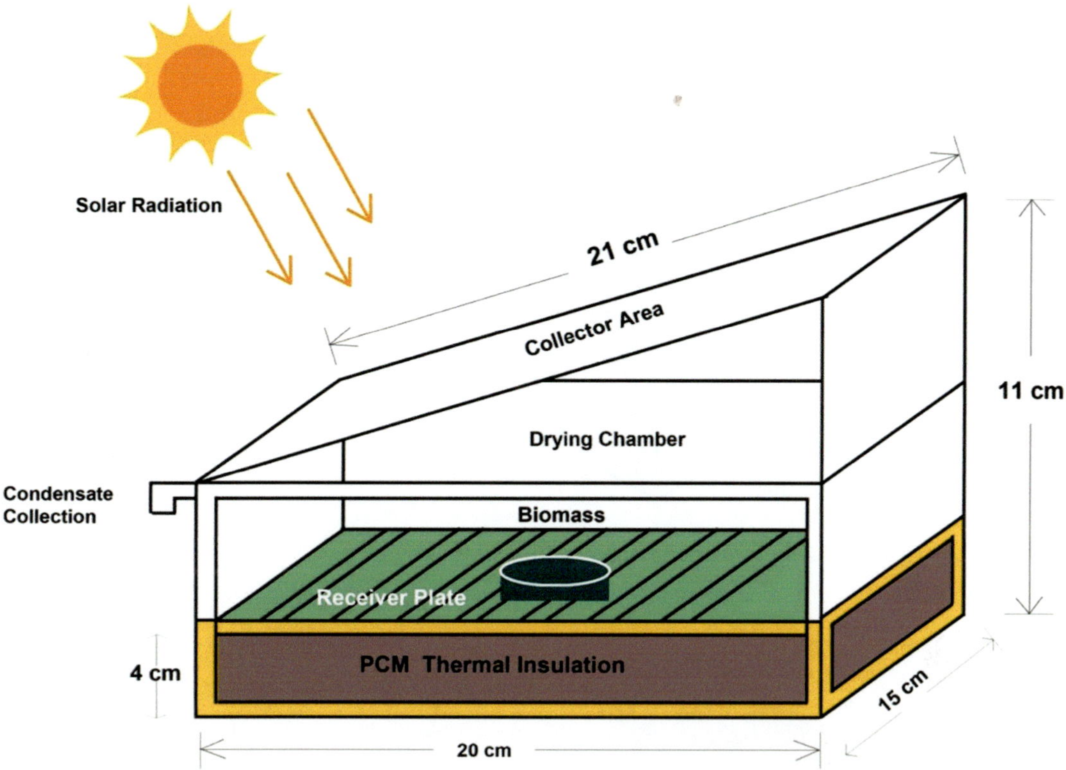

FIGURE 9.2 Passive solar dryer (Box type) for direct method of drying. (Kalaiselvan and Mathimani, 2022.)

2. *Hybrid Solar Dryers:* In this type of dryer, two different modes (single and mixed) of direct or indirect are there to operate the system. The product is dried by either direct radiation or stored energy during absence of sunlight or through applying both the methods. For stored energy, various sources (electricity, photovoltaic [PV], liquefied petroleum gas [LPG]) can be used to preheat the air. An enhanced hybrid tunnel dryer with PV system and flat plate solar dryer is also developed by Eltawil et al. (2018), which exhibited a drying efficiency of 31% with 32 tons of CO_2 emission. Microbial contamination is reduced in this solar system.

Apart from these, solar dryers can be categorized according to their structure also, for instance, box solar dryer (Figure 9.2), parabolic greenhouse solar dryer, hybrid box solar dryer (Figure 9.3), tunnel solar dryer (Figure 9.4), and greenhouse solar dryer.

9.6 ADVANTAGES

In view of food security, solar drying is very important as it curbs the postharvest losses prior to the product's packaging and marketing. In developing countries, about 30% of meat produce gets wasted during postharvest phase. Solar drying is a safe, viable, and cost-effective method for highly perishable food products such as meat, and the advantage related to this method cannot be overlooked.

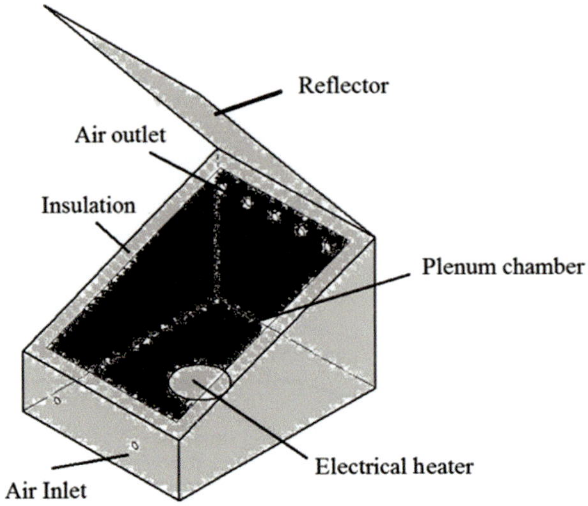

FIGURE 9.3 Hybrid solar dryer. (Chaudhari et al., 2018.)

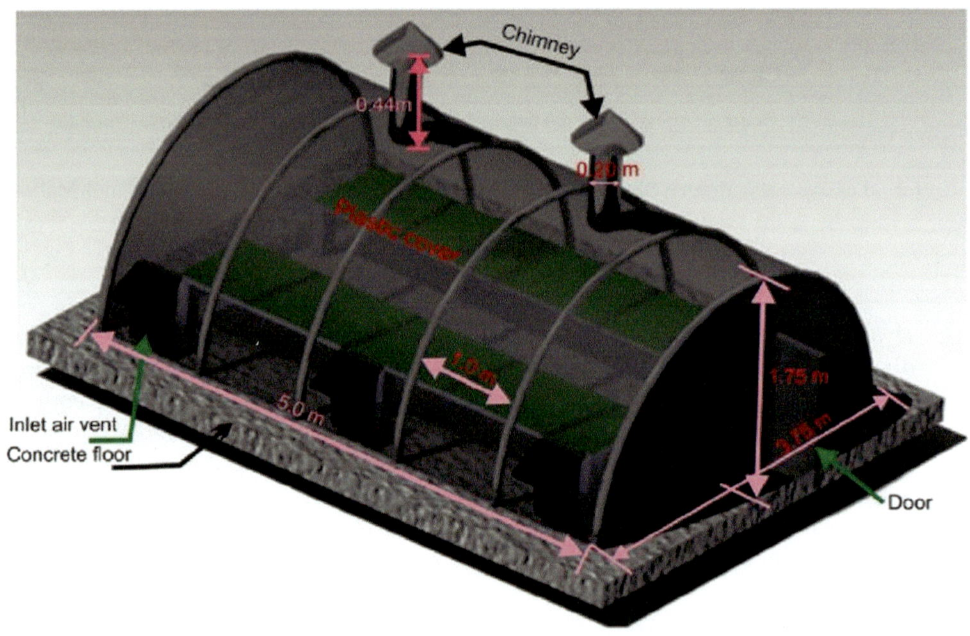

FIGURE 9.4 Solar dryer (tunnel type) for mixed-method of drying. (Panwar, 2014.)

Advantages of solar dried meat

1. Solar-dried meat has extended shelf life, so it can be consumed during off-season.
2. Solar-dried meat retains essential meat components such as nutrients and antioxidants in it. Therefore, it is nutritionally good to consume.
3. Solar drying process is hygienic as performed in a climatized chamber, and any physical and microbial contamination in the solar-dried meat does not occur, hence it is safe for human consumption.

4. In view of above advantage, no product damage can be there, so processing losses do not take place apart from moisture loss.
5. It improves market demand of the dried meat products as it ensures product's hygiene and extended shelf life.
6. Solar-dried meat has the potential to combat food and nutritional insecurity as well as to augment the income of farmer by increasing profit over investment.
7. Direct sun dryer are less expensive, therefore low installation cost is required. Also, these are simple to operate, thus untrained manpower can operate this.
8. The product is protected against ultraviolet radiation when dried in indirect mode solar dryer.
9. With mixed-mode solar dryer, temperature damage to dried meat can be minimized.
10. Hybrid solar dryer can be operated without solar radiation. It provides better temperature control and faster drying than other dryers.
11. Solar drying protects environment as it emits less pollutants (greenhouse gases) than electricity/fuel-operated dryers.
12. Risk of respiratory, allergic ailment and cancer morbidity can be trimmed down with solar drying practice.

9.7 DISADVANTAGES

1. In direct sun dryer, food can be damaged with UV rays.
2. Hybrid mode solar dryers are fuel-dependent.
3. Expensive structure.
4. Skillful operation is also the primary requirement.

Though solar drying technology is expensive, nevertheless solar dryers can be designed as per the location, budget, and capacity requirements.

9.8 RETROSPECTIVE RESEARCH/CASE STUDIES OF SOLAR DRYING TECHNOLOGIES FOR MEAT PRODUCTS

Meat contains good amounts of vitamin B complex along with some minerals (iron), and it is easily digestible after processing applying appropriate method. This food commodity is highly perishable; therefore, since ages, meat drying is into practice to prolong its shelf life (Ahmad et al., 2018). Initially, open-sun drying was exercised to preserve meat, which leads to contamination. For this reason, cabinet solar drying of meat was introduced (Figure 9.5). Research evidences based on solar drying technologies of meat are plentiful throughout the globe. Numerous studies have discussed the design and advantages of solar dryers for meat. Among all, environment and economics are the decisive factors in designing of an apt solar drying technology. A solar cabinet-type dryer was developed and found suitable for meat drying in it. Further, at the temperature between 45.75°C and 48.63°C, tray drying of seasoned raw pork is apt for two days, with highest and lowest temperature in upper and lower trays, respectively. Sensory quality of solar-dried meat was also acceptable since the uniform temperature was sustained within the dryer (Jangsawang, 2017). Also, 1.9% more weight loss was recorded in solar-dried pork than open-sun drying. A PSD with natural convection (indirect mode) with attached collector unit (70 × 60 × 30 cm) and drying chamber (60 × 60 × 70 cm) was developed in Nigeria by Modibbo Adama University of Technology for meat drying. The materials used in construction were mild steel sheet, plywood, wire net, glass, and paint. Drying capacity was 4 kg with two trays. The average temperature can be 370°C, with diverse temperature component-wise as 420°C, 400°C, and 350°C in drying chamber, collector, and product's temperature, respectively, at 2:00 p.m. The drying efficiency of 27% was noted, with a decline of 0.029 kg in meat sample at the mean relative humidity and drying rate as 26.4% and 0.032 kg/h, in that order

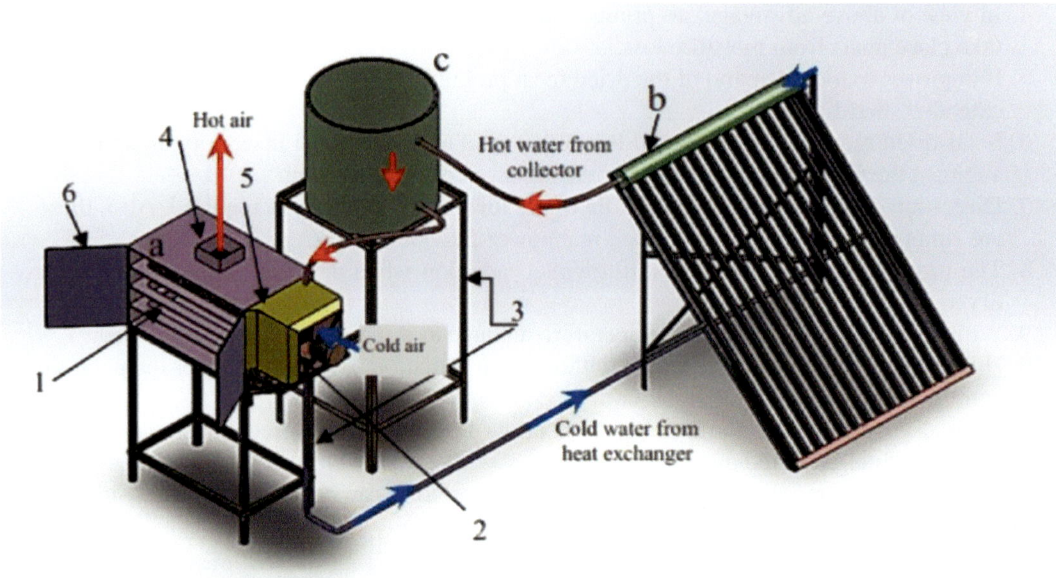

FIGURE 9.5 a: Solar dryer (cabinet type) for indirect method of drying. b: Evacuated tube solar collector. c: Tank for storage and container with phase change material is available inside. 1: Cabinet and trays; 2: fan; 3: chassis; 4: channel for exhaust; 5: heat exchanger; 6: doors. (Getahun et al., 2021.)

(Dasin et al., 2019). Aiming at hygienic meat processing, a PSD with capacity of 5 kg was designed in the Philippines. This device had four components—solar heat collector, combustion chamber, drying chamber, and chimney. It was used to dry smoked meat called "etag." In this solar dryer, air is heated in solar heat collector and this hot air then absorbs moisture from the product and get exhausted via chimney (Molintas, 2017). Besides, at Basrah University, a solar dryer using indirect mode was developed which consists of three main components—solar collector with a drying tube (iron), chimney, and tubular met. It was designed for drying of beef. This investigation reported a significant ($p < 0.05$) increase in solar radiation energy with 810.55 W/m^2 and 680.1 W/m^2 as maximum and minimum level of energy at 12:00 noon and 4:00 p.m., in that order. At 60°C, the drying efficiency was 65.91%. Further, it is reduced when drying temperature increased. In this study, at different storage period, i.e. 0, 14, 28, 56, and 70, both salted and unsalted beef were dried and found an augmented ash, fat, and protein content, while decreased microbial content in both type of beef (AL-RUBAI et al., 2021).

Another investigation on sensory attributes of spent hen meat powder suggests that scores obtained for sensory parameters of spent hen meat powder were higher when dried in solar dryer as compared to oven-dried spent hen meat powder (Sarkar et al., 2021). Further, diverse drying methods (tunnel solar dryer and air-oven dryer) were compared in view of physicochemical characteristics of shrimp meat. It was observed that drying rate was more in solar tunnel dryer. There was a change in color as solar-dried shrimp meat had darker color than that of air-oven dried meat. A significant lesser (1.5%) fat content was recorded in solar-dried meat when compared to the air-oven dried (2.1%), although with no significant difference between ash and protein content (Akonor et al., 2016). Further, two different solar methods (indirect and mixed modes) were assessed in view of physicochemical characteristics of chicken breast meat (López-Vidaña et al., 2022). This investigation concluded that mixed-mode solar dryer had faster drying than indirect mode solar dryer. However, a slight color difference was seen in the product; nevertheless, content of protein and fat were higher than the cooked meat. Similarly, Orhevba and Moru (2015) compared the open-sun-dried and solar-dried meat for organoleptic and nutritional properties after 8 hours of drying.

10 Energy and Exergy Analyses in Various Solar Energy Applications

Poonam Rani
Food Chemistry & Technology Department, Teagasc Food Research Centre, Fermoy, Cork, Ireland

Rohit Biswas and P. P. Tripathy
Agricultural and Food Engineering Department, Indian Institute of Technology Kharagpur, Kharagpur, West Bengal, India

ABBREVIATIONS

α	absorptivity
A	area
β	thermal expansion
C	solar cooker concentration ratio
c	specific heat capacity
COR	cooker opto-thermal ratio
d	diameter
D	thermohydraulic factor
ε	emissivity
η	efficiency
Ex	exergy
f	factor
F′	collector efficiency factor
F_1	first figure of merit
F_2	second figure of merit
G	global solar radiation
γ	reflectivity
g	gravitational acceleration
Gr	Grashof number
h	heat transfer coefficient
I	incident solar radiation
J	heat flux
k	thermal conductivity
L	length
m	mass of fluid
m	meter
ṁ	mass flow rate
μ	viscosity
Nu	Nusselt number
ν	kinematic viscosity

DOI: 10.1201/9781003344087-10

p	pressure
Pr	Prandlt number
Q	heat gained
$\dot{q}$	rate of heat transfer
R	resistance
Ra	Rayleigh number
RH	relative humidity
ρ	density
σ	Stefan–Boltzmann constant
τ	transmissivity
t	thickness
T	temperature
U	heat loss coefficient
V	velocity
W	watt
x	distance

SUBSCRIPT

a	absorber plate
air	fluid
amb	ambient
b	back heat loss
b	bottom of the absorber plate
bottom	bottom flow below absorber plate
c	convection
con	conduction
cover	top glazing
eff	effectively
f	fluid
g1	glass on the top
g2	second glass for double-glaze system
i	insulation
inst	instantaneous
k	conductivity
L	total heat loss
m	fluid
pv	photovoltaic
r	radiation
ref	reference temperature
s	side of the solar collector
st	structure top
T	top heat loss
top	top flow above absorber plate
u	total heat transfer
ut	cooking utensil/pot
w	wind

10.1 INTRODUCTION

The demand for employing renewable energy sources is increasing in importance since fossil fuels like coal and oil are slowly being exhausted. As renewable energy technologies become more sustainable economically as well as environmentally, support for this approach grows. Among other methods of restricting nonrenewable sources, use of solar energy is outstanding as it is clean, abundantly available, sustainable, and environment-friendly. Through the use of solar energy, the sunrays can be used to heat water, air, or other fluids, dry agro-commodities, cook meals, charge batteries, and generate electricity. The predicted energy crisis for the future emphasizes the conservation of fuel resources along with the better analysis and improvement of power systems thermodynamically. Now, solar thermal energy systems are evaluated on the basis of exergy analysis in addition to energy analysis. The basis of this new approach is the concurrent use of the first and second laws of thermodynamics to both research and design. Thermodynamic analysis employing energy and exergy can be utilized during the design process to discover patterns and the presence of possibilities for optimization with the goal to minimize costs. Therefore, in this chapter, concepts of energy and exergy analysis are clearly explained. Further, a comprehensive review of various solar energy applications, including solar collector, dryer, and cooker, has been conducted on energy- and exergy-based performance evaluation.

10.2 ENERGY AND EXERGY ANALYSIS: THEORETICAL CONCEPT

The principles of thermodynamics are utilized to describe the execution, performance, as well as effectiveness of systems for transferring one kind of energy to other. The first rule of thermodynamics forms the primary focus of conventional thermal analysis and describes the concept of energy conservation. An examination of the energy entering and leaving a system of transforming energy is the basis of energy evaluation. The output energy can be divided into useful products and waste. Efficiency ratings are widely used to assess the similarities and differences between various systems and are frequently calculated as energy quantity ratios. Thermal losses that occur inside a system are typically not accurately identified and assessed when using energy analysis.

Exergy is a term used to define how much theoretical work a system can produce before it achieves thermodynamic equilibrium with respect to the surroundings (Akbulut and Durmuş, 2010). The identification of the sources, locations, and magnitude of losses that have developed as a result of various irreversibilities along with the creation of entropy in diverse processes has led to the development of exergy flow calculation as an important technique to conduct a detailed and precise perspective of energy use. Exergy assessment is described by the second rule of thermodynamics. This law describes the way a part of input exergy used by the thermodynamic system is destroyed as a result of the associated irreversible processes (Shukuya and Hammache, 2002). Any energy system's exergy analysis is essential for its optimization and performance assessment. The primary goal of this analysis is to increase system performance by identifying the causes and points of exergy loss.

10.2.1 RELATIONSHIP BETWEEN ENERGY AND EXERGY

The exergy of a substance/energy is an indicator of its use, value, or ability for transformation. The link between energy and exergy is presented in Figure 10.1. The two components that combined to make "energy" are

i. Exergy, i.e., usable components
ii. Anergy, i.e., nonusable components

Exergy describes the capacity of any system to carry out work, and it represents the work composition of the many different streams (mass, heat, and work) that move through the system.

Anergy refers to energy that has no potential for use and is discarded to the surroundings. The whole sum of anergy and exergy in a process remains constant (Tiwari et al., 2016). In any real process, exergy declines and anergy rises at the same rate as entropy production. The anergy cannot be turned into exergy since it has no work potential. Exergy is gradually converted into anergy rather than being conserved (Tiwari et al., 2016).

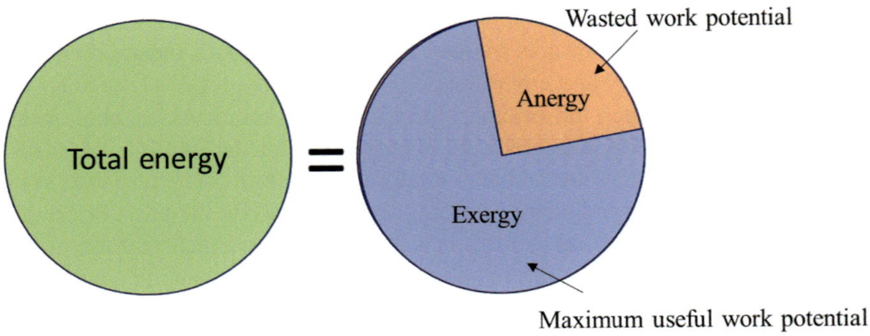

FIGURE 10.1 Relationship between energy, exergy, and anergy.

Figure 10.2 illustrates a comparison of exergy, entropy, and exergy transfer. Energy transfer happens at the cost of the thermal gradient when heat is transferred between two bodies from the warmer side to the cold side, as seen in Figure 10.2. Despite the possibility of achieving a 100% energy utilization between two bodies, the temperature of the resulting bodies might not match that of the heat source. This indicates that energy was degraded throughout the heat transfer process, which is frequently described as the creation of entropy. Entropy generation lowers the quality of energy transported from the source to the sink, and shows up in the sink's availability of energy. Exergy loss (availability loss) is the term used to describe this deterioration in energy quality. Another name for the energy loss is irreversibility. Entropy production and consumption are inversely correlated (Shukuya and Hammache, 2002).

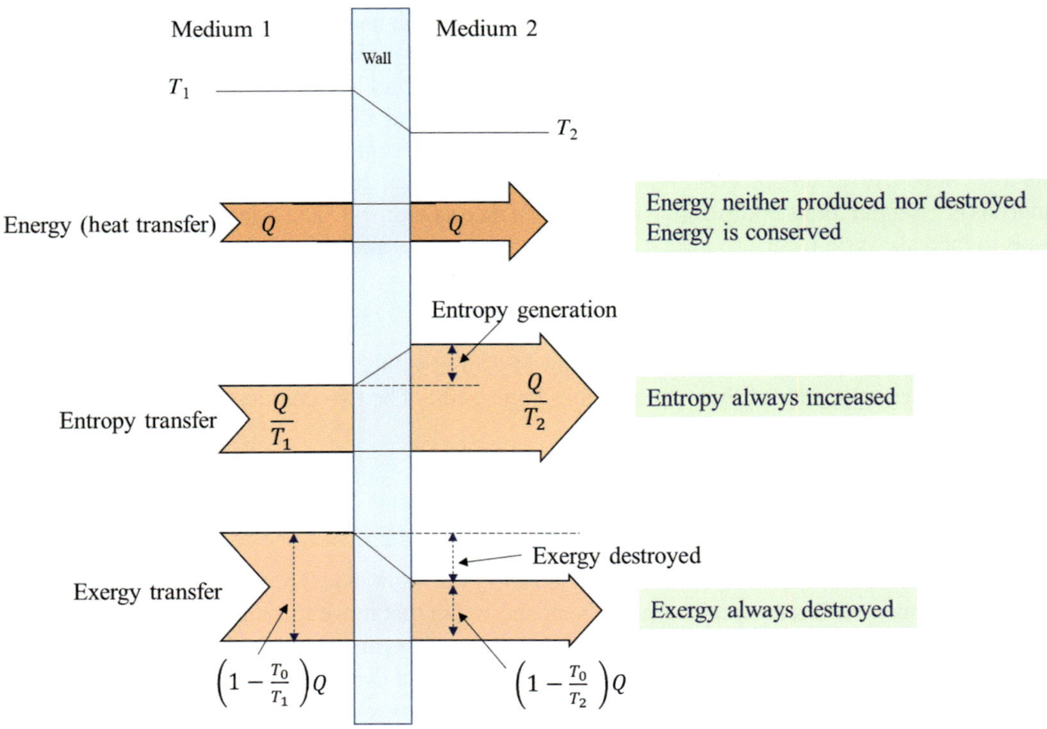

FIGURE 10.2 Concepts of energy, entropy, and exergy transfer.

Energy and exergy differ primarily in that energy is a measure of energy quality or work potential and cannot be consumed, whereas exergy may be consumed in a process. Exergy is only conserved in reversible processes, but energy is always conserved. Anergy destruction is another name for anergy. The created entropy is directly related to the exergy that is destroyed. Exergy is always lost, whether completely or partially. The actual thermodynamic efficiency of any system is usually lower-than-theoretical thermodynamic efficiency and this is because of the exergy destruction or the entropy that is created. The quality of energy, i.e., exergy, decreases when there is higher entropy creation inside the system (Shukuya and Hammache, 2002).

The main benefits of exergy analysis are as follows:

1. It offers a thorough and precise perspective of energy use by identifying the origins, locations, and sizes of losses that have developed as a result of various irreversibilities and entropy formation in various phenomena inside a solar thermal system.
2. By addressing the primary cause of losses and irreversibilities, identifying and describing exergy destruction aids in improving system performance.
3. Exergy analysis has become a more valuable tool for energy system design, monitoring, effectiveness, and advancement.
4. It can be applied to reduce losses to increase system effectiveness.
5. Exergy might help people comprehend ideas like energy scarcity and energy conservation.

10.3 METHODOLOGIES FOR ENERGY ANALYSIS OF SOLAR ENERGY SYSTEMS

10.3.1 Energy Analysis of Solar Collector

Solar air/water collector is a type of equipment that traps the incident solar energy and transfers it to a fluid medium for any thermal operation (Ramani et al., 2010). Various design changes and modification are considered to enhance the thermal performance of solar collector. The energy analysis of solar energy system aims towards the behavior of the solar collector in different scenarios such as terrain, application, and constraints (Wenceslas Koholé et al., 2022). There are a variety of design modification and adaptation in the conventional solar collector design to achieve an improved overall thermal performance. In this chapter, two most common designs used for water (Figure 10.3)

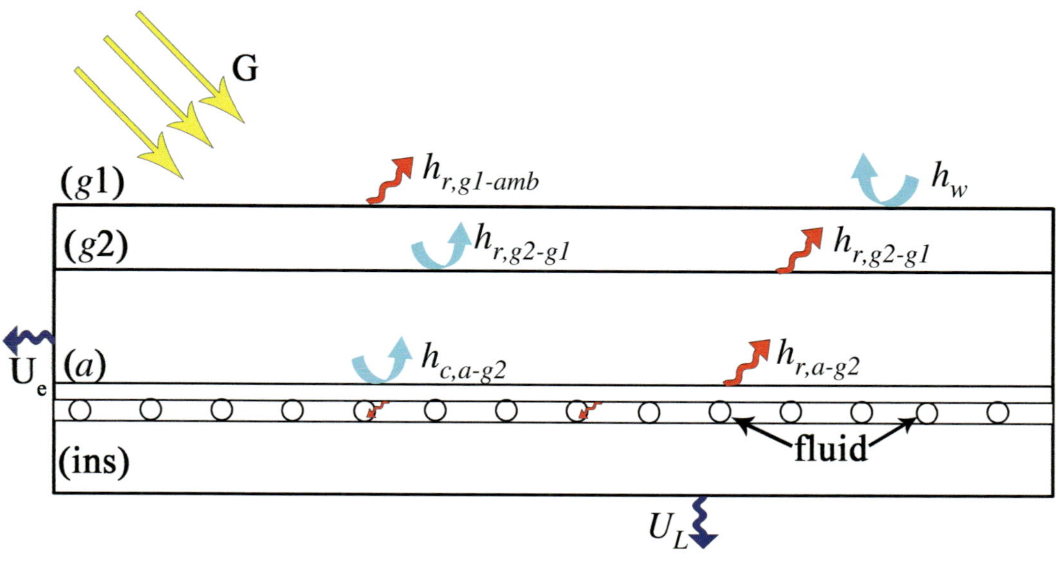

FIGURE 10.3 Solar collector with double glazing and fluid flowing within pipe.

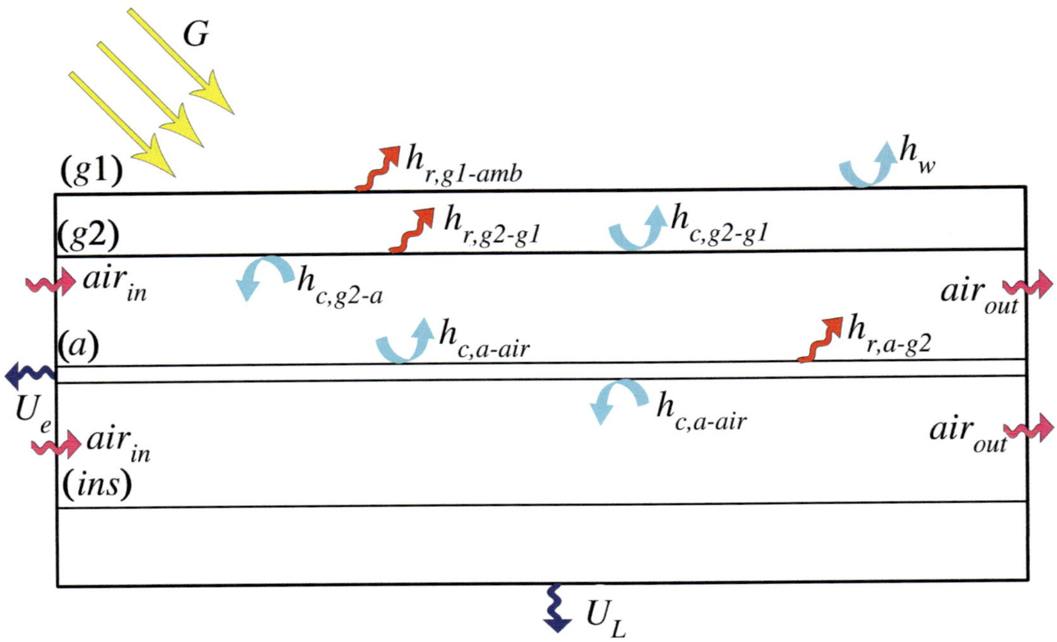

FIGURE 10.4 Solar collector with double glazing and fluid flowing over and under the absorber plate.

and air (Figure 10.4) heating will be discussed along with various heat transfer coefficients and loss constraints. The solar collector as shown in Figure 10.3 mainly deals with a water-heating solar collector into which different design modifications can be incorporated, so that the overall efficiency of collector can be attained. Certain assumptions are made for simplification of thermal analysis:

1. The connection between absorber plate and absorber tube has negligible thermal resistance.
2. The fluid flow rate inside the absorber tube is identical.
3. Negligible temperature gradient and pressure drop is observed within the absorber tube.
4. Thermos-physical properties of different materials employed in the system are independent of temperature.
5. The ambient condition around the collector is assumed to be same for all sides.

The energy balance for solar collector as shown in Figure 10.1 can be estimated using Equation 10.1 (Joseph and Thomas, 2022):

$$G_i = \left(G_{ab-g1} + G_{ref-g1} + G_{ab-g2} + G_{ref-g2} + G_{ab-a} + G_{ref-a} + G_{fluid/air}\right) \times A \quad (10.1)$$

The energy absorber by each of the top (g1) and bottom (g2) glass cover can be estimated using Equation 10.2:

$$G_{ab-g} = \alpha_g I \times \gamma_{g1} A \quad (10.2)$$

Next, the absorber plate (*a*) is calculated using Equations 10.3–10.6 (Wu et al., 2022):

$$Q_a = Q_L + Q_u \quad (10.3)$$

$$Q_a = (\tau\alpha) IA \quad (10.4)$$

$$Q_L = U_L\left(T_a - T_{amb}\right) \tag{10.5}$$

$$Q_a = A\left((\tau\alpha)I - U_L\left(T_a - T_{amb}\right)\right) \tag{10.6}$$

The thermal load or performance of the solar collector can be estimated mainly by the heat transfer coefficient within the system. The heat lost to the surrounding from the solar collector glass, at the top of the collector, back of the solar collector through insulation, and edge of the solar collector can be estimated in order to effectively optimize the solar collector for different applications.

The heat transfer between top and bottom glass due to convection is expressed as Equation 10.7 (Alawi et al., 2022):

$$h_{c,g2-g1} = Nu_{g2-g1}\ K_{air}/d \tag{10.7}$$

The Nusselt number in Equation 10.7 can be obtained using Hollands et al. (1976), with air being the medium in between two surfaces as given in Equation 10.8:

$$Nu = 1 + 1.44\left[1 - \frac{1708}{Ra\ cos\beta}\right]^{+}\left(1 - \frac{\sin(1.8\beta)^{1.6}\,1708}{Ra\cos\beta}\right) + \left[\left\{\frac{Ra\cos\beta}{5830}\right\}^{1/3} - 1\right]^{+} \tag{10.8}$$

In Equation 10.8, "+" sign denotes the absolute positive value of the bracket to be taken into consideration, else for negative value zero is to be used in place and for Rayleigh number Equation 10.9 can be used as given:

$$Ra = Gr\ Pr = \frac{g\ \beta'\Delta T\ L^3}{\nu\alpha} \tag{10.9}$$

Equations 10.7–10.9 can also be employed to find the convective heat transfer coefficient $\left(h_{c,a-g2}\right)$ between the absorber plate (a) and bottom glass ($g2$) (Tiwari and Shyam, 2016).

The radiative heat loss coefficient from the top glass can be expressed using Equation 10.10:

$$h_{r,g1-amb} = \sigma\varepsilon_{g1}\left(\frac{T_{g1}^4 - T_s^4}{T_{g1} - T_{amb}}\right) \tag{10.10}$$

T_s can be estimated for hourly (t) basis using Equation 10.11 (Bennour and Mzad, 2022):

$$T_s = T_{amb}\left(0.711 + 0.0056T_{dew} + 0.000073T_{dew}^2 + 0.013\cos(15t)\right)^{1/4} \tag{10.11}$$

Equation 10.12 can be employed in order to estimate the dew point temperature T_{dp} for 0–93°C (Duffie et al., 2020):

$$T_{dew} = 6.54 + 14.526\ \ln(p_w) + 0.7389\ \ln(p_w)^2 + 0.09486\ \ln(p_w)^3 + 0.4569p_w^{0.1984} \tag{10.12}$$

where $RH = \frac{p_w}{p_{sw}}$

Equation 10.13 can be used to estimate the vapor saturation pressure for the ambient air, as proposed by Buck (1981).

$$p_{sw}(T) = 0.61121\exp\left(\left(18.678 - \frac{T}{235.5}\right)\left(\frac{T}{257.14 + T}\right)\right) \tag{10.13}$$

Apart from Equation 10.13, an average temperature of sky can be estimated using Equation 10.14 (Gao et al., 2022):

$$T_{sky} = 0.0552\ T_{amb}^{1.5} \tag{10.14}$$

The top glass convective heat transfer (h_w) can be attributed to the wind blowing above the solar collector using Equations 10.15 and 10.16 (Tiwari and Shyam, 2016):

$$h_w = 2.8 + 3\ V_w,\ \text{if } V_w \leq 5 \text{ m/s} \tag{10.15}$$

$$h_w = 6.15 V_w^{0.8},\ \text{if } V_w \geq 5 \text{ m/s} \tag{10.16}$$

The radiative heat transfer coefficient between top and bottom glass can be estimated using Equation 10.17:

$$h_{r,g1-g2} = \frac{\sigma\left(\left(T_{g1} + T_{g2}\right)\left(T_{g1}^2 + T_{g1}^2\right)\right)}{\left(\frac{1}{\varepsilon_{g1}}\right) + \left(\frac{1}{\varepsilon_{g2}}\right) - 1} \tag{10.17}$$

The radiative heat transfer coefficient between absorber plate and bottom glass is given by Equation 10.18 (Bennour and Mzad, 2022):

$$h_{r,a-g1} = \frac{\sigma\left(\left(T_a + T_{g1}\right)\left(T_a^2 + T_{g1}^2\right)\right)}{\left(\frac{1}{\varepsilon_a}\right) + \left(\frac{1}{\varepsilon_{g1}}\right) - 1} \tag{10.18}$$

The energy conservation equation for working fluid flowing inside the pipe can be given by Equation 10.19 (Wenceslas Koholé et al., 2022):

$$h_{c,a-fluid}\ \pi d_i \left(T_a - T_{fluid}\right) = \rho_{fluid}\ c_{a-fluid} \frac{\pi d_p^2}{4} \frac{dT_{fluid}}{dt} + \frac{\dot{m}_{fluid} c_{a-fluid}}{n_t} \frac{\partial T_{fluid}}{\partial y} \tag{10.19}$$

Finally, the overall loss from the system can be given by Equation 10.20.

$$U_o = U_b + U_e + U_t \tag{10.20}$$

The loss from the back (U_b) and edge (U_e) of the solar collector can be estimated using Equations 10.21–10.22.

$$U_b = \frac{k_b}{x_b} \tag{10.21}$$

$$U_e = \frac{k_e p x}{L_e A_c} \tag{10.22}$$

The top loss coefficient is the parallel stacking of the layers, including glass and absorber plate, as given in Equations 10.23–10.26 (Duffie et al., 2020; Murugan et al., 2022):

$$U_t = \frac{1}{R_1 + R_2 + R_3} \tag{10.23}$$

$$R_1 = \frac{1}{h_w + h_{r,g2-amb}} \tag{10.24}$$

$$R_2 = \frac{1}{h_{c,a-g1} + h_{r,a-g1}} \tag{10.25}$$

$$R_3 = \frac{1}{h_{c,g1-g2} + h_{r,g1-g2}} \tag{10.26}$$

To evaluate the fluid heating performance inside the solar collector, Equations 10.27–10.29 are utilized for fluid flowing inside the pipe (Alawi et al., 2022):

$$Re = \frac{\rho U_{in} D_h}{\mu} \tag{10.27}$$

$$Pr = \frac{C_p \mu}{k} \tag{10.28}$$

$$\Delta P = \left(\bar{P}_{out} - \bar{P}_{in}\right) = f \frac{L}{D_h} \frac{\rho U_{in}^2}{2} \tag{10.29}$$

$$\eta = \frac{Q_{gain}}{AI} = \frac{\dot{m} c_p \left(T_{out} - T_{in}\right)}{AI} \tag{10.30}$$

Following the solar water heater energy calculation, different applications can be employed with the water heated using the solar collector. A similar type of solar collector design is also employed for heating of air that is mainly applied for room heating, space heating, drying, and heat exchangers. A schematic design with various heat transfer coefficients are illustrated in Figure 10.4 with air flowing above and below the absorber plate.

The configuration presented in Figure 10.4 can be modified as per specification, a wide variety of design changes have been made in recent years to increase the overall performance of the solar collector, mainly by creating diversion and adding fins for either increasing the residence time within the solar collector or increasing the surface area for the fluid contact, respectively. Also, since the air has low specific heat and thermal conductivity which is mainly used as fluid inside the solar air heater collector, improvement is necessary for the overall achievement of the desired temperature for any specific physical operation to take place.

The heat transfer coefficients in Figure 10.4, mainly $h_{r,g2-amb}$, h_w, $h_{r,g1-g2}$, and $h_{c,g1-g2}$, are 10, 14–15, 17, and 7, respectively.

For the air flowing above the absorber plate, the heat transfer coefficient $h_{c,g2-air}$ and $h_{c,a-air}$ can be estimated using Equations 10.31 and 10.32:

$$h_{r,a-g1}\left(T_a - T_c\right) + h_{c,g1-a}\left(T_{air} - T_{g1}\right) = h_{a-g1}\left(T_{g1} - T_{air}\right) \tag{10.31}$$

$$\dot{q}_u = h_{c,a-air}\left(T_a - T_{air}\right) + h_{c,air-g2}\left(T_{air} - T_{g2}\right) \tag{10.32}$$

Equations 10.31 and 10.32 can be reduced to Equation 10.33:

$$\dot{q}_u = F'\left[(\alpha\tau)I(t) - U_L\left(T_{air} - T_a\right)\right] \tag{10.33}$$

F' is the collector efficiency factor that can be estimated using Equation 10.34:

$$F' = B/\left(\left[h_{c,g1-air}\left(h_{c,a-air} + U_b\right) + h_{c,a-air}h_{a-g1}\left(h_{c,g1-air} + h_{c,a-air} + U_b + h_{a-g1}\right) + U_b h_{a-g1}\right]\right. \tag{10.34}$$

The heat loss factor U_L can be also be estimated using Equation 10.35:

$$U_L = \left[h_{c,g1-air}\left(h_{c,a-air}U_b + h_{c,a-air}h_{a-g1} + U_b h_{a-g1}\right) + h_{c,a-air}U_b h_{a-g1}\right.$$
$$\left. + h_{r,a-g1}\left(U_b h_{c,g1-air} + h_{c,g1-air}h_{a-g1} + h_{c,a-air}U_b + h_{c,a-air}h_{a-g1}\right)\right] \tag{10.35}$$

where

$$B = h_{c,g1-air}h_{c,a-air} + h_{c,a-air}h_{a-g1} + h_{r,a-g1}\left(h_{c,g1-air} + h_{c,a-air}\right) \tag{10.36}$$

Further, the instantaneous thermal efficiency for solar air heater can be estimated using Equation 10.37 (Tiwari and Shyam, 2016).

$$\eta_{inst} = \frac{\dot{q}_u}{I(t)} = F'[(\alpha\tau) - \frac{U_L\left(T_f - T_a\right)}{I(t)} \tag{10.37}$$

In another case, when the air is flowing from both sides of the absorber plate as shown in Figure 10.2, the steady-state heat gain for the fluid flowing can be given by Equation 10.38 (Ramani et al., 2010a; S. Tiwari and Tiwari, 2017):

$$\dot{q}_u = h_{c,a-air}\left(T_a - T_{air,top}\right) + h_{c,a-air}\left(T_a - T_{air,bottom}\right) - h_{c,g1-air}\left(T_{ari,top} - T_{g1}\right) + h_{c,g1-air}\left(T_{air,bottom} - T_b\right) \tag{10.38}$$

where $T_{air,top}$ and $T_{air,bottom}$ are air flowing above and below the absorber plate, respectively.

It is considered that the heat loss from the top of the abuser plate is higher as compared to the bottom of the absorber plate due to which the air flowing from the bottom of the absorber plate has slightly higher temperature as compared to the air flowing from the above. The air flowing on both sides of the absorber plate has similar air flowrate and T_{air} is the average temperature of $T_{air,top}$ and $T_{air,bottom}$, accordingly, the deviation of T_{air} from both the temperature must also be equal which can be given by Equations 10.39 and 10.40:

$$\left(T_{air} - T_a\right)(1-n) = \left(T_{air,top} - T_a\right) \tag{10.39}$$

$$\left(T_{air} - T_a\right)(1-n) = \left(T_{air,bottom} - T_a\right) \tag{10.40}$$

On solving Equation 10.38 using Equations 10.39 and 10.40, the expression of F' and U_L can be obtained as given in Equation 10.41:

$$F' = D/\left\{2h_{c,g1-air}h_{c,a-air}P + 2h_{c,a-air}U_b h_{a-g1} + h_{r,a-g1}\left(h_{c,g1-air} + h_{r,a-b}\right)\left(P + 2h_{c,a-air}\right)\right.$$
$$\left. + U_b\left(2h_{c,a-air} + h_{a-g1}\right) + \left(h_{g1-air}h_{r,a-b}\right) + h_{r,a-b}\left(h_{c,g1-air}\left[p + 2h_{a-air}\right] + 2h_{c,a-air}h_{a-g1} + U_b h_{a-g1}\right)\right\} \tag{10.41}$$

$$
\begin{aligned}
U_L = {} & 4h_{c,g1-air}h_{c,a-air} + 2h_{r,a-g1}\left(\left[h_{c,g1-air} + h_{c,a-air}\right]\left[h_{r,a-b}h_{a-g1} + U_b h_{a-g1} + h_{r,a-b}U_b\right] + h_{c,g1-air}\right) \\
& + 2h_{r,p-b}U_b\left(h_{c,g1-air}h_{c,a-air} + h_{c,g1-air}h_{a-g1} + h_{c,a-air\,a-g1}\right) + (1-n)\,h \\
& + (h_{c.g1-air}h_{a-g1}(h_{c,g1-air}[2h_{\left(h_{c,a-air} + h_{r,a-g1} + (1+n)h_{c,g1-air}U_b\left(h_{c,g1-air}\left[h_{r,a-b} + 2h_{c,a-air}\right] + h_{r,a-g1}Q\right)\right]}/D
\end{aligned} \tag{10.42}
$$

$$
\begin{aligned}
\text{where } D = {} & \left[2h_{c,g1-air}\,h_{c,a-air}P + 2h_{c,a-air}U_b h_{a-g1} + h_{r,a-g1}\left(Q\left\{h_{c,g1-air} + h_{r,a-b} + U_b\right\} + h_{c,g1-air}h_{r,a-g1}\right.\right. \\
& \left.\left. + h_{r,a-b}\,Q\left(h_{c.g1-air} + h_{a-g1}\right)\right)\right]
\end{aligned} \tag{10.43}
$$

$$P = h_{c,g1-air} + U_b + h_{a-g1} \tag{10.44}$$

$$Q = h_{c,g1-air} + 2h_{c,a-air} \tag{10.45}$$

It is worth noting that the F' and U_L are calculated using the same method as above for air flowing above absorber plate, but due to increase in the configuration of the air within the collector, the complication of the expression increases in order to accommodate the different configurations. Also, the error can generate in case of omission of deviation within the temperature of air flowing above and below the absorber plate.

The above-described configuration can be taken as base of various other configurations such as double-pass flow and triple-pass flow, and the same base expression can be modified depending upon the configurations. Several design modifications such as addition of baffle and fins can also be incorporated within the design and the expression for the same can be modified according to the increase in the variables.

10.3.2 Energy Analysis of Solar Cooker

Various types of solar cooker designs are being researched and different design changes are taken into consideration in order to improve the overall thermal efficiency and achievement of the desired temperature. In general, three types of solar cooker are there: (i) solar panel cooker, (ii) solar parabolic cooker, and (iii) solar box cooker. Among all these, solar box cooker is the simplest and most widely used. Due to its simple design and construction ease, it has been widely accepted and is being modified in order to improve the thermal efficiency. In box-type solar cooker, reflector is the most common type of modification that is added to increase the amount of incident solar radiation to reach higher temperature. Another type of modification includes use of thermal energy storage materials inside the box so as to increase the amount of heat trapped as well as increasing the duration of heating even during off-sunshine period or to provide a more constant temperature in case of fluctuation in solar radiation (Aquilanti et al., 2023; Cuce and Cuce, 2013; Goyal and Eswaramoorthy, 2023b, 2023a; Koshti et al., 2023).

The thermal analysis of the box-type solar cooker with single reflector involves the calculation of overall heat loss coefficient using Equation 10.46 (Goyal and Eswaramoorthy, 2023b):

$$\mathrm{U_L} = U_T + U_B + U_S \tag{10.46}$$

The top loss coefficient can be estimated using Equation 10.47:

$$\frac{1}{U_T}=\left[\frac{5.78\left\{\left(T_a-T_{g1}\right)\cos\beta\right\}^{0.27}}{\left(\dfrac{T_a+T_{g1}}{2}\right)^{0.31}L_{a-g1}^{0.21}}+\frac{\sigma\left(T_a^2+T_{g1}^2\right)\left(T_a+T_{g1}\right)}{\dfrac{1}{\varepsilon_a}+\dfrac{1}{\varepsilon_{g1}}-1}\right]^{-1}$$
$$+\left[\frac{5.78\left\{\left(T_{g1}-T_{ref}\right)\cos\beta\right\}^{0.27}}{\left(\dfrac{T_{g1}+T_{ref}}{2}\right)^{0.31}L_{g1-ref}^{0.21}}+\frac{\sigma\left(T_{g1}^2+T_{ref}^2\right)\left(T_{g1}+T_{ref}\right)}{\dfrac{1}{\varepsilon_{g1}}+\dfrac{1}{\varepsilon_{ref}}-1}\right]^{-1}$$
$$+\left[h_w+\sigma\varepsilon_{ref}\left(\left(T_{ref}^2+T_{amb}^2\right)\left(T_{ref}+T_{amb}\right)\right)\right]^{-1}+\frac{2t_{ref}}{k_{ref}} \tag{10.47}$$

where h_w can be estimated using Equations 10.15 and 10.16.

The heat loss coefficient from the bottom of the solar cooker is estimated using Equation 10.48:

$$U_b=\frac{t_i}{k_i}+\frac{t_{st}}{k_{st}} \tag{10.48}$$

The heat loss from the side of the solar cooker mainly depends upon the insulation and the frame material for which Equation 10.49 can be used for estimation:

$$\frac{1}{U_s}=\frac{1}{4L_{a-g1}}\left[\frac{t_i}{k_i}+\frac{t_{st}}{k_{st}}\right] \tag{10.49}$$

The energy balance for the top glass cover $\left(T_{g1}\right)$ of the box solar cooker can be expressed as Equation 10.50 (Goyal and Eswaramoorthy, 2023a):

$$\left(mc_p\right)_{g1}\frac{dT_{g1}}{dt}=A_{g1}G_{g1}-A_{g1}h_{c,g1-i}\left(T_{g1}-T_i\right)-A_{g1}h_{r,g1-s}\left(T_{g1}-T_s\right)-A_{g1}h_{c,g1-amb}\left(T_{g1}-T_{amb}\right)$$
$$+A_{g1}h_{c,g1-a}\left(T_{g1}-T_a\right)+A_ah_{r,g1-a}\left(T_{g1}-T_a\right)-A_{ut}h_{r,g1-ut}\left(T_{g1}-T_{ut}\right) \tag{10.50}$$

The energy balance at the insulation (T_i) of the box is given by Equation 10.51.

$$\left(mc_p\right)_{air,inside}\frac{dT_{air,inside}}{dt}=A_{g1}h_{c,air,inside-g1}\left(T_{air,inside}-T_{g1}\right)+A_{g1}h_{c,air,inside-a}\left(T_{air,inside}-T_a\right)$$
$$-A_{ut}h_{c,air,inside-ut}\left(T_{ut}-T_{air,inside}\right) \tag{10.51}$$

The inside of the box is the absorber plate (T_a) and the energy balance is given by Equation 10.52:

$$(mc_p)_a \frac{dT_a}{dt} = G_a A_a - A_a h_{c,a-g1}(T_a - T_{g1}) - A_a h_{c,a-ut}(T_a - T_{ut}) - A_a h_{c,a-air,inside}(T_a - T_{air,inside})$$
$$- A_a h_{a-amb}(T_a - T_{amb}) - A_a h_{r,g1-a}(T_{g1} - T_a) - A_a \frac{\varepsilon_a}{1-\varepsilon_a}(\sigma T_a^4 - J_a) - A_a h_{con,a-ut}(T_a - T_{ut})$$
(10.52)

The energy balance at the surface of the cooking pot placed inside the solar box cooker is given by Equation 10.53.

$$(mc_p)_{ut} \frac{dT_{ut}}{dt} = A_a h_{con,a-ut}(T_a - T_{ut}) + A_{ut}(h_{c,ut-a} + h_{r,ut-a})(T_{ut} - T_a) + A_a \frac{\varepsilon_a}{1-\varepsilon_a}(\sigma T_{ut}^4 - J_{ut}) \quad (10.53)$$

Inside the utensil, the energy balance is given by Equation 10.54:

$$(mc_p)_{ut,inside} \frac{dT_{ut,inside}}{dt} = A_{ut} h_{c,ut,inside-ut}(T_{ut,inside} - T_{ut}) \quad (10.54)$$

Similarly, radiative heat flux for the cooking pot and absorber plate can be estimated using Equations 10.55 and 10.56:

$$J_a - (1-\varepsilon_a) F_{a-ut} J_{ut} = \varepsilon_a \sigma T_a^4 \quad (10.55)$$

$$J_{ut} + (\varepsilon_{ut} - 1)[F_{ut-a} J_a] = \varepsilon_{ut} \sigma T_{ut}^4 \quad (10.56)$$

The radiative heat transfer from glass to sky $h_{r,g1-s}$ and from absorber plate to glass $(h_{r,a-g1})$ can be estimated using Equations 10.10 and 10.18, respectively. Also, the convective heat transfer coefficient from glass to ambient $(h_{c,g1-amb})$ can be estimated using Equations 10.15 and 10.16 (Duffie et al., 2020; Tiwari and Shyam, 2016).

For the convective heat transfer coefficient between the absorber plate inside the solar cooker box with air inside $(T_{air,inside})$ and the top glass cover, Equation 10.57 is utilized:

$$Nu = \left[0.825 + \frac{0.387 Ra^{\frac{1}{6}}}{\left[1 + \left(\frac{0.492}{\Pr} \right)^{\frac{9}{16}} \right]^{\frac{8}{27}}} \right]^2 \quad (10.57)$$

Assuming the heat transfer between the cooking pot and inside air is similar to that of a hot air flowing over heated vertical plate and also assuming cooking pot is a vertical cylinder with uniform heating all over the surface, the heat transfer coefficient can be estimated using relation given by Churchill and Chu (1975). Also, assuming the pot lid is flat and uniform, the heat transfer coefficient for the lid can be estimated using Equations 10.58 and 10.59 (Goyal and Eswaramoorthy, 2023a).

$$Nu = 0.54 Ra^{\frac{1}{4}} \quad 10^4 \leq Ra \leq 10^7 \tag{10.58}$$

$$Nu = 0.15 Ra^{\frac{1}{3}} \quad 10^7 \leq Ra \leq 10^{11} \tag{10.59}$$

Convective heat transfer coefficient between cooking pot and air inside the solar cooker can be estimated using Equation 10.60:

$$h_{c,air,inside-ut} = A_{env} h_{c,air,inside-env} \frac{A_{cover} h_{c,air,inside-cover}}{A_{cover+env}} \tag{10.60}$$

Also, the conductive heat transfer coefficient between the absorber plate and cooking pot can be estimated using Equation 10.61.

$$h_{k,ut-a} = \frac{k_a}{t_a} \tag{10.61}$$

The absorbed solar radiation can be estimated using Equations 10.62 and 10.63.

$$G_{g1} = I(t) \rho_m^n \tau_{g1} \left[\alpha_{g1} + \alpha_{g1} \rho_{g1} + \alpha_{g1} \tau_{g1} \rho_a \right] C \tag{10.62}$$

$$G_a = I(t) \rho_m^n \tau_{g1} \left[\alpha_a + \alpha_a \rho_{g1} \rho_a \right] C \tag{10.63}$$

where C is the parabolic curve aided solar cooker concentration ratio and $n = 0.5 + 0.07C$.

The solar cooker overall utilizable efficiency can be estimated using Equation 10.64 (Cuce and Cuce, 2013):

$$\eta_u = \frac{Q_F}{Q_{input}} \tag{10.64}$$

For the characterization of solar cooker, certain parameters are taken into consideration such as first figure of merit (F_1), which is the ratio of optical efficiency and overall thermal loss coefficient, as given in Equation 10.65, and second figure of merit (F_2) that is defined as the ability of the solar cooker to convert the incident solar energy into useful thermal energy as given in Equation 10.66 (Aquilanti et al., 2023; Koshti et al., 2023):

$$F_1 = \frac{T_{a,max} - T_{amb}}{G_n} \tag{10.65}$$

$$F_2 = \frac{F_1 m_f c_f}{A_a \Delta t_h} \ln \left[\frac{1 - \frac{1}{F_1}\left(T_1 - T_{amb,avg}\right) / G_{n,avg}}{1 - \frac{1}{F_1}\left(T_2 - T_{amb,avg}\right) / G_{n,avg}} \right] \tag{10.66}$$

Further, the overall efficiency (η_{avg}) and cooker opto-thermal ratio can be estimated using Equations 10.67 and 10.68 (Aquilanti et al., 2023):

$$\eta_{avg} = \frac{m_f c_f (T_2 - T_1)}{G_{n,avg} A_a \Delta t_h} \tag{10.67}$$

$$COR = \frac{\eta_0 C}{U_1} \tag{10.68}$$

10.3.3 PV-T Solar Collector

The energy balance equation for PV-T solar collector can be given by Equation 10.69:

$$Q_{in} = \eta_{PV} I(t) + \eta_u GA \tag{10.69}$$

The temperature of solar cell in the PV panel can be expressed as Equation 10.70:

$$T_{pv} = \frac{(\alpha\tau)_{eff} I + U_t T_a + U_b T_a}{U_t + U_b} \tag{10.70}$$

where

$$(\alpha\tau)_{eff} I = (\alpha_c p + \alpha_T (1-p) - p\eta_e) \tag{10.71}$$

The cell efficiency (η_e) can be estimated using Equation 10.71:

$$\eta_e = \eta_0 \left(1 - \beta\left(T_{pv} - T_{ref}\right)\right) \tag{10.72}$$

Apart from this, other heat transfer coefficients, mainly radiative and convective from glass to photovoltaic module, photovoltaic module to absorber plate in case of sandwich solar collector, or single photovoltaic module replacing the glass, can be estimated using similar expression as that of single- and double-pass air solar collectors as discussed above.

10.4 METHODOLOGIES FOR EXERGY ANALYSIS

Analysis of exergy of any system mainly involves the assessment of the total exergy input, output, and losses within the system as well as evaluation of exergy parameters at various steady-state points in the system. Figure 10.5 presents the distribution of input exergy to exergy losses and useful exergy output. Exergy losses fall into two categories:

i. Internal exergy losses
ii. External exergy losses

Internal exergy losses occur within the system boundary and may be caused by a variety of factors, including heat transfer, uncontrolled combustion, friction, etc. These losses are because of internal irreversibilities, and contribute to exergy destruction/consumption inside the system.

External exergy losses are external to the system because of external irreversibilities and include emissions of waste heat/energy output and gases into the environment (Tahir et al., 2021).

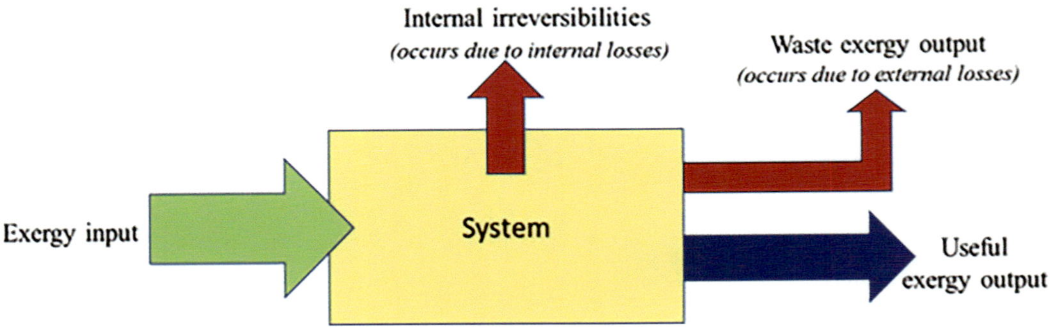

FIGURE 10.5 Exergy flow in a system.

The validity of the mass and energy balances should be established before doing an exergy analysis. The thermodynamic analysis should be assessed to determine whether a combined energy and entropy balance, known as an exergy balance, can be obtained.

Based on the energy balance of air, the basic expression for estimating the working medium's exergy at steady-state flow for a thermal process is utilized (Chowdhury et al., 2011).

$$\text{Exergy} = \text{Internal energy} - \text{Entropy} + \text{Work} + \text{Gravity} + \text{Chemical energy} \quad (10.73)$$

$$Ex = (u_i - u_a) - T_{amb} \cdot (S - S_a) + \left(\frac{P_a}{J}\right) . (V - V_a) + \left(\frac{v^2}{2gJ}\right) + (z - z_a) \cdot \left(\frac{g}{g_c J}\right) + \Sigma(\mu_c - \mu_a) \cdot N_c \quad (10.74)$$

where u_i is the internal energy (kJ/s), T is the ambient temperature (°C), S is the specific entropy (kJ/kg·K), P is the ambient pressure (kPa), J is the Joule constant, V is the specific volume (m^3/kg), v is the velocity (m/s), g is the acceleration due to gravity (m/s^2), g_c is the Newton's constant, z is the altitude (m), μ_c is the chemical potential (kJ/kg), and N_c is the number of species. The subscript a indicates the ambient conditions (Chowdhury et al., 2011).

The exergy of fluid (Ex_{fluid}) is most frequently expressed by Equation 10.75, with gravitational, momentum, and potential energy being ignored:

$$Ex_{fluid} = \dot{m}c_p \left[(T_{fluid} - T_a) - T_a \ln\left(\frac{T_{fluid}}{T_a}\right) \right] \quad (10.75)$$

10.4.1 Exergy Analysis of a Solar Collector

The general exergy flow in a conventional solar collector is presented in Figure 10.6a. The following assumptions can be considered for conducting the assessment of exergy flow of a solar collector:

1. It is pre-summated that the flow is a steady-state wand and there is no influence of kinetic and potential energy.
2. The effect of ambient humidity is disregarded.
3. The ambient temperature and pressure are used as baseline conditions.

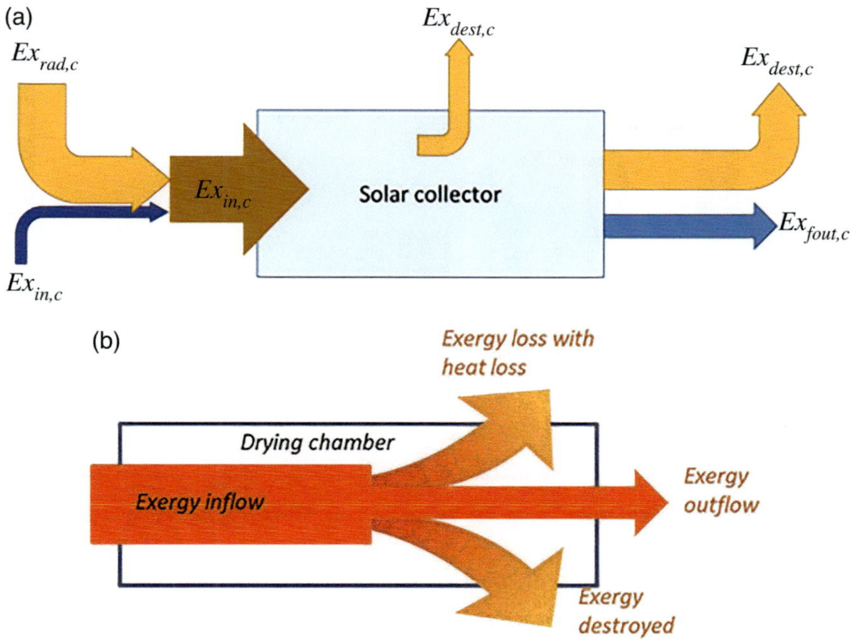

FIGURE 10.6 Exergy flow in (a) solar collector and (b) drying chamber.

By utilizing Equation 10.76, where T_{sky} is the sky's temperature, the exergy of the sun's radiation that hits a solar collector ($Ex_{rad,c}$) is calculated (Petela, 2003):

$$Ex_{rad,c} = A_c I\left[1-\frac{4}{3}\left(\frac{T_{amb}}{T_{sky}}\right)+\frac{1}{3}\left(\frac{T_{amb}}{T_{sky}}\right)^4\right] \tag{10.76}$$

The exergy of fluid at the inlet ($Ex_{fin,c}$) and the outlet ($Ex_{fout,c}$) of collector are computed using Equations 10.77 and 10.78. Fluid medium gains exergy ($Ex_{gain,c}$) from the absorber plate and this exergy is measured using Equation 10.79. The summation of exergy of fluid entering the collector and the solar energy exergy hitting the solar collector forms the total exergy input to the solar collector, as indicated by Equation 10.80.

$$Ex_{fin,c} = \dot{m}c_p\left[(T_i - T_{amb}) - T_{amb}\ln\left(\frac{T_i}{T_{amb}}\right)\right] \tag{10.77}$$

$$Ex_{fout,c} = \dot{m}c_p\left[(T_o - T_{amb}) - T_{amb}\ln\left(\frac{T_o}{T_{amb}}\right)\right] \tag{10.78}$$

$$Ex_{gain,c} = Ex_{fout,c} - Ex_{fin,c} \tag{10.79}$$

$$Ex_{in,c} = Ex_{fin,c} + Ex_{rad,c} \tag{10.80}$$

Equation 10.81 provides the exergy of the absorber plate (Ex_a) (Sahu and Prasad, 2016).

$$Ex_a = \eta_o A_s I\left(1-\frac{T_{amb}}{T_a}\right) \tag{10.81}$$

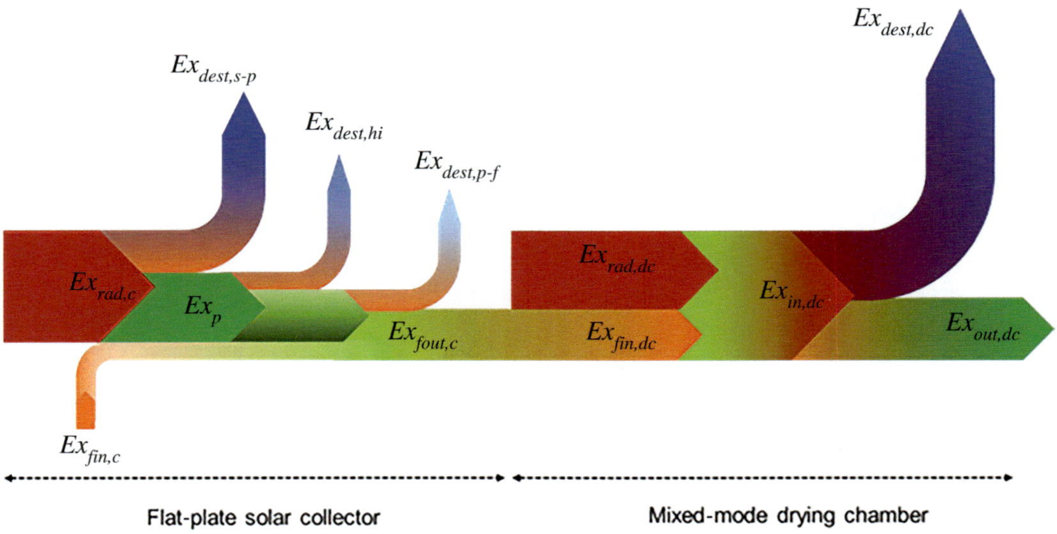

FIGURE 10.7 Exergy loss and exergy destruction in an air-based flat-plate solar collector and drying chamber.

Equation 10.82 can be used to express the exergy balance within the collection:

$$\sum Ex_{in,c} - \sum Ex_{out,c} = \sum Ex_{dest,c} \tag{10.82}$$

where $Ex_{dest,c}$ is the entire amount of energy that has been destroyed inside the solar collector.

Three mechanisms in particular are responsible for the exergy destruction in a solar collector. Figure 10.7 is a schematic Grassmann diagram showing exergy flow, exergy loss, and exergy destruction in different parts of a mixed-mode solar dryer (Rani and Tripathy, 2021). Exergy is lost during the absorber plate's conversion of solar energy into heat ($Ex_{dest,s\text{-}a}$), heat loss from the collector ($Ex_{dest,hl}$), and energy transfer from the absorber plate to fluid (*Exdest, p-f*).

Exergy destruction refers to the exergy loss within the system boundary due to irreversibility, whereas exergy losses refer to exergy flowing to the surroundings. Exergy destruction is defined as the actual alteration in exergy for the irreversible process less the change in exergy for reversible process. Exergy destruction results from irreversible processes such as chemical reaction, heat transfer, pressure drop, and mixing (Bayrak et al., 2013). Equations 10.83–10.85 are used to calculate the mentioned exergy losses in a solar collector (Sahu and Prasad, 2016):

$$Ex_{dest,s-a} = Ex_{rad} - Ex_a \tag{10.83}$$

$$Ex_{dest,hl} = U_L A_c \left(T_a - T_{amb}\right)\left(1 - \frac{T_{amb}}{T_a}\right) \tag{10.84}$$

$$Ex_{dest,a-f} = Ex_{a-f} - Ex_{gain,c} \tag{10.85}$$

The exergy transferred from the absorber plate to the fluid medium, indicated by $Ex_{p\text{-}f}$, can be estimated using Equation 10.86:

$$Ex_{a-f} = \dot{m}c_p\left[\left(T_o - T_i\right)\left(1 - \frac{T_{amb}}{T_a}\right)\right] \tag{10.86}$$

Exergy efficiency is a crucial factor employed for assessment and improvement of thermal energy systems. The exergy efficiency of the solar collector is defined as the ratio of total exergy gain by the fluid in the solar collector to the total exergy going into the collector, as represented by Equation 10.87:

$$Ex_{eff,c} = \frac{Ex_{gain,c}}{Ex_{in,c}} \tag{10.87}$$

10.4.2 Exergy Analysis of Drying Chamber

Using the generalized exergy equation, fluid exergy at dryer outlet section ($Ex_{out,dc}$) is given by Equation 10.88, where $T_{o,d}$ is the air temperature at the dryer outlet:

$$Ex_{out,dc} = \dot{m}c_p\left[\left(T_{o,d} - T_{amb}\right) - T_{amb}\ \ln\left(\frac{T_{o,d}}{T_{amb}}\right)\right] \tag{10.88}$$

The internal exergy loss of drying chamber section ($Ex_{dest,dc}$) is calculated using Equation 10.89:

$$\Sigma Ex_{in,dc} - \Sigma Ex_{out,dc} = \Sigma Ex_{dest,dc} \tag{10.89}$$

As stated, exergy efficiency can also be defined as the drying chamber's exergy outflow to inflow ratio. The ratio of the exergy output from the drying chamber to the exergy input to dry the food sample is known as the exergy efficiency of the drying chamber ($Ex_{eff,dc}$), shown by Equation 10.90 (Karthikeyan and Murugavelh, 2018):

$$Ex_{eff,dc} = \frac{Ex_{out,dc}}{Ex_{in,dc}} = 1 - \frac{Ex_{dest,dc}}{Ex_{in,dc}} \tag{10.90}$$

The exergy input to the drying cabinet of a mixed-mode solar dryer consists of fluid exergy output from the solar collector, which acts as the fluid exergy input to drying chamber ($Ex_{fin,dc}$) and exergy of the solar power received by the drying chamber ($Ex_{rad,dc}$).

$$Ex_{in,dc} = Ex_{fin,dc} + Ex_{rad,dc} \tag{10.91}$$

In case of indirect solar dryer, the exergy input to drying cabinet is equivalent to exergy output from the collector, as there is no solar radiation falling on the drying chamber in indirect mode of solar dryer.

10.5 ENERGY AND EXERGY PERFORMANCE OF SOLAR THERMAL SYSTEMS

Exergy evaluation is a helpful technique for formulating plans for the designing and carrying out various in-factory operations concerning to crucial need of the best utilization of energy. According to researchers, this information is useful for calculating costs of plant operation, flexibility of fuel in addition to energy-saving measures, and pollution level controls (Yahya et al., 2017). Examining the exergy has been a popular tool for assessing the efficiency of thermal systems in the recent past. Therefore, exergy analysis is employed for many solar-powered systems such as solar collectors, solar dryer, solar refrigeration cycle, solar cookers, etc. for performance analysis. The goal of increasing provided exergy is similar to lowering entropy generating rates.

10.5.1 Performance Analysis of Solar Collectors

Generally, the efficiency figures for the energy analysis have been demonstrated to be greater than exergy values (Bayrak et al., 2013). Further, it was also witnessed that a reciprocal link between energy loss and collector efficiency also exist and higher irreversibility tends to lower the collector efficiency. Therefore, it is necessary to decrease energy losses for improvement of the energetic and exergetic performance of solar collector system. The higher temperature gradient between the surface of the absorber plate and solar radiation contributes to majority of the energy losses occurring in the flat-plate solar collector. Because of the lower heat losses from the collector's surface, double-glass cover collectors have higher absorber plate temperatures than single-cover collectors, which lowers energy loss (Gholami et al., 2019). At average absorber plate temperature, the absorber surface's potential to absorb radiation contributes a most significant part to the system loss. The exergy losses caused by the absorption are reduced by a high absorber temperature, but the heat losses to the surroundings are increased. (Rani and Tripathy, 2021).

The pressure drop and heat transfer both are greatly influenced by air mass flow rate, which also affect the thermal characteristics of solar collectors. It is a well-known fact that increase of air mass flow rates contribute to the improvement of thermal efficiency of solar collector (Abi Mathew and Thangavel, 2021). Studies have revealed that at larger flow rates, more usable energy can be collected from the system; as a result, exergy efficiency also get enhanced, thus improving the overall performance of solar air heater (Raj et al., 2019). The increase in exergy gain with higher air velocity is also responsible for the collector's improved exergy efficiency (Rani and Tripathy, 2021). But, contrarily, Abuska and Sevik (2017) reported that increases in Reynolds number and mass flow rate result in a exergy efficiency decrement due to increased fan power. On the other hand, Chamoli and Thakur (2014) stated that exergetic efficiency starts increasing with a rise in Reynolds number and achieves a maximum value before it starts to decline. They observed highest effective efficiency for the Reynolds number ranges between 4,000 and 12,000; therefore, it is recommended to run the solar collector around the mentioned range to attain the maximum exergy. For better optimization of any system, the exergy analysis should be taken into account by integrating the exergy parameter as an objective function. Depending on the flow rate, the value of exergy efficiency rises as solar radiation value increases. However, because of the involvement of numerous interconnected variables, the efficiency of exergy does not observe to be improved linearly (Abuska and Sevik, 2017).

Chamoli and Thakur (2014) provided evidence of the exergetic efficiency-based optimization of the roughness geometrical characteristics of V down baffles with perforations. A mathematical model and an experimental research were used by Farhan et al. (2021) to investigate the impact of the twisted tape insertion on the energy and exergetic efficiencies of the V corrugated solar air heater. They showed that adding more twisted tapes reduced exergy loss, increasing the thermal and exergy efficiency of the collector. Energy and exergy performance characteristics of the solar collector having fins or baffles are found to be a strong function of fins geometrical parameters and ambient parameters along with air mass flow rate (Raj et al., 2019). Studies summarized in Table 10.1 for different collectors under various operational parameters, climacteric factors, collector design and configurations, etc. indicate that research now give consideration to exergy analysis in addition to energy analysis for performance analysis.

10.5.2 Performance Analysis of Solar Dryers

Utilizing the first rule of thermodynamics, energy analysis illustrates the idea of energy conservation during the drying process. The natural and forced convection PV-assisted direct solar dryer during green onion drying presented the average drying efficiencies of 32.3% and 38.3%, respectively, while specific energy consumption (SEC) was 18.3 and 16.4 kWh/kg (Hidalgo et al., 2021). The SEC of the black ginger drying in a mixed-mode horizontal solar dryer at forced convection

TABLE 10.1
Recent Works Analyzing the Performance of Different Configurations of Solar Collectors Using Energy and Exergy Approach

Authors and Year	Type of Solar Collector	Energy and Exergy Analysis
Haloui et al. (2022)	PVT collector based on polymer and Si	Estimated thermal efficiency of 85.77% and 68.71% for polymer and Si, respectively
Ramani et al. (2010b)	Double-pass solar collector with and without porous materials	Estimated thermal efficiency in the range of 39.3–78.2% with 25% higher for double-pass solar collector and 35% higher for single-pass with porous material than without
Bennour and Mzad (2022)	Flat-plate solar air heater with inert gas in between double-glazing layer	Heat transfer for trapped gas reached up to 14 W/m^2 for a gap of less than 10 mm between glazing, with an exergy of 667, 664, 630, and 574 W for xenon, krypton, argon, and air, respectively
Joseph and Thomas (2022)	C-dot nanofluid influence onto the solar absorption in parabolic solar collector	The maximum efficiency of C-dot nanofluid flowing solar collector was found to be 73.41%, which was 57.62% above than that of the distilled water flowing. An exergy destruction of 922.3 W was observed at 90 L/h flow rate condition of nanofluid
Alawi et al. (2022)	Effect of CuO nanofluid morphology on the thermal efficiency	The base distilled water showed a thermal efficiency of 40.79–47.64%, whereas CuO (20 nm) of shape spherical, blade, brick, cylindrical, and platelets illustrated 64.74%, 69.32%, 69.86%, 84.95%, and 90.74%, respectively
Said et al. (2022)	Parabolic trough solar collector using silicon oil–based nanofluid	Illustrated improved thermal conductivity with 0.1 wt% nanofluid by 69.67% and 88.70% for 25°C and 150°C, respectively. The convective heat transfer coefficient improvement was 23.42%, 61.10%, and 93.75% for 0.05, 0.08 and 0.1 wt%, respectively. An exergy improvement of 4.52% was observed
Gao et al. (2022)	Glazed solar preheating collector wall integrated with nonuniform perforated corrugated plate	Illustrated thermal and exergy efficiency decrease of 20% and 25%, respectively, for air volume having low porosity. Change in air porosity showed improved thermal efficiency from 64.7% to 76.7% and exergy reduction from 9.4% to 1.6%
Wu et al. (2022)	Evacuated tube solar collector with phase change material	Observed increase in thermal performance with increase in density of thermal storage material till 1,700 kg/m^3. Also observed thermal conductivity beyond 1 W/(m·K) is not favorable. Increase in specific heat capacity of the thermal storage material beyond 4,000 J/(kg·K) is not significant for thermal efficiency improvement
Wenceslas Koholé et al. (2022)	Flat-plate solar collector	Increase in solar insolation from 100 W/m^2 to 1,000 W/m^2 decreased the thermal efficiency from 57% to 53.37%, while the exergy improved from 12.58% to 32.69%. Also, increase in ambient temperature from 4°C to 50°C lead to reduction in exergy by 41.41%
Abozoor et al. (2022)	Solar still with evacuated flat-plate collector	Energy efficiency of 39.74% was observed solar still, whereas evacuated tube solar still showed thermal efficiency of 45.85–46.64%. Exergy of the solar still was found to be 2.69%, while evacuated solar collector integrated solar still depicted exergy of 7.95–8.21%
Lamrani et al. (2022)	Mixed-mode solar dryer with V-groove integrated in thermal storage	About 78.19% of thermal energy was contributed by solar radiation and granite heat storage material provided 21.82% of thermal energy along with a collector efficiency of 60.08%

(Continued)

TABLE 10.1 ***(Continued)***

Recent Works Analyzing the Performance of Different Configurations of Solar Collectors Using Energy and Exergy Approach

Authors and Year	Type of Solar Collector	Energy and Exergy Analysis
Henein et al. (2023)	Evacuated tube solar collector using hybrid nanofluid	Highest heat energy gained was observed for multiwalled carbon nanotube and magnesium oxide for a flow rate of 3 L/min at 495 and 479 W. Energy yield factor was 27.09 for lifetime in case of magnesium oxide
Rani and Tripathy (2021)	Flat-plate solar collector	The exergy efficiency of flat-plate solar collector was observed at natural convection as 0.38%, and increased to 1.46%, 1.52%, 1.65% and 1.76% because of rise in air mass flow rate to 0.006, 0.01, 0.015, and 0.02 kg/s, respectively.
Raj et al. (2019)	Double-pass solar air heater system with asymmetric channel flow passages	The exergy efficiency varied in-between 5.6–18.04% and enhanced with increased mass flow rate condition
Gholami et al. (2019)	Solar air heaters with arcuate-shaped obstacles	When the mass flow rate is increased, the exergy efficiency continues to rise significantly until it reaches a maximum of 18% at 0.0177 kg/s, at which point it begins to decline
Sivakumar et al. (2019)	Heater using absorber plate with pin-fins	The solar collector's exergy efficiency was varied in the range of 5–17% for flat absorber plates and 7–26% for pin-fin absorber plates
Matheswaran et al. (2019)	Parallel pass jet plate solar air heater with artificial roughness	Reduced temperature of the absorber plate minimizes the exergy loss brought on by the temperature gradient between the absorber plate and air. Reduction of loss of heat to ambient on both sides of the absorber plate led to lower the exergy destruction and improvement of exergy enhancement ratio is improved
Farhan et al. (2021)	V-corrugated solar air heater integrated with twisted tape inserts	The exergy varies with the strength of the hourly solar radiation. The energy rises with the sun's irradiance during the day for three months until noon, after which it falls until sunset

was 1.07 and 0.56 kWh/kg at airflow of 0.062 and 0.018 kg/s, respectively (Ekka et al., 2020). A comparison was made among the efficiency of modified solar collector–integrated hybrid solar dryer, traditional open-sun drying, and active-mode solar PV and electric dryer during the drying of pineapple (Mohammed et al., 2020). Results showed that drying efficiency of hybrid solar dryer improved with integration of multiple metallic solar concentrators was near to PV and electric solar dryer, but 18% greater than the open-sun drying. Research by Kareem et al. (2017) indicated that Roselle leaves dried and picked up moisture with an efficiency of 36.22% and 66.95%, respectively, in a solar drier with a multi-pass solar air collector under forced convective. The drying efficiency and moisture pickup efficiency of Roselle leaves were 36.22% and 66.95%, respectively, at air velocity of 0.81 m/s.

Numerous studies (Mahapatra and Tripathy, 2019; Morad et al., 2017) have investigated the relationships between the energy balances inside the dryer and assessed its thermal performance on the basis of solar collector effectiveness, dryer efficiency, SEC, coefficients of heat transfer due to convection, conduction, and radiation, temperature of different components of dryer in addition to drying characteristics of food materials. However, the quality of heat/thermal energy and irreversibility observed during different energy conversion processes cannot be well expressed by energy analysis. A more effective drying system may be possible with the help of information from the exergy analysis, which might help food industry engineers and dryer manufacturers to distinguish the effectiveness of each solar dryer component parts.

In a study, due to the exergy's dependence on solar radiation, the incoming, outgoing, and loss of energy in the drying chamber attained the peak value within the initial 5 hours of experiment before steadily declining until the end of the day (Vijayan et al., 2020). The exergy loss also reached its maximum value when solar radiation attained the peak, and the exergy efficiency steadily rises as the time of day increases. Exergy efficiency demonstrated a rise with advancement of the drying process, due to a decrease in exergy degradation throughout the drying process, the exergy efficiency augmented with increment of time of drying experiment. Considering the mass flow rates change from 0.0141 kg/s to 0.0872 kg/s, there was 11.939% escalation in the values of average exergy efficiency.

Figure 10.7 illustrated the exergy flow of a solar dryer particularly of mixed-mode type, using the schematic Grassmann diagram (Rani and Tripathy, 2021). According to the study, 82.6–90% of the solar radiation exergy was lost because of the absorption process of solar radiation by the collector. When the solar dryer's operating mode was changed from natural to forced convection, energy destruction related to absorber plate thermal losses reduced. Positive impact of increased air velocity on reduced exergy destruction is justified by the lower absorber plate temperature. Exergy exchange between absorber and flowing air led to an increase in exergy in the air, known as exergy gain by the air. Some of the exergy linked with process irreversibility was also destroyed at the same time. By assuming equivalent conditions at the outlet of solar collector and inlet of chamber, the collector output exergy serves as exergy of air entering the drying chamber. At every point in the experiment, the natural convection exergy efficiency was much lower than that observed at forced convection mode. It is evident that the energy efficiency of dryers first displayed a progressive increasing pattern with some variation and gave a diminishing pattern toward the end of the experiment.

The mixed-mode solar tunnel dryer operating at forced convection was used for turmeric drying and process was studied for energy and exergy analysis (Karthikeyan and Murugavelh, 2018). The investigation led to the findings that energy utilization factor varies in the range of 9.73–33.98% while the tunnel dryer's energy efficiency was between 23.25% and 73.31%. A higher air mass flow rate indicates better performance from a solar dryer since it reduces drying time and energy waste while also improving the energy gain of air and energy utilization factor (Akbulut and Durmuş, 2010). The research conducted by Rabha et al. (2017) revealed lowest, maximum, and average values of drying chamber exergy efficiency as 21%, 98% and 63% for drying ghost peppers and 4%, 96%, and 47% for drying ginger. When drying of jackfruit leather was performed in a solar tunnel drier, increase of exergy input and exergy loss was proportional to rise in solar radiation intensity (Chowdhury et al., 2011). When solar radiation fluctuated between 100 W/m^2 and 600 W/m^2, it was discovered that the exergy efficiencies of the collector and solar tunnel dryer were in the range of 27.45–42.50% and 32.34–65.30%, respectively. The stevia leaves drying using a solar dryer attaining solar radiation on collector and drying chamber resulted in an overall dryer efficiency of 33.5% and average exergy efficiency was found to be 59.1% (Lakshmi et al., 2019). During the drying of banana pulp, Lingayat et al. (2020) performed an exergy analysis for an indirect solar drier. According to their findings, exergy loss ranged between 3.36 kJ/kg and 25.21 kJ/kg, and exergy efficiency was revealed to be in the range of 7.4% and 45.32%.

10.5.3 Performance Analysis of Other Solar Energy Applications

Solar thermal technology is a sustainable way to satisfy the world's rising energy needs and the peak of summer's cooling season also happens to be when solar radiation is at its strongest. The solar thermal collectors connected to a sorption chiller make up the majority of the solar thermal cooling system. Solar collectors, a heat retention tank, a thermal-driven chilling equipment, an indoor air-conditioning system, and an additional (backup) component are the key elements of such a system. The generator, after the condenser, was found to be the component with the second-highest energy loss in the model ammonia–water absorption cycle solar thermal systems. Lower condenser temperature levels result in the highest system performance and energy efficiency. The overall energy loss also rises sharply as the condenser temperature rises. The exergy study reveals that the generator and the absorbing process have the highest exergy losses (about 76%). The absorber should receive the most attention in order to increase cycle efficiency, with the generator perhaps coming in second. The outcomes can also be applied to absorption system thermoeconomic optimization (Aman et al., 2013).

The performance analysis of solar cooker studied by various researchers using energy and exergy concept is given in Table 10.2.

TABLE 10.2
Exergy and Energy Analysis of a Solar Cooker

S. No.	Application	Energy and Exergy Calculation	Authors
1	Energy and exergy assessment of compressed air–based energy storage system and solar collector	The hybrid system depicted an energy and exergy efficiency of 86.96% and 67.57%. The exergy destruction flow was about 649.26 kW	Li et al., 2023
2	Studied thermal efficiency and coefficient of performance for four different types of solar cooker	The values of average thermal efficiency were 5–9%, 7–15%, 5–9%, and 8–14% for Kimono, Funnel, DPSC, and cookit, respectively. The second figure of merit F_2 was lying in the ranges of 0.1–0.17, 0.12–0.23, 0.15–0.16, and 0.16–0.24, respectively	Aquilanti et al., 2023
3	Thermal efficiency analysis of triangular finned incorporated box-type solar cooker over conventional cooking pot	The first and second figures of merit were 0.139 and 0.36 for 1 kg and 0.56 for 1.5 kg indicating grade-1 solar cooker. Cooking power of 53.51 W and 77.06 W were found for 1 kg and 1.5 kg load	Goyal and Eswaramoorthy, 2023b
4	Energy estimation of thermal storage integrated indirect solar cooker with and without reflectors	The energy and exergy were obtained to be 79% and 48% for solar cooker with reflector and 63% and 42% without reflector. The utilization efficiency was found to be 22.3% for solar cooker with reflector and total power consumption was 458 W for 3 kg of water to boil	Getnet et al., 2023
5	Exergy and energy analysis for solar still integrated with box-type solar cooker	The energy efficiency of solar still with solar cooker was found to be 27.6% higher than that without solar cooker and annual exergy production of 55.3 and 35.3 kWh for active and passive solar still, respectively	Angappan et al., 2023
6	Comparative study of different box-type solar cookers	The figures of merit for multipurpose box-type solar cooker were 0.117 and 0.467, 0.11, and 0.33 for novel combined solar cooker, 0.11 and 0.31 for double-glazed solar cooker, and 0.117 and 0.154 for box solar cooker with aluminum cooker and reflector	Koshti et al., 2023

10.6 CONCLUSIONS

Solar thermal applications harness the power of the sun to generate heat, which can be utilized for various purposes such as space heating, water heating, and industrial processes. Understanding the energy and exergy flows in solar thermal systems is essential for assessing their performance, efficiency, and potential for improvement. Energy flow in solar thermal applications involves the conversion of solar radiation into thermal energy, which can be utilized for various purposes. Exergy flow, on the other hand, considers the quality of energy and its ability to do useful work. Exergy is a measure of the maximum useful work that can be obtained from a given energy flow when it comes into equilibrium with the environment. In solar thermal applications, exergy losses occur due to various factors such as heat transfer irreversibility, temperature differences, and system inefficiencies. To minimize exergy losses and enhance system performance, several strategies can be employed. One approach is to optimize the design and operation of the solar collector to maximize solar absorption and minimize heat losses. Insulation and glazing materials with low thermal conductivity can reduce heat losses from the collector. Energy and exergy analyses are important to precisely assess the performance of solar collectors, solar dryer, solar cooker, solar refrigeration, etc. to identify the unforeseen opportunities of system improvement.

REFERENCES

Abi Mathew, A., Thangavel, V., 2021. A novel thermal storage integrated evacuated tube heat pipe solar air heater: Energy, exergy, economic and environmental impact analysis. Solar Energy, 220, 828–842. https://doi.org/10.1016/j.solener.2021.03.057

Abozoor, M.K.S., Meraj, M., Azhar, M., Khan, M.E., Seraj, M., Ahsan, M., Ahmed, S.A., Bani Hani, E.H., 2022. Energy and exergy analyses of active solar still integrated with evacuated flat plate collector for New Delhi. Groundwater for Sustainable Development, 19, 100833. https://doi.org/10.1016/J.GSD.2022.100833

Abuska, M., Sevik, S., 2017. Energy, exergy, economic and environmental (4E) analyses of flat-plate and V-groove solar air collectors based on aluminium and copper. Solar Energy, 158, 259–277.

Akbulut, A., Durmuş, A., 2010. Energy and exergy analyses of thin layer drying of mulberry in a forced solar dryer. Energy, 35, 1754–1763.

Alawi, O.A., Kamar, H.M., Abdelrazek, A.H., Mallah, A.R., Mohammed, H.A., Ibrahim Abdulla, A., Alwan Gatea, H., Khiadani, M., Kazi, S.N., Yaseen, Z.M., 2022. Hydrothermal and energy analysis of flat plate solar collector using copper oxide nanomaterials with different morphologies: Economic performance. Sustainable Energy Technologies and Assessments, 49, 101772. https://doi.org/10.1016/J.SETA.2021.101772

Aman, J., Ting, D.S.K., Henshaw, P., 2013. Residential solar air conditioning: Energy and exergy analyses of an ammonia–water absorption cooling system. Applied Thermal Engineering, 61, 424–432.

Angappan, G., Pandiaraj, S., Alruabie, A.J., Muthusamy, S., Said, Z., Panchal, H., Katekar, V.P., Shoeibi, S., Kabeel, A.E., 2023. Investigation on solar still with integration of solar cooker to enhance productivity: Experimental, exergy, and economic analysis. Journal of Water Process Engineering, 51, 103470. https://doi.org/10.1016/J.JWPE.2022.103470

Aquilanti, A., Tomassetti, S., Coccia, G., Muccioli, M., Di Nicola, G., 2023. Experimental characterization and performance comparison of four prototypes of panel solar cooker for low to high sun elevations. Journal of Cleaner Production, 390, 136158. https://doi.org/10.1016/J.JCLEPRO.2023.136158

Bayrak, F., Oztop, H.F., Hepbasli, A., 2013. Energy and exergy analyses of porous baffles inserted solar air heaters for building applications. Energy Building, 57, 338–345. https://doi.org/10.1016/j.enbuild.2012.10.055

Bennour, F., Mzad, H., 2022. A numerical study of the double-glazed solar air heater (DG-SAH) performance using optimized gas-filled interspaces. Case Studies in Thermal Engineering, 40, 102492. https://doi.org/10.1016/J.CSITE.2022.102492

Buck, A.L., 1981. New equations for computing vapor pressure and enhancement factor. Journal of Applied Meteorology and Climatology, 20(12), 1527–1532. https://doi.org/10.1175/1520-0450(1981)020<1527:NEFCVP>2.0.CO;2

Chamoli, S., Thakur, N.S., 2014. Exergetic performance evaluation of solar air heater having V-down perforated baffles on the absorber plate. Journal of Thermal Analysis and Calorimetry, 117, 909–923.

Chowdhury, M.M.I., Bala, B.K., Haque, M.A., 2011. Energy and exergy analysis of the solar drying of jackfruit leather. Biosystem Engineering 110, 222–229.

Churchill, S.W., Chu, H.H.S., 1975. Correlating equations for laminar and turbulent free convection from a vertical plate. International Journal of Heat and Mass Transfer, 18(11), 1323–1329. https://doi.org/10.1016/0017-9310(75)90243-4

Cuce, E., Cuce, P.M., 2013. A comprehensive review on solar cookers. Applied Energy, 102, 1399–1421. https://doi.org/10.1016/J.APENERGY.2012.09.002

Duffie, J.A., Beckman, W.A., Blair, N. 2020. Solar Engineering of Thermal Processes, Photovoltaics and Wind. John Wiley & Sons, Inc.

Ekka, J.P., Bala, K., Muthukumar, P., Kanaujiya, D.K., 2020. Performance analysis of a forced convection mixed mode horizontal solar cabinet dryer for drying of black ginger (*Kaempferia parviflora*) using two successive air mass flow rates. Renewable Energy, 152, 55–66.

Farhan, A.A., Issam M. Ali, A., Ahmed, H.E., 2021. Energetic and exergetic efficiency analysis of a v-corrugated solar air heater integrated with twisted tape inserts. Renewable Energy, 169, 1373–1385.

Gao, M., Fan, J., Furbo, S., Xiang, Y., 2022. Energy and exergy analysis of a glazed solar preheating collector wall with non-uniform perforated corrugated plate. Renewable Energy, 196, 1048–1063. https://doi.org/10.1016/J.RENENE.2022.07.026

Getnet, M.Y., Gunjo, D.G., Sinha, D.K., 2023. Experimental investigation of thermal storage integrated indirect solar cooker with and without reflectors. Results in Engineering, 18, 101022. https://doi.org/10.1016/j.rineng.2023.101022

Gholami, A., Ajabshirchi, Y., Ranjbar, S.F., 2019. Thermo-economic optimization of solar air heaters with arcuate-shaped obstacles. Journal of Thermal Analysis and Calorimetry, 138, 1395–1403. https://doi.org/10.1007/s10973-019-08273-x

Goyal, R.K., Eswaramoorthy, M., 2023a. Modeling and analysis of compound parabolic concentrator integrated box type solar cooker. Materials Today: Proceedings, 72, 1047–1055. https://doi.org/10.1016/J.MATPR.2022.09.160

Goyal, R.K., Eswaramoorthy, M., 2023b. Thermal performance enhancement on a box-type solar cooker using a triangular fin over a conventional cooking pot. Solar Energy, 258, 339–350. https://doi.org/10.1016/j.solener.2023.03.053

Haloui, H., Touafek, K., Khelifa, A., 2022. Modeling of a hybrid thermal photovoltaic (PVT) collector based on thin film organic solar cells. Electric Power Systems Research, 210, 108131. https://doi.org/10.1016/J.EPSR.2022.108131

Henein, S.M., Abdel-Rehim, A.A., El-Nagar, K., 2023. Energy, economic and environmental analysis of an evacuated tube solar collector using hybrid nanofluid. Applied Thermal Engineering, 219, 119671. https://doi.org/10.1016/J.APPLTHERMALENG.2022.119671

Hidalgo, L.F., Candido, M.N., Nishioka, K., Freire, J.T., Vieira, G.N.A., 2021. Natural and forced air convection operation in a direct solar dryer assisted by photovoltaic module for drying of green onion. Solar Energy, 220, 24–34.

Hollands, K.G.T., Unny, T.E., Raithby, G.D., Konicek, L. 1976. Free convective heat transfer across inclined air layers. International Journal of Heat and Mass Transfer, 39, 3925–3930.

Joseph, A., Thomas, S., 2022. Energy, exergy and corrosion analysis of direct absorption solar collector employed with ultra-high stable carbon quantum dot nanofluid. Renewable Energy, 181, 725–737. https://doi.org/10.1016/J.RENENE.2021.09.079

Kareem, M.W., Habib, K., Sopian, K., Ruslan, M.H., 2017. Multi-pass solar air heating collector system for drying of screw-pine leaf (*Pandanus tectorius*). Renewable Energy, 112, 413–424. https://doi.org/10.1016/j.renene.2017.04.069

Karthikeyan, A.K., Murugavelh, S., 2018. Thin layer drying kinetics and exergy analysis of turmeric (*Curcuma longa*) in a mixed mode forced convection solar tunnel dryer. Renewable Energy, 128, 305–312.

Koshti, B., Dev, R., Bharti, A., Narayan, A., 2023. Comparative performance evaluation of modified solar cookers for subtropical climate conditions. Renewable Energy, 209, 505–515. https://doi.org/10.1016/j.renene.2023.04.021

Lakshmi, D.V.N., Muthukumar, P., Layek, A., Kumar, P., 2019. Performance analyses of mixed mode forced convection solar dryer for drying of stevia leaves. Solar Energy, 188, 507–518.

Lamrani, B., Elmrabet, Y., Mathew, I., Bekkioui, N., Etim, P., Chahboun, A., Draoui, A., Ndukwu, M.C., 2022. Energy, economic analysis and mathematical modelling of mixed-mode solar drying of potato slices with thermal storage loaded v-groove collector: Application to Maghreb region. Renewable Energy, 200, 48–58. https://doi.org/10.1016/J.RENENE.2022.09.119

Li, Y., Teng, S., Xi, H., 2023. 3E analyses of a cogeneration system based on compressed air energy storage system, solar collector and organic Rankine cycle. Case Studies in Thermal Engineering, 42, 102753. https://doi.org/10.1016/J.CSITE.2023.102753

Lingayat, A., Chandramohan, V.P., Raju, V.R.K., 2020. Energy and exergy analysis on drying of banana using indirect type natural convection solar dryer. Heat Transfer Engineering, 41, 551–561.

Mahapatra, A., Tripathy, P.P., 2019. Thermal performance analysis of natural convection solar dryers under no load condition: Experimental investigation and numerical simulation. International Journal of Green Energy, 16, 1448–1464.

Matheswaran, M.M., Arjunan, T.V., Somasundaram, D., 2019. Energetic, exergetic and enviro-economic analysis of parallel pass jet plate solar air heater with artificial roughness. Journal of Thermal Analysis and Calorimetry, 136, 5–19. https://doi.org/10.1007/s10973-018-7727-4

Mohammed, S., Edna, M., Siraj, K., 2020. The effect of traditional and improved solar drying methods on the sensory quality and nutritional composition of fruits: A case of mangoes and pineapples. Heliyon, 6, e04163.

Morad, M.M., El-Shazly, M.A., Wasfy, K.I., El-Maghawry, H.A.M., 2017. Thermal analysis and performance evaluation of a solar tunnel greenhouse dryer for drying peppermint plants. Renewable Energy, 101, 992–1004.

Murugan, M., Saravanan, A., Elumalai, P.V., Kumar, P., Ahamed Saleel, C., Samuel, O.D., Setiyo, M., Enweremadu, C.C., Afzal, A., 2022. An overview on energy and exergy analysis of solar thermal collectors with passive performance enhancers. Alexandria Engineering Journal, 61(10), 8123–8147. https://doi.org/10.1016/J.AEJ.2022.01.052

Petela, R., 2003. Exergy of undiluted thermal radiation. Solar Energy, 74, 469–488.

Rabha, D.K., Muthukumar, P., Somayaji, C., 2017. Energy and exergy analyses of the solar drying processes of ghost chilli pepper and ginger. Renewable Energy, 105, 764–773.

Raj, A.K., Kunal, G., Srinivas, M., Jayaraj, S., 2019. Performance analysis of a double-pass solar air heater system with asymmetric channel flow passages. Journal of Thermal Analysis and Calorimetry, 136, 21–38. https://doi.org/10.1007/s10973-018-7762-1

Ramani, B.M., Gupta, A., Kumar, R., 2010a. Performance of a double pass solar air collector. Solar Energy, 84(11), 1929–1937. https://doi.org/10.1016/j.solener.2010.07.007

Ramani, B.M., Gupta, A., Kumar, R., 2010b. Performance of a double pass solar air collector. Solar Energy, 84(11), 1929–1937. https://doi.org/10.1016/J.SOLENER.2010.07.007

Rani, P., Tripathy, P.P., 2021. Drying characteristics, energetic and exergetic investigation during mixed-mode solar drying of pineapple slices at varied air mass flow rates. Renewable Energy, 167, 508–519. https://doi.org/10.1016/j.renene.2020.11.107

Sahu, M.K., Prasad, R.K., 2016. Exergy based performance evaluation of solar air heater with arc-shaped wire roughened absorber plate. Renewable Energy, 96, 233–243.

Said, Z., Ghodbane, M., Boumeddane, B., Tiwari, A.K., Sundar, L.S., Li, C., Aslfattahi, N., Bellos, E., 2022. Energy, exergy, economic and environmental (4E) analysis of a parabolic trough solar collector using MXene based silicone oil nanofluids. Solar Energy Materials and Solar Cells, 239, 111633. https://doi.org/10.1016/J.SOLMAT.2022.111633

Shukuya, M., Hammache, A., 2002. Introduction to the concept of exergy. Low Exergy Systems for Heating and Cooling of Buildings. IEA Annex37, pp. 41–44.

Sivakumar, S., Siva, K., Mohanraj, M., 2019. Experimental thermodynamic analysis of a forced convection solar air heater using absorber plate with pin-fins. Journal of Thermal Analysis and Calorimetry, 136, 39–47.

Tahir, M.F., Haoyong, C., Guangze, H., 2021. A comprehensive review of 4E analysis of thermal power plants, intermittent renewable energy and integrated energy systems. Energy Reports, 7, 3517–3534. https://doi.org/10.1016/j.egyr.2021.06.006

Tiwari, G.N., Shyam, A.T. 2016. Energy Systems in Electrical Engineering Handbook of Solar Energy: Theory, Analysis and Applications, Springer, Singapore.

Tiwari, G.N., Tiwari, A., Shyam, 2016. Solar Water-Heating Systems.

Tiwari, S., Tiwari, G.N., 2017. Energy and exergy analysis of a mixed-mode greenhouse-type solar dryer, integrated with partially covered *n*-PVT air collector. Energy, 128, 183–195. https://doi.org/10.1016/j.energy.2017.04.022

Vijayan, S., Arjunan, T.V., Kumar, A., 2020. Exergo-environmental analysis of an indirect forced convection solar dryer for drying bitter gourd slices. Renewable Energy, 146, 2210–2223.

Wenceslas Koholé, Y., Cyrille Vincelas Fohagui, F., Tchuen, G., 2022. Flat-plate solar collector thermal performance assessment via energy, exergy and irreversibility analysis. Energy Conversion and Management: X, 15, 100247. https://doi.org/10.1016/J.ECMX.2022.100247

Wu, Y., Tong, X., Li, D., Arıcı, M., Liu, C., Liu, Y., Yang, R., Yu, Y., 2022. Energy analysis of evacuated tube solar collector integrating phase change material in northeast China. Journal of Energy Storage, 55, 105772. https://doi.org/10.1016/J.EST.2022.105772

Yahya, M., Fudholi, A., Sopian, K., 2017. Energy and exergy analyses of solar-assisted fluidized bed drying integrated with biomass furnace. Renewable Energy, 105, 22–29. https://doi.org/10.1016/j.renene.2016.12.049

11 Numerical Modeling and Simulations of Renewable Energy-Based Food-Drying Technologies

Eshetu Getahun
Faculty of Chemical and Food Engineering, Bahir Dar Institute of Technology, Bahir Dar University, Po. Box 26, Bahir Dar, Ethiopia
Bahir Dar Energy Center, Bahir Dar Technology Institute, Bahir Dar University, Po. Box 26, Bahir Dar, Ethiopia

Mulugeta A. Delele
Faculty of Chemical and Food Engineering, Bahir Dar Institute of Technology, Bahir Dar University, Po. Box 26, Bahir Dar, Ethiopia
Department of System Process Engineering, Leibniz Institute for Agricultural Engineering and Bioeconomy, Potsdam, Germany

11.1 INTRODUCTION

One of the challenges in the globe is the continuous increase of human population growth over the optimum utilization of inadequate energy and natural resources. Food security is one of the ultimate prominent inspirations driven by one primitive endurance instinct. Thus, a full understanding of global population growth and the expected energy and food demands to come is critical for sustainable development. According to the United Nations population projection, the world population is expected to reach 10.9 billion in 2100 with high natural resource demand such as energy and food (United Nations, 2019). Energy and food are the important driving factors for the survival of human beings. To sustain the balance between food supply and population growth, food losses during postharvesting should be reduced by the application of appropriate technologies. The amount and quality of food products are highly affected by poor processing methods and preservation mechanisms. One-third of the food produced globally is lost in the postharvest stage (Joardder and Masud, 2019). Concerning this, many reports indicated that the postharvest losses of fruits and vegetables in developing countries are 30–40% (El-Sebaii and Shalaby, 2012). Thus, the preservation of food products is very essential for extending the shelf life by retarding quality deterioration of the products.

Cold storage is the best way to conserve the nutritional value of fresh fruit and vegetable products amid food preservation technologies; however, most cold storage techniques require low temperatures, which are very difficult to sustain throughout the delivery chain of the products (Sagar and Suresh Kumar, 2010). On the other way, drying is the most appropriate option to preserve food products for longer time, minimize the postharvest loss, preserve quality attributes, increase shelf life, and reduce transport weight and cost. Removing moisture leads to slowing down the activity of bacteria, enzymes, yeasts, and molds that belittle the quality of products. Table 11.1 presents the

DOI: 10.1201/9781003344087-11

TABLE 11.1
Details of Moisture Content of the Various Food Commodities

Food Commodities	Initial Moisture Content (wb%)	Final Moisture Content (wb%)	Maximum Allowable Temperature (°C)
Paddy, raw	22–24	11	50
Paddy, parboiled	30–35	13	50
Maize	35	15	60
Wheat	20	16	45
Corn	24	14	50
Rice	24	11	50
Pulse	20–22	9–10	40–60
Oilseed	20–25	7–9	40–60
Green peas	80	5	65
Cauliflower	80	6	65
Carrot	70	5	75
Green beans	70	5	75
Onion	80	4	55
Garlic	80	4	55
Cabbage	80	4	55
Sweet potato	75	7	75
Chilies	80	5	65
Potatoes	75	7	75
Apricot	85	18	65
Apple	80	24	70
Grapes	80	15–20	70
Bananas	80	15	70
Guavas	80	7	65
Okra	80	20	65
Pineapple	80	10	65
Tomatoes	96	10	60
Brinjal	95	6	60

Source: Getahun et al. (2021a).

drying temperature, moisture content, and the safe limit of the various food commodities (Getahun et al., 2021a; Prakash and Kumar, 2014).

Drying is one of the appropriate preservation techniques for food products which avoids the growth and replica of microorganisms and results in the decline of many moisture-mediated deteriorative reactions (Calado et al., 2014; Dincer and Hussain, 2002; Xiao et al., 2015). The cause of food quality deterioration and health risks is the growth of microorganisms in the food. Drying also brings about a considerable drop in weight and volume and reduces the cost of packaging, storage, and transportation.

11.1.1 Principles of Drying

Drying is a complex phenomenon that takes place through simultaneous and coupled momentum, heat, and mass transfer; for instance, for convective-type dryers, moisture is taken away from a solid product by the surrounding hot air flowing around it. These two processes (mass and heat transfer) can be divided into an internal phase and an external phase. Heat is provided from the energy source to the food product surface in the first phase, and moisture is transported from the product to the external environment. The second phase comprises heat transfer from the lateral surface to the core of the food product, and internal mass/moisture transfer from the core to the external boundary. Various heat transfer modes may occur during the drying process, namely, convection, conduction, and radiation. Convective and conductive drying is the most widely implemented type in both traditional food and industrial sectors. Radiation is a cutting-edge method for producing high-quality dry food products, but it remains costly and inapplicable in emergent countries.

The drying of wet porous materials behaves differently and depends on moisture content, physical nature, and operating conditions. Internal moisture is found in the form of either bound or unbound. The liquid, which is found in the microcapillaries on solid surfaces is bound to moisture, and it has lower vapor pressure compared to water at the same temperature. The surplus of bound water is defined as unbounded moisture. During the initial drying stage, unbound or free water on the solid surface vaporizes rapidly, resulting in rapid moisture removal, as shown in Figure 11.1. After the first stage, shrinkage might start, and the drying rate becomes constant.

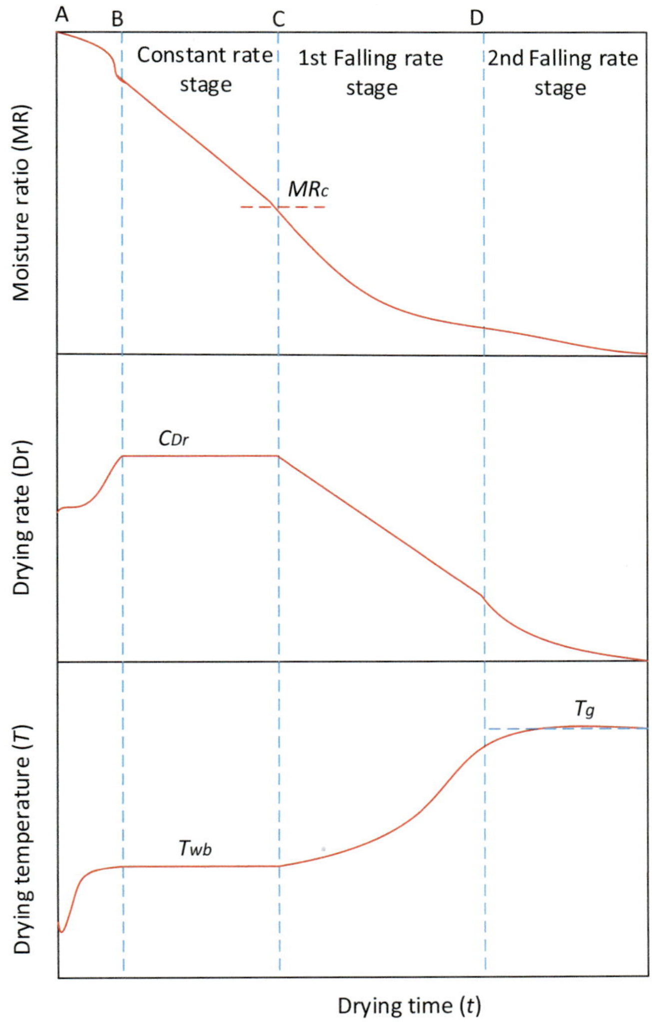

FIGURE 11.1 Drying curve of food products that shows the constant rate and falling rate periods. (Getahun et al., 2021a.)

This stage's rate-controlling steps are water vapor diffusion through the air–moisture interface and surface diffusion removal (Hashim et al., 2014). The cellular-bound water is transferred to the surface by capillary forces (Erbay and Icier, 2010). The next stage is the falling rate period, which influences the drying time. Applying additional heat accelerates the drying process. However, low humidity and high temperatures during drying may form hard shells on the surface, damaging the dried food, particularly soft and heat-sensitive food products. In general, the change in the moisture ratio, wet bulb temperature, and drying rate versus time is portrayed in Figure 11.1 as a main drying principle.

When the temperature of food increases, the drying rate also increases (A to B) until the surface temperature sustains equilibrium (B to C). The falling rate period is highly controlled through diffusion of moisture from the interior to the surface. Furthermore, convective drying operations rely on external parameters, mainly air velocity, temperature, relative humidity, and pressure. All these parameters must be controlled and optimized to enhance final dried food quality. Moreover, different chemical and physical changes, such as crystallization, shrinkage, nutritional compound loss, and puffing, occur during the drying process and may affect product quality. Thus, the main concern in the drying operation is the determination of appropriate drying conditions without minimal change in the quality of the dried products. Accurate investigation of moisture transfer parameters such as moisture diffusivity, drying rate, and mass transfer coefficient through appropriate numerical drying models plays a vital role to get high-quality dried food products.

Most of the drying studies mostly focused on experimental studies and analyzing the dryer performance. There are also enormous research outputs using numerical methods based on heat and mass transfer at a microlevel. The numerical modeling based on microlevel indicates that the modeling and simulations are performed on product level, and gave little attention to the performance of the overall drying chamber. However, the overall performance of a dryer has a great impact on the drying rate and kinetics, consequently determining the product quality (Curcio et al., 2016, Getahun et al., 2021a). The dried food product quality is determined indirectly through the sustaining of a uniform distribution of drying conditions in the dryer chamber. The quality of the food product is very sensitive to drying parameters such as drying air velocity, temperature, and other drying conditions. Thus, the optimization of these parameters through rigorous numerical modeling and simulation methods especially using the CFD model is very important. The CFD modeling and simulation methods are tremendously significant to develop efficient sustainable dryers, by analyzing and predicting the performance of drying systems that preserve the quality of food products. Since drying is considered as a combined multiphysics, multiscale, and multiphase problem, which is a very complex process, rigorous numerical modeling is very indispensable to explicitly define the drying process and behavior in terms of drying rate, drying uniformity, drying time, dryer design, and quality attributes preservation. Therefore, this chapter emphasizes on the application of numerical modeling and simulation in the design and operation of sustainable food product drying systems.

11.1.2 Conventional Drying Technologies

Different types of drying technologies are utilized for the drying of food products. Food products are commonly dried using convective-type dryers that use hot air as a drying medium. Sun and solar drying are the other types of drying techniques that are practiced in the drying of significant amount of food products. The advancement of novel fruit and vegetable drying technologies, which preserve product quality at low energy consumption, has progressively increased. Such novel dryers mainly used osmotic, vacuum, microwave, ultrasound, spray, freeze, and fluidized bed drying techniques (Getahun et al., 2021a).

Nonthermal drying processes such as pulsed electric, ultrasound, and ultraviolet radiation may cause a temperature change inside the product without generating heat within the product. This process results in lower food quality deterioration due to low drying temperature as compared to the thermal drying process (Raso and Barbosa-Cánovas, 2003). The nonthermal drying processes have the advantage of reduced drying time, increased processing efficiency, and enhanced quality attributes. However, some of the advanced drying technologies have their limitations. Khin et al. (2005) showed that freeze-drying, vacuum drying, osmosis, ultrasound, and dehydration are highly expensive for the large-scale drying process, especially in developing countries.

The sustainable provision of energy for food product drying technology is very important. There are several sources of energy such as fossil fuel (natural gas, coal, and oil), and renewable energy (solar, biomass, geothermal, and others), which can be used for drying food products. Drying is a simultaneous heat and mass transfer phenomenon that removes water molecules from food cells structure through the use of heat (Getahun et al., 2021a). However, the drying process needs a huge amount of thermal energy due to the high latent heat needed for water evaporation. The convective drying techniques that are widely used around the world require both electrical energy and mechanical input (Masud et al., 2020). The existing industrial drying operation consumes about 20–25% of the energy utilized in the food processing operation since drying 1 kg of food needs about 14.53–90 MJ energy (Ananno et al., 2020). Food drying energy should account for almost 15% of the overall industrial energy consumption (Catorze et al., 2022). Other reports indicated that drying requires about 40–50% of the energy out of total energy needed for the processing and production of food products (Bennett et al., 2007). Hence, drying is an energy-intensive operation, which results in an increase in greenhouse gas emissions (GHG) and evolves huge environmental concerns. Drying operations cause significant environmental pollution since the energy source is mostly coming from fossil fuel.

Thus, it is imperative to utilize technological advancement in line with numerical modeling to develop drying systems that integrate renewable and sustainable energy sources to minimize the postharvest loss of food products.

11.2 MATHEMATICAL MODELING APPROACHES OF SUSTAINABLE DRYER TECHNOLOGIES

Nowadays, a wide variety of worthwhile mathematical modeling techniques have been developed to design, evaluate, and control different drying systems. Mathematical modeling is an alternative to experimental techniques to predict the performance of different types of food product drying processes. Mathematical modeling is contributing a lot in improving the drying systems by maximizing the drying performance and improving the final product quality. Commonly, drying operation is modeled using two approaches:

- Mechanistic models (based on fundamental principles of physics, chemistry, biology, and mathematics)
- Empirical and semiempirical models (based on experimental data and dimensionless analysis)

Mechanistic models integrated fundamental physics involved in the drying operations, mostly expressed through the coupling of fundamental transport equations with phase equilibrium expressions and thermodynamically interactive fluxes. Mechanistic models involve a number of parameters and high mathematical complexity. Advanced mechanistic models mainly used theoretical models to understand fully the food drying behaviors by involving several coupled mechanisms and source terms in the drying process. Computational fluid dynamics (CFD) of dryers is one of the mechanistic models. Empirical and semiempirical models are models that are developed based on experimentally collected data.

Mathematical modeling approach comprises several steps, as summarized in Figure 11.2, beginning with conceptualization and establishing the transport phenomena equations, thermophysical properties, solver type, and discretization techniques to illustrate actual operating systems.

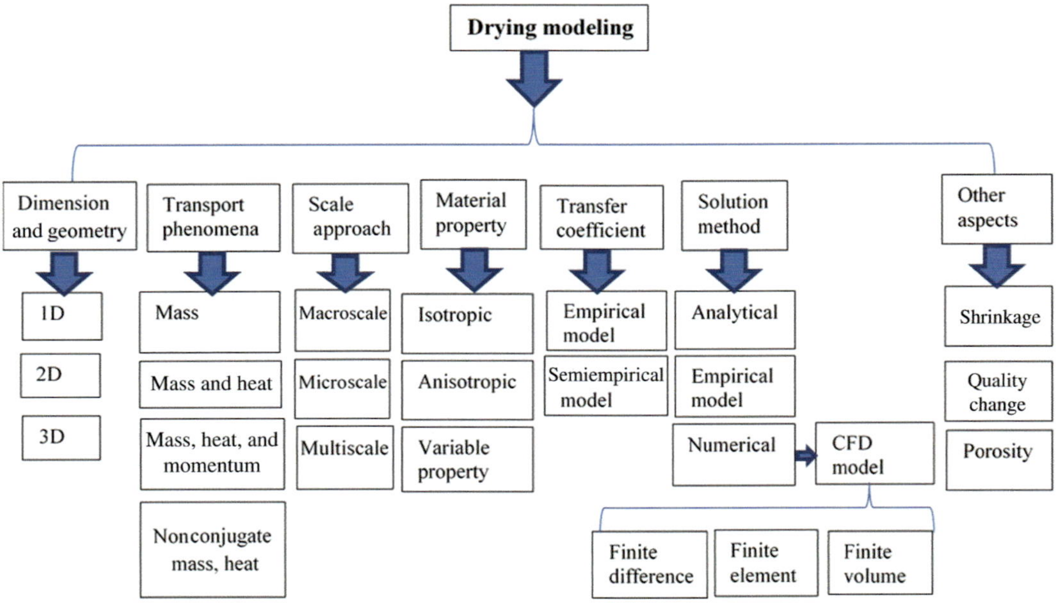

FIGURE 11.2 Common approaches in the modeling of sustainable food products drying systems.

11.2.1 Empirical and Semi-empirical Models

Empirical models for food product drying are essential to understand better the behavior of mass and heat transfer mechanisms that govern the drying operation. It also helps to perform further research to optimize the drying design and operating conditions and maximize the preservation of product quality attributes.

11.2.1.1 Moisture Sorption Isotherm Model

A full understanding of the behavior of the moisture sorption isotherm of food products is important for the design and optimization of the drying process, packaging, storage, and shelf life of the product and to control moisture migration of the products (Getahun et al., 2020a, Fasina et al., 1997, Getahun et al., 2021a). The important parameters such as texture, color, drying rate, storage life, and structure of dried food products also depend on the sorption characteristics (Mehta and Singh, 2006). There are different types of sorption models to describe the sorption behavior of food products. For example, different chili pepper varieties throughout the world have been described using different moisture sorption isotherm models (Table 11.2).

11.2.1.2 Thin-Layer Drying Model

Thin-layer drying model is expressed as the drying of a sample particle of one layer or slices. Due to its thin structure, the distribution of drying conditions like temperature can be easily expected as uniform throughout the product, which is very appropriate for lumped heat transfer models. The use of simple mathematical drying model equations to examine the water transport phenomenon is an important alternative to experimental methods. The importance of thin-layer drying models is very significant due to their easiness of use and require fewer data compared to complex distributed models such as coupling and phenomenological coefficients (Getahun et al., 2021c). There are empirical and semi-empirical thin-layer drying models that have been developed based on diffusion and Newton's law of cooling (Getahun, et al, 2021a), as shown in Table 11.3. These simple models are mostly considered lumped parametric models and explain the heat

TABLE 11.2
Chili Pepper Varieties, Their Sorption Model, and Type for Different Countries

Isotherm	Chili Pepper Varieties	Best Model	Sorption Type	Monolayer Moisture	Country
Desorption	Lamuyo	BET	Type II	0.072–0.100	Spain
	Black peppercorns	GAB	Type III	0.030–0.050	Sri Lanka
	Sweet peppers	OSWIN	Type II	0.030–0.040	India
	Polyester	Modified OSWIN	Type II	0.190	India
	Bursa	Halsey	Type II	0.040–0.101	Turkey
	Aji	Andrade	Type II	0.270	Brazil
	Bico	GAB	Type II	0.060–0.080	Brazil
Adsorption	Lamuyo	BET	Type II	0.070–0.10	Spain
	Black peppercorns	GAB, Peleg	Type III	0.040–0.050	Sri Lanka
	Sweet peppers	OSWIN, Halsey	Type II	0.100–0.110	India
	Polyester	Modified Halsey	Type II	0.290	India
	Bursa	Halsey	Type II	0.050–0.080	Turkey
	CH-1	Modified Oswin	Type II	0.210	India

Source: Getahun et al. (2021a).

and mass transfer rate for the whole product ignoring the internal resistance for mass and heat transfer (Getahun et al., 2021c). They strongly depend on the experimental conditions and give insufficient data about the drying behaviors of the food produce (Erbay and Icier, 2010; Khoshhal et al., 2010).

Empirical models are simple and useful for describing drying kinetics. They simulate the drying curve and therefore predict the drying results, but they have little physical significance than theoretical models (Getahun et al., 2021c). These models neglect food mass diffusivity/conductivity

TABLE 11.3
Different Thin-Layer Mathematical Models for the Drying of Fruits and Vegetables

Model Name	Model Equation
Newton	$MR = \exp(-kt)$
Page	$MR = \exp(-kt^n)$
Modified page	$MR = \exp(-(kt)^n)$
Henderson and Pabis	$MR = \alpha \exp(-kt)$
Modified Henderson and Pabis	$MR = \alpha \exp(-kt) + b\exp(-gt) + c\exp(-ht)$
Logarithmic	$MR = \alpha \exp(-kt) + c$
Two-term	$MR = \alpha \exp(-k_o t) + b\exp(-k_1 t)$
Two-term exponential	$MR = \alpha \exp(-kt) + (1-\alpha)\exp(-k\alpha t)$
Wang and Singh	$MR = 1 + \alpha t + bt^2$
Diffusion approach	$MR = \alpha \exp(-kt) + (1-\alpha)\exp(-kbt)$
Midilli and Kucuk	$MR = \alpha \exp(-kt^n) + bt$

Source: Getahun et al. (2021a); Getahun et al. (2021c).

shrinkage and geometry. To characterize the dynamic drying behavior of food materials, it is vital to choose a model that can grasp the moisture diffusion mechanism throughout the drying operation and properly fit the experimental moisture ratio curve. The moisture content data are transformed into the expressions of the moisture content ratio (*MR*), and then mathematical models are subjected to a curve-fitting calculation with the drying time (t). The other parameters (k, n, α, b, and c) are model constants.

Theoretical models are developed using the fundamental physics principle in the drying operations, primarily expressed through the coupling of fundamental transport equations with thermodynamically interactive fluxes and then phase equilibrium expressions. Commonly, these models involve an excessive number of parameters and a high mathematical complexity.

In general, the Page model is most suitable for describing the drying behavior of various vegetables and fruits as compared to the other thin-layer drying models (Getahun et al., 2021c). Detail information about the pros and cons of each model and the derivation mechanism of each empirical model is found in Erbay and Icier's (2010) and Ertekin and Firat's (2017) studies.

11.2.2 Mechanistic Drying Model

Drying is a very complex phenomenon and is affected by different external factors, in addition to the internal factors. For this reason, the mechanistic drying model is very essential to integrate the external and internal mass transfer resistances.

11.2.2.1 Biot Number–Based Model

The mechanistic drying models simultaneously involve both internal and external diffusion mechanisms using appropriate correlations mechanism such as Bi-lag (Dincer and Hussain, 2004), Biot-Reynolds (Bi-Re) (Dincer et al., 2002), and Biot-Dincer (Bi-Di) (Dincer and Hussain, 2002).

The mechanistic drying models efficiently describe the moisture transfer mechanisms for the drying of food products as compared to theoretical and empirical thin-layer drying models (Getahun et al., 2021b). The mechanistic drying model uses the drying coefficient (S) that relates the drying capability of a food product to drying time and it is highly affected by relative humidity and the lag factor (G) that relates the internal resistance of a food product to the moisture transfer (Getahun et al., 2021b). The dimensionless moisture content (MR) in terms of an exponential form using the drying coefficient (S) and the lag factor (G) is given as follows:

$$\mathrm{MR} = G\exp(-St) \tag{11.1}$$

where t is the drying time.

11.2.2.2 CFD Modeling

CFD is the discipline of science and engineering that is devoted to predicting the characteristics of heat transfer, mass transfer, fluid flow distribution, chemical reactions, and related phenomena by solving the mathematical models that govern these processes using numerical computations. The user-friendly interface of the computer simulation tool allows for evaluating the performance of the drying operation with a visualization of flow throughout the process. Data from CFD simulation is analyzed and applied to improve the drying efficiency and product quality and also used to design a new optimized drying system. CFD results give better reliability, higher productivity, more assured scale-up, and the ability to investigate hazardous conditions with a reduced cost, workload, and drying space compared to laboratory experiments. CFD modeling is a numerical computation and the governing equations are mainly the continuity (conservation of mass), Navier–Stokes (balance of momentum), and energy (conservation of energy) and mass transfer equations, as indicated in Equations 11.2–11.6 (Defraeye and Radu, 2017, Getahun et al., 2021a). Numerical methods deliver an approximate solution for the equations by discretizing the computational domain into several finite elements or grids, by transforming complex governing equations into simple polynomial or linear equations. The suitable solution method may differ depending on the information and the situation that the researchers want to get into a specific problem. The choice includes the features of mathematical models, such as the domain geometry determination, initial and boundary conditions, and the categorization of the differential equation. The most conventional numerical methods that have been effectively employed to solve those complex equations in convective drying are the finite volume method, finite differences method, and finite elements method (Demissie et al., 2019, Chandramohan, 2020; Getahun et al., 2021a).

11.2.2.2.1 Continuity Equation

$$\frac{\partial \rho}{\partial t} = \nabla(\rho \vec{u}) = 0 \tag{11.2}$$

where $\vec{u}$ and ρ are velocity and density of air.

11.2.2.2.2 Momentum Equation

$$\frac{\partial(\rho \vec{u})}{\partial t} = \nabla.(\rho \vec{u}\vec{u}) = -\nabla P + \bar{\bar{\tau}} + \rho \vec{g} + S_m \tag{11.3}$$

where $\bar{\bar{\tau}}$ and S_m are Reynolds shear stress and momentum source terms, respectively.

11.2.2.2.3 Energy Equation

$$\frac{\partial(\rho E)}{\partial t} = \nabla.(\vec{u}(\rho E + P)) = \nabla.(-\vec{q} + \bar{\bar{\tau}}\vec{u}) + S_h \tag{11.4}$$

where, S_h, $\vec{q}$, and E are source terms for heat, heat flux vector, and total energy, respectively. The total energy is the summation of internal energy and kinetic energy, which is given by

$$E = h - \frac{P}{\rho} + \frac{u^2}{2}$$

where h is the sensible enthalpy of air.

11.2.2.2.4 Mass Transfer

Water vapor is a scalar property and does not disturb the solutions of the momentum, mass, and energy of the drying air (Thorpe, 2008). The mass transport of the scalar property is, thus, solved independently with a general conservation equation:

$$\frac{\partial(\rho w)}{\partial t} = \nabla.(\rho \vec{u} w) = \nabla.(\rho D) + S_w \tag{11.5}$$

where w is the humidity ratio (kg water vapor/kg dry air) and D is the water vapor diffusion coefficient which depends on temperature, as defined by Sanghi et al. (2017). The moisture and heat source terms S_w and S_h are determined as follows:

$$S_w = \rho_p \frac{\partial M}{\partial t}, \quad S_h = \rho_p \frac{\partial M}{\partial t} L_v \tag{11.6}$$

where L_v is the water latent heat and M is the moisture content.

Moreover, the modeling of the convective mass and heat transfer of porous food products with their surroundings during the drying operation is differentiated into two main groups: conjugated and nonconjugated models. The nonconjugated method treats mass and heat transfer in the air and inside the food product independently by supposing that airflow properties are unchanged by heat and moisture transfer within the product, so the influence of internal vaporization of water on the outside fluid flow is ignored. The momentum transfer usually starts with determining velocity and temperature profiles to evaluate convective transfer coefficients (CTCs) at the food product–air interface and then simulating mass and heat transport inside the product and CTCs values are inputs to this model. Thus, a proper estimation of those coefficients is needed. The nonconjugate model study indicated that the CTCs could be considered inconsistent along the surface during the simulation of the drying process (Esfahani et al., 2014). While nonconjugate models are simple to formulate, flexible, and generally require less computational cost, conjugate models are complex and engage profound knowledge of strong computational tools and mathematical equations. In the conjugate approach, the mass and heat transport phenomena in the surrounding airflow and the porous food material region are determined simultaneously in a transient way.

The fluid flow can be considered as either turbulent or laminar flow. Several wall functions and turbulence models are usually used when the flow is turbulent, which is usually the case in modeling of food drying operations. Thus, the requirement of an accurate flow-field expectation and identifying an appropriate model is crucial and significantly impacts the CFD performance. The most familiar turbulence models stated in the open literature are k–ω, k–ε, and Reynolds stress models.

11.2.2.3 Quality Modeling Using CFD

Although technological developments in CFD have substantially contributed to modeling the drying kinetics of food products, few efforts have been made to predict the food quality evolution of dried products. Nowadays, very few studies reported on the application of kinetics of CFD modeling for food quality evaluation during the drying process.

The open literature reports indicated that the kinetic-rate-law model is widely used to determine the generic quality attribute (C) of fruits and vegetables in drying process (Defraeye, 2016; Defraeye and Radu, 2018), as shown in Equation 11.7:

$$\frac{dC}{dt} = KC^{n} \tag{11.7}$$

where n is the order of the reaction, k is the rate constant, and t is the time.

Solving this differential equation at constant temperature results in a linear reduction (for zero order) or an exponential reduction (for the first order and second order) of the quality attribute over drying time as shown in Equations 11.8–11.10:

$$C = C_o - kt \quad (\text{zero order}) \tag{11.8}$$

$$C = C_o e^{-kt} \quad (\text{first order}) \tag{11.9}$$

$$\frac{1}{C} = kt + \frac{1}{C_o} \quad (\text{second order}) \tag{11.10}$$

where C_0 is the quality attribute at the start of the drying process (t = 0 s) and C is the quality attribute at the final of the drying process. Food products quality deterioration depends on drying temperature and in this case, the rate constant k is determined as a function of temperature in the form of an Arrhenius relationship, as indicated in Equation 11.11:

$$k(T) = k_o e^{-E_A/RT} \tag{11.11}$$

where k_0 is a constant (s^{-1}), E_A is the activation energy (J/mol), R is the ideal gas constant (8.314 J/mol·K), and T is the absolute temperature (K).

11.2.3 Other Modeling Approaches

The nonlinearity, complexity, and multivariable characteristics of the drying process have triggered researchers into the involvement of advanced soft-computing technologies such as artificial intelligence (AI) methods (Sinha and Bhargav, 2022). As AI is developing quickly, considerable efforts are being used to introduce this powerful approach with a promising future. The application of AI is based on the combination of science and technology, using an intelligent, robust, and stable machine with high learning and analytical capabilities, which includes big data and cloud computing to deal with complexities and nonlinearities of ill-defined operations such as the drying of food products. AI technologies have been widely used to solve problems of nonlinear function approximation, pattern detection, data interpretation, optimization, simulation, diagnosis, control, data sorting, clustering, and noise reduction in different food drying technologies due to the advantages of self-learning ability, adaptive ability, strong fault tolerance, and high degree robustness to map the nonlinear structures of arbitrarily complex and dynamic phenomena (Sun et al., 2018). AI could be an extremely useful tool to develop "soft sensors" and for control of complex drying process (Martynenko, 2017). AI systems could be used for the calculation, design, and selection of drying equipment (Menshutina et al., 2019). The AI modeling techniques include the following:

- Artificial neural network (ANN) modeling
- Adaptive-network-based fuzzy inference system (ANFIS) modeling
- Fuzzy modeling

These AI models can control drying parameters, optimize the process, improve the quality of dried products, and perform energy analysis. ANN modeling is a part of intelligence and statistical paradigms inspired by the functional behavior of the human brain neural network. It has been reasonably applied to establish relationships between input and output data to solve complex problems of growing interrelated variables due to its excellent information-processing capabilities. These include clustering, nonlinear function approximation, noise reduction, high parallelism, pattern recognition, and control (Liu et al., 2020). It has been reported that ANN approach can efficiently estimate the moisture distribution along the depth in a fixed-bed dryer for drying of grains and hay (Farkas, 2013). Fuzzy modeling converts linguistic description into an algorithm whose result is an action. In this model, different variables could describe fuzzy subsets of the system, which results in necessitating the division of the system into parts. The modeling methodology has been well-known widely in numerous real-world applications to describe the complex nonlinear system with an uncertain situation and imprecision (Novák et al., 2016). ANFIS combines the computational capabilities of Takagi–Sugeno-type fuzzy inference systems and ANN, approaching highly accurate and efficient (Gill et al., 2020).

11.3 APPLICATION OF NUMERICAL MODELING IN DIFFERENT RENEWABLE ENERGY–BASED DRYER TECHNOLOGIES

Supplying sustainable, environment-friendly, and low-cost energy resources is a global challenge that directly influences economic and industrial growth. Clean energy sources have recently become a major concern due to belittling fossil fuel reserves, volatile crude oil prices, and high environmental pollution associated with human health risks and global warming. Hence, renewable energy–based food product dryers are exceptional alternatives. Some of the renewable energy-based dryer technologies and the application of numerical modeling on those dryers are discussed in detail in the subsequent sections.

11.3.1 Solar Energy–Based Dryer

Sun drying has been used since ancient times for the dehydration of food produces. Open sun drying is the simplest and cheapest method of drying food products. Although open-sun drying requires little workforce and capital, it has an enormous limitation on the quality of the products (Sahdev, 2014). Therefore, researchers have improved open-sun drying through the application of solar drying technologies using the same abundant energy source. In solar drying, the important heat to remove moisture from the product is mostly obtained from solar energy. The air mass flow can be either natural or forced convection. The heating of the product occurs by either the direct absorption of solar energy by the product or through the passage of hot air within the product or a combination of both (Ekechukwu and Norton, 1999). Solar dryers may be categorized as direct solar dryers, indirect solar dryers, and hybrid or mixed solar dryers (Getahun et al., 2021a).

Several advanced numerical modeling and simulation approaches for food solar dryer systems have been examined so far by different researchers. Those advanced modeling and simulation approaches were based on the microscale level excluding the overall solar dryer modeling and simulation and mainly focused on the food item drying. In this scenario, there were various studies which were conducted to investigate the optimum drying parameters (Getahun et al., 2021a). However, the quality of dried food depends also on the type of dryer, humidity, temperature, pressure, and velocity distribution at each position in the solar dryer chamber. Uncontrolled solar drying deteriorates the nutritional value of food products, thus advanced numerical modeling and simulation of solar dryers is an important approach to optimize the drying conditions and product quality.

Amanlou and Zomorodian (2010) developed a new cabinet fruit dryer with a side-mounted plenum chamber using seven geometrical configurations (Figure 11.3) with the application of CFD to examine the distribution of velocity and temperature of drying air inside the dryer chamber. The standard k–ϵ model was implemented in the CFD simulation. The report indicated that design type (Figure 11.3b) was the most suitable option for obtaining uniform air temperature and velocity distributions in the drying chamber among the other alternative geometries. The experimental and simulation results showed good correlation coefficients of 99.9% and 86.5% for drying air temperature and air velocity in the drying chamber, respectively.

Sanghi et al. (2018) carried out a CFD simulation of the drying of corn using natural convection solar dryer at fair and overcast weather conditions to visualize humidity, temperature, and air velocity distributions of the dryer considering dual-band spectrums. The model was validated with experimental results, which indicated an overprediction of humidity (21.4%) and temperature (8.5%). The experimental humidity profile clearly showed that there was stagnation in the airflow (Figure 11.4) of the dryer which displayed a perfect prediction of the model and the overcast weather predicted moisture removal was 32% less than the simulated fair-weather case. Other reports indicated the distribution of air velocity and temperature inside cabinet solar dryers using CFD (Abhay et al., 2018; Alqadhi et al., 2017; Darabi et al., 2015; Ghaffari and Mehdipour, 2015, Mutabilwa and Nwaigwe, 2020; Sanghi et al., 2018; Tegenaw et al., 2019).

Tegenaw et al. (2019) carried out a marching heat transfer research in a solar food (sweet potato) dryer by using both CFD and lumped capacitance modeling techniques. The CFD model was used to simulate the transient airflow and heat transfer within the dryer. The result showed that the two models calculated a steady-state temperature rise of 40°C within the drying cabinet. The transient heat transfer phenomenon from CFD modeling (Figure 11.5) also indicated that the spatial distributions of temperature were uniform after 30 min, which validates the assumption of lumped capacitance modeling.

Capossio et al. (2022) research a natural convection solar dryer (NCSD) and conventional electric convective dryer (CECD) for the dehydration process of local microbrewers' spent grains drying using artificial neural networks and CFD modeling (Figure 11.6) and assess the environmental advantages and disadvantages of the NCSD. The result showed that the NCSD has a short drying time and huge environmental benefit.

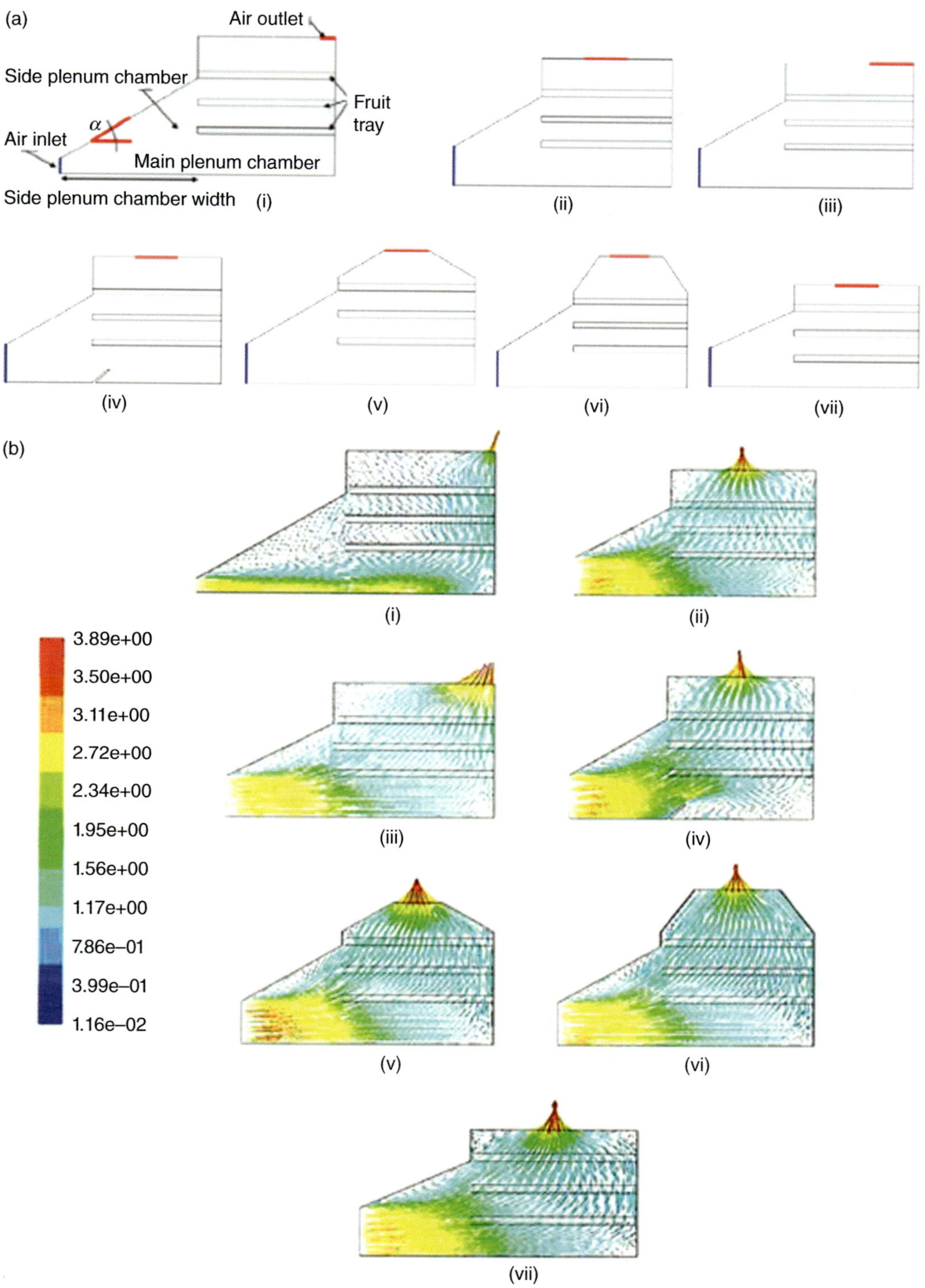

FIGURE 11.3 Air velocity distribution profiles for the seven designs (a) and CFD simulation results (b) at inlet air velocity = 3 m/s (Amanlou and Zomorodian, 2010; Getahun et al., 2021a).

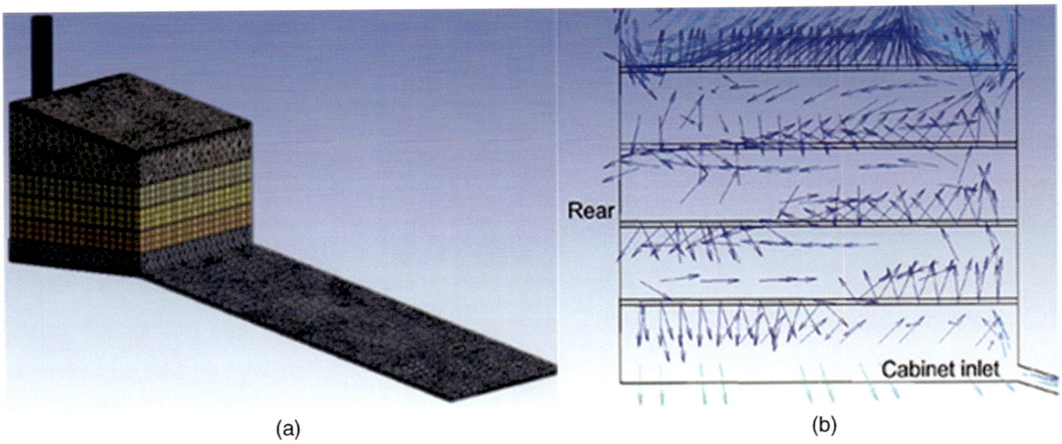

FIGURE 11.4 Geometry of a cabinet dryer (a) and airflow profile at the center plane (b) of a cabinet dryer (Getahun et al., 2021a; Sanghi et al., 2018).

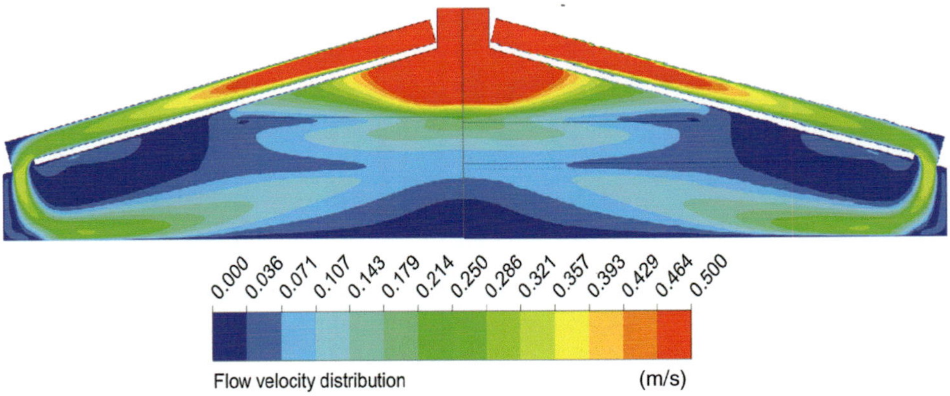

FIGURE 11.5 Contour plot of the velocity distribution of the solar sweet potato drying cabinet (Getahun et al., 2021a; Tegenaw et al., 2019).

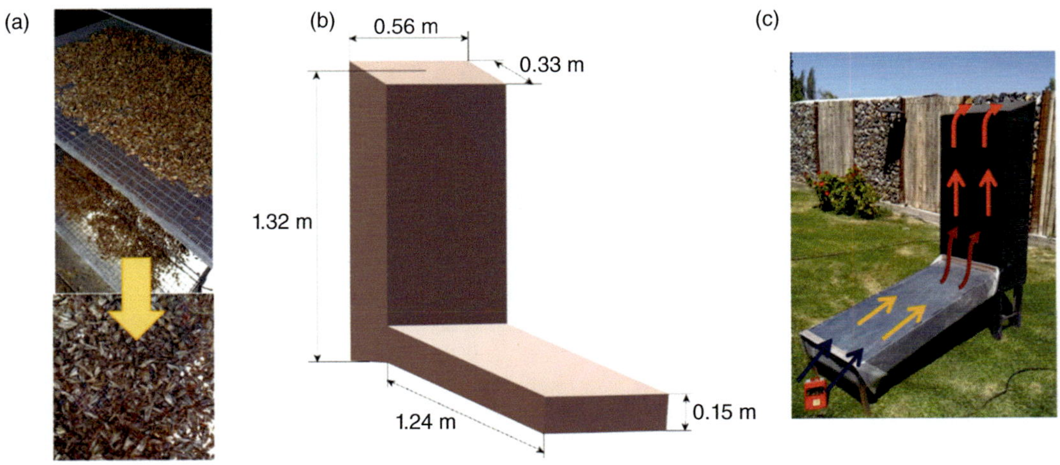

FIGURE 11.6 (a) Placement of the biomass on trays inside the solar drying chamber. (b) Natural convection solar dryer (NCSD) dimensions. (c) Heating and air circulation inside the NCSD. (Capossio et al., 2022.)

Chaudhari et al. (2021) developed an empirical and a CFD model to determine the optimal air temperature in a solar drying room. The results revealed that optimal performance was achieved at 45°C, where, the total energy consumption was 97.3 kWh per day, and solar energy utilization was maximized. The results of the CFD simulation were in good agreement with the findings of the mathematical model. Besides, the numerical study indicated that the dried product should be put far from the wall boundary since the air temperature was higher near the boundaries.

A sustainable solar tunnel dryer was designed by Afshari et al. (2021) based on an experimental and numerical study. For various process conditions and three different structures, a 3D steady-state CFD simulation was performed to estimate the suitable geometry for the tunnel dryer. The CFD simulation was effectively carried out, and the simulation results were in close agreement with the experimental results. The study proved that a Quonset-type dryer with fins had better thermal performance than the rectangular configuration in terms of homogeneity in velocity and temperature distribution, energy efficiency, and drying time.

Zewdie et al. (2022) developed a prototype solar energy-driven onion curing system using CFD modeling. The developed CFD model was validated by comparing the predicted temperature, air velocity, and mass loss to measured values on a prototype curing system operated on a field in Ethiopia. A good agreement between the experimental and model value was observed for the temperature profiles. The onions were cured to a desirable level for long-term storage within 48 hours.

Getahun et al. (2020b) investigated the potential of a solar cabinet chili dryer through rigorous CFD modeling based on the porous media using a double-pass solar collector and the simulation result indicated that the drying chamber had uniform drying air velocity distribution. Abebe, M., and Getahun, E. (2025), developed a detailed computational fluid dynamics (CFD) and experimental model for a solar bubble parboiled rice drier.The CFD model's ability to precisely forecast the airflow, heat, and mass transfer inside a solar bubble rice drier is demonstrated by the findings.Singh et al. (2021) carried out a CFD modeling and simulation of an indirect forced convection solar dryer to obtain the dynamic and thermal performance of the dryer at different operating conditions. The predicted results are validated using experimental results. The average velocity at the exit of the collector at 0.0872 kg/s mass flow rate was 2.89 m/s, while for a mass flow rate of 0.0447 kg/s it was 1.36 m/s. The maximum absorber plate temperature was 69.57°C and the maximum temperature at the collector outlet was 44.6°C for a magnitude of 0.0872 kg/s mass flow rate.

Iranmanesh et al. (2020) developed a CFD model and evaluated the performance of a solar cabinet apple slices dryer equipped with a thermal storage system and evacuated tube solar collector. The performance of the dryer was simulated and validated by experimental results. The result indicated that the maximum overall drying efficiency coupled with energy storage material (PCM) at the airflow rate of 0.025 kg/s was 39.9%. The CFD simulation of the storage system and the dryer showed that there is a good agreement between the simulated and experimental data.

Comprehensive modeling of solar energy systems using ANN including solar dryers is found in the work of Elsheikh et al. (2019). In this report, an effort has been made to examine the applications of ANN as an intelligent system–based approach for optimizing and predicting different sustainable energy systems performance, such as solar-assisted heat pumps, solar collectors, solar water and air heaters, solar stills, photovoltaic/thermal (PV/T) systems, solar cookers, and solar dryers. The frequently used artificial neural network categories and architectures are multilayer perceptron neural network, neural network using wavelet transform, Elman neural network, and radial basis functions.

Moghimi et al. (2021) studied an experimental and numerical optimal design of a household solar vegetable and fruit dryer (Figure 11.7). Several well-designed experiments and CFD simulations using the finite volume method were carried out. It was concluded that the proposed conversion procedure could be achieved using a blackened absorber plate, a honeycomb air manifold, and an air-circulating unit. In comparison with the direct type, the drying capacity of the dryer could be increased by more than 50% by adding a new tray to the optimized solar dryer. The thermal efficiency of the suggested solar dryer was 16.4%. Moreover, it was observed that the suggested numerical method was able to produce a harmonious quantitative and qualitative insight into both the fluid and thermal behavior of the dryer.

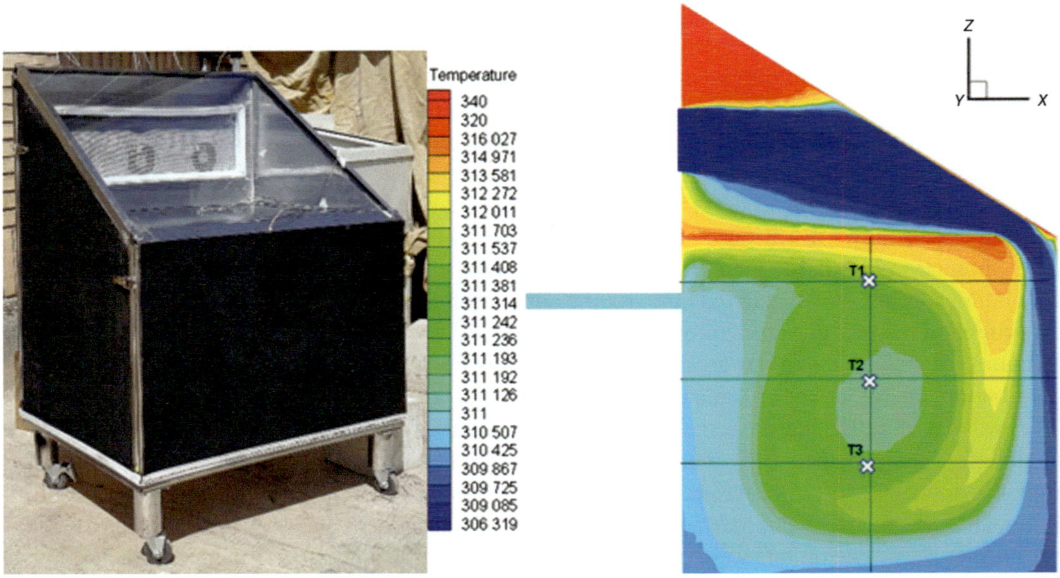

FIGURE 11.7 The cabinet dryer setup (*left*) and predicted temperature contour in the middle section of the dryer (*right*). (Moghimi et al., 2021.)

11.3.2 Biomass Energy–Based Dryer

Biomass can be a favorable way to get sustainable energy for the drying of food produces. It has been known that biomass is such a kind of renewable energy source that brings sustainable development and economic growth to developing countries. Over 90% of households in rural areas and 15% of households in urban areas use biomass to meet their primary energy demands (Sen et al., 2016).

A food drying system fueled with biomass energy looks to be mainly suitable for rural areas since it is a locally accessible and cheap source of energy. Biomass fuel is considered a "carbon neutral process" since carbon generated through the combustion process is used during its production. Biomass energy has been used for centuries as a renewable energy source for food product drying functions. The biomass-fired dryer comprises a combustion chamber, a drying chamber, drying air inlets, a passage for flue gases, trays, a drying air outlet chimney, a flue gas chimney, etc.

Efficiency is very important in biomass-based food dryers. The heat transfer efficiency of the biomass-fired dryer is the ratio of heat gained into the drying chamber to the summation of energy supplied by the fuel and enthalpy of air for the combustion of fuel. However, low efficiency is the major concern for biomass energy–based food dryers as a large amount of energy is wasted usually due to the poor design of the dryer system. Thus, an advanced design of biomass dryer that minimizes the loss of heat and improves mass and heat transfer rate in line with good product quality may overcome the limitations of traditional biomass energy-based dryer systems. Biomass dryers can also be integrated with solar collectors to improve the performance of the dryer. Continuous direct-heat rotary dryer, flash dryer, and fluidized bed dryers are some of the suggested improved biomass dryers (Qu et al., 2022). A schematic diagram of an improved and continuous fluidized bed dryer based on biomass fuel is presented in Figure 11.8, where two major components of the improved dryer are a biomass combustion chamber and a continuous feeding mechanism.

Francik et al. (2018) developed an artificial neural network models to describe the process of convective drying of willow woodchips. They obtained highly precise neural models describing changes in the temperature and the mass alterations of the water content of the chips.

Dhanushkodi et al. (2017) used empirical models to examine the drying behavior of cashew in a solar biomass hybrid dryer. Fifteen different empirical drying models were used to explain the

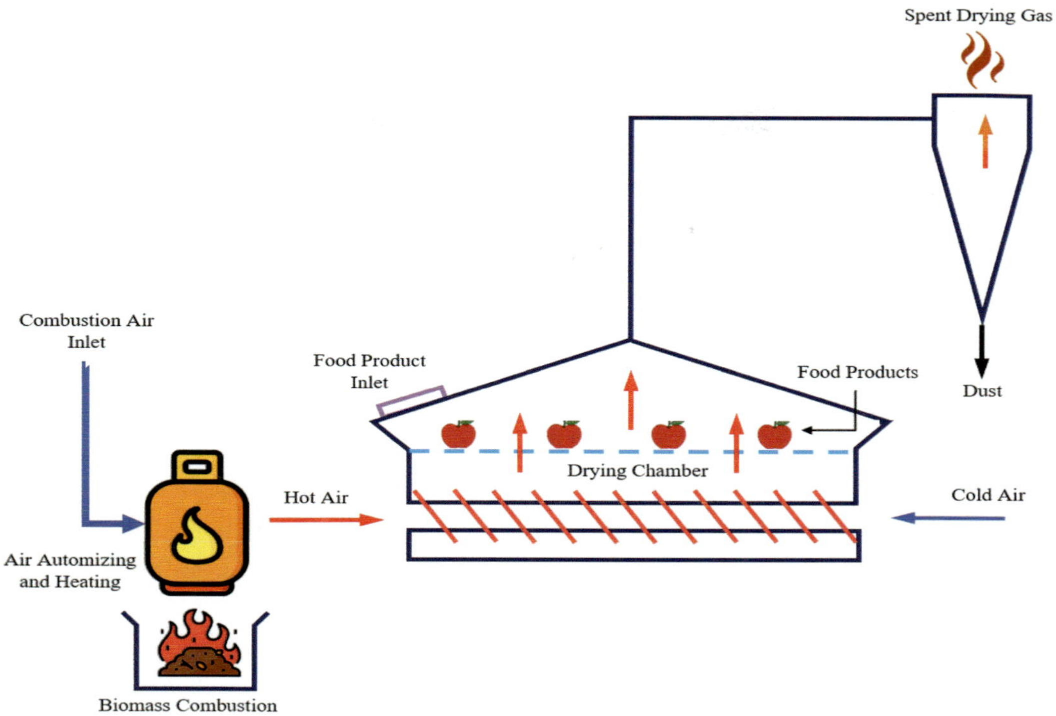

FIGURE 11.8 Fluidized bed dryers use biomass as an energy source. (Qu et al., 2022.)

drying behaviors of cashew. The result indicated that the Page model was found to be the best model to fit the thin-layer drying behavior of cashew for hybrid drying and biomass drying.

Anand et al. (2020) developed a CFD model to evaluate the performance of an innovative biomass energy–based dryer. The effect of tray length, tray number, hole arrangement, and tray arrangement was examined in detail. A drying system with good heat transfer performance and almost no addition to pressure drop was obtained from this CFD modeling and simulation.

11.3.3 Geothermal Energy–Based Dryer

Geothermal energy exploitation can be categorized as near ground and deep underground by a conceptual boundary, and deep geothermal is further categorized as low enthalpy and high enthalpy reservoirs (Qu et al., 2022). Geothermal energy, which is environment-friendly, reliable, green, and sustainable, is considered one of the vital kinds of renewable energy. Geothermal energy is a plentiful and sustainable energy source that produces lower carbon emissions to the environment. Geothermal energy mostly refers to the part of the Earth's heat that can be found from the shallow ground and used for various industrial applications such as space heating/cooling, electricity generation, aquaculture produce drying, and domestic and therapeutic applications such as swimming and bathing. Nowadays, it has already been integrated with food drying technology. Furthermore, the 24 h availability of geothermal energy makes it more feasible than wind energy (wind speed dependent) and solar energy (only in daylight). The temperature deviation of the geothermal energy depends on the reservoir well depth. A reservoir well around 1.5 km deep can create hot water to the surface that can be transformed directly or indirectly as useful energy.

A geothermal well with low enthalpy at a temperature below 150°C is appropriate for drying fruits and vegetables (Qu et al., 2022). The general application of geothermal energy in the agriculture and food industries is presented in Figure 11.9.

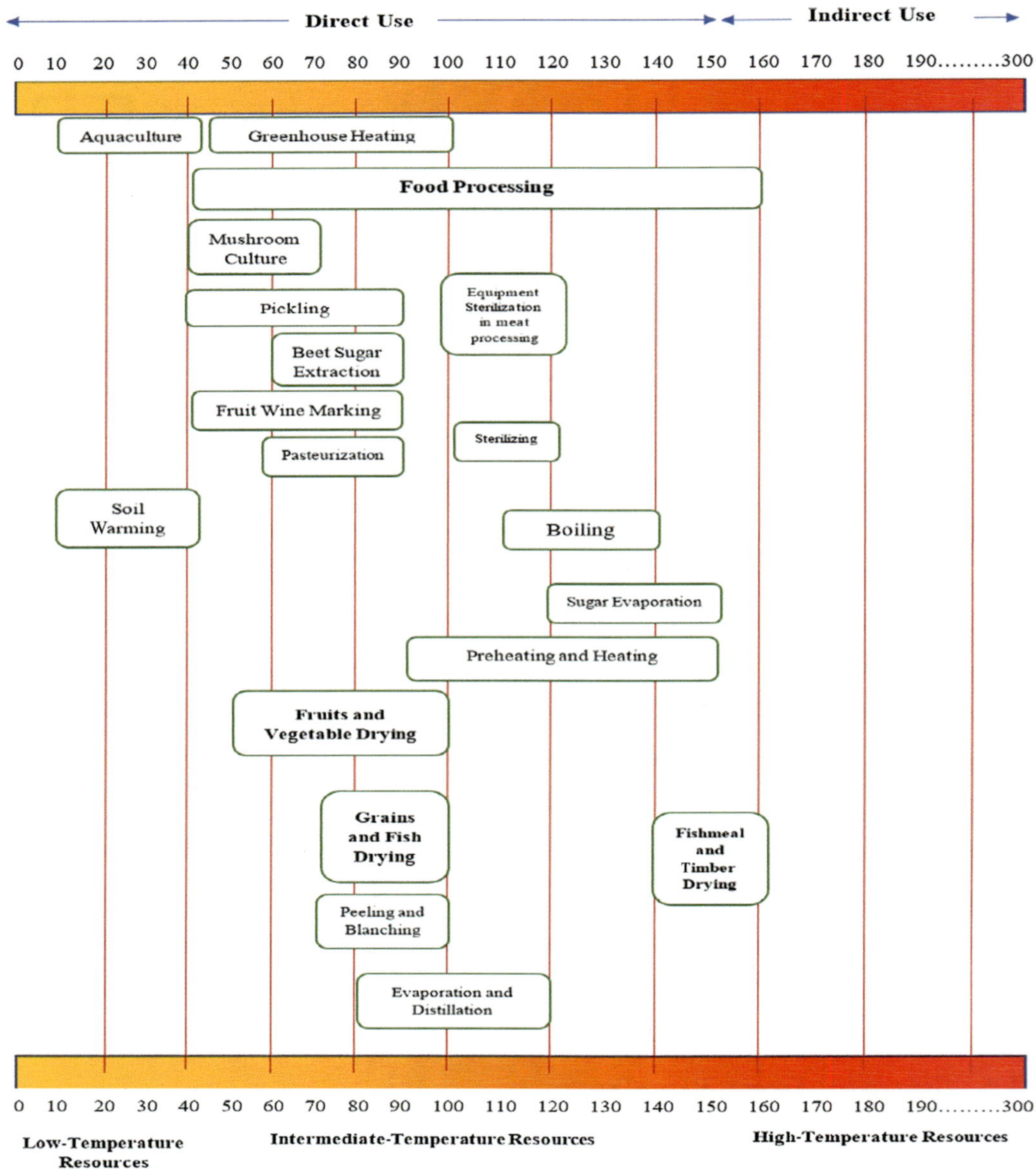

FIGURE 11.9 Lindal diagram for geothermal energy use in agriculture and food industries. (Ananno et al., 2020.)

Low-temperature geothermal hot waters can be used effectively in drying various vegetables and fruits, substituting sun drying or drying using conventional fuels. Since geothermal driers are generally conducted at lower temperatures than driers using conventional fuels, the food products do not expose to discoloration or contamination, which results in a higher quality of final dried product (Kostoglou et al., 2013). The typical schematic of a geothermal food dryer is presented in Figure 11.10. As can be seen in the figure, underground hot water is pumped through a heat exchanger to heat the air as the drying medium. The hot air is then passed through the trays in the drying chamber to pick up the moisture from the food products.

Numerical modeling is an important approach in geothermal energy–based food product drying to optimize the drying parameters. Kostoglou et al. (2013) developed a tomato dehydration model

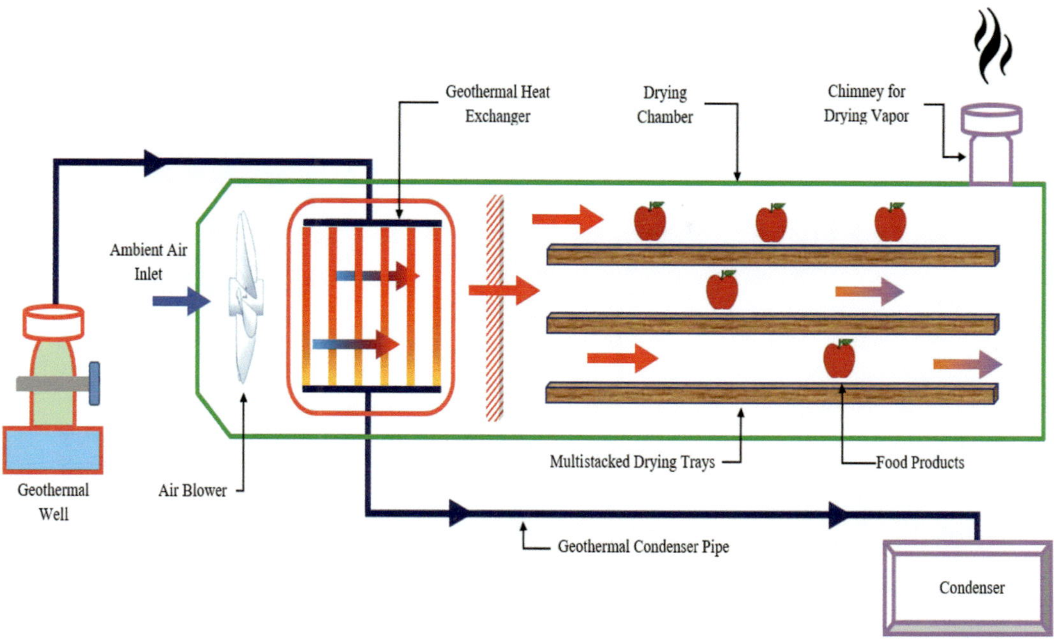

FIGURE 11.10 Schematic of multistacked tray geothermal dryer. (Qu et al., 2022.)

in a tunnel dryer using low-temperature geothermal energy. The model was based on the empirical model conducted using the diffusion model. The effect of flow configuration, such as countercurrent or co-current operation, on the drying characteristics, was also examined and good results were observed. The model can be used for the design and optimization of continuous tomato drying operations.

Sandali et al. (2019) studied the improvement of a direct solar dryer performance using a geothermal water heat exchanger as a supplementary energetic supply through experimental investigation and CFD simulation study. This integration of solar and geothermal energy in the dryer ensured the continuity of the drying process on cloudy days and even at nighttime with high performance.

Kulundu et al. (2022) developed a CFD model of a modified geothermal grain dryer at Menengai Well 3 in Kenya to scrutinize the airflow distribution and its effect on the drying performance of maize. The model was simulated using ANSYS 21 and experimental data were used to validate the simulation results. The model was discretized using finite volume to get the required mesh size. The pressure-based segregated solver was implemented in the CFD simulation and K-Omega turbulent model was employed for improving wall treatment. The simulation result indicated that nonuniform hot-air distribution in the drying chamber causes uneven drying. After implementing filleted flow guides, the distribution of drying parameters such as drying velocity, temperature, and turbulence kinetic relatively became uniform. The average temperature and velocity magnitudes in lower compartments improved by 0.25% and 153.3%, respectively, for the improved dryer (Figure 11.11). In the upper compartments, the temperature and velocity improvements were 0.22% and 176.5%, respectively.

11.3.4 Wind Energy–Based Dryer

Wind energy systems are clean and environment-friendly sources of power like solar energy but can serve at all times of the day as long as the wind blows exist. In developing countries in rural areas where the electricity grid is far from the existence of unreliable power and high cost of generation

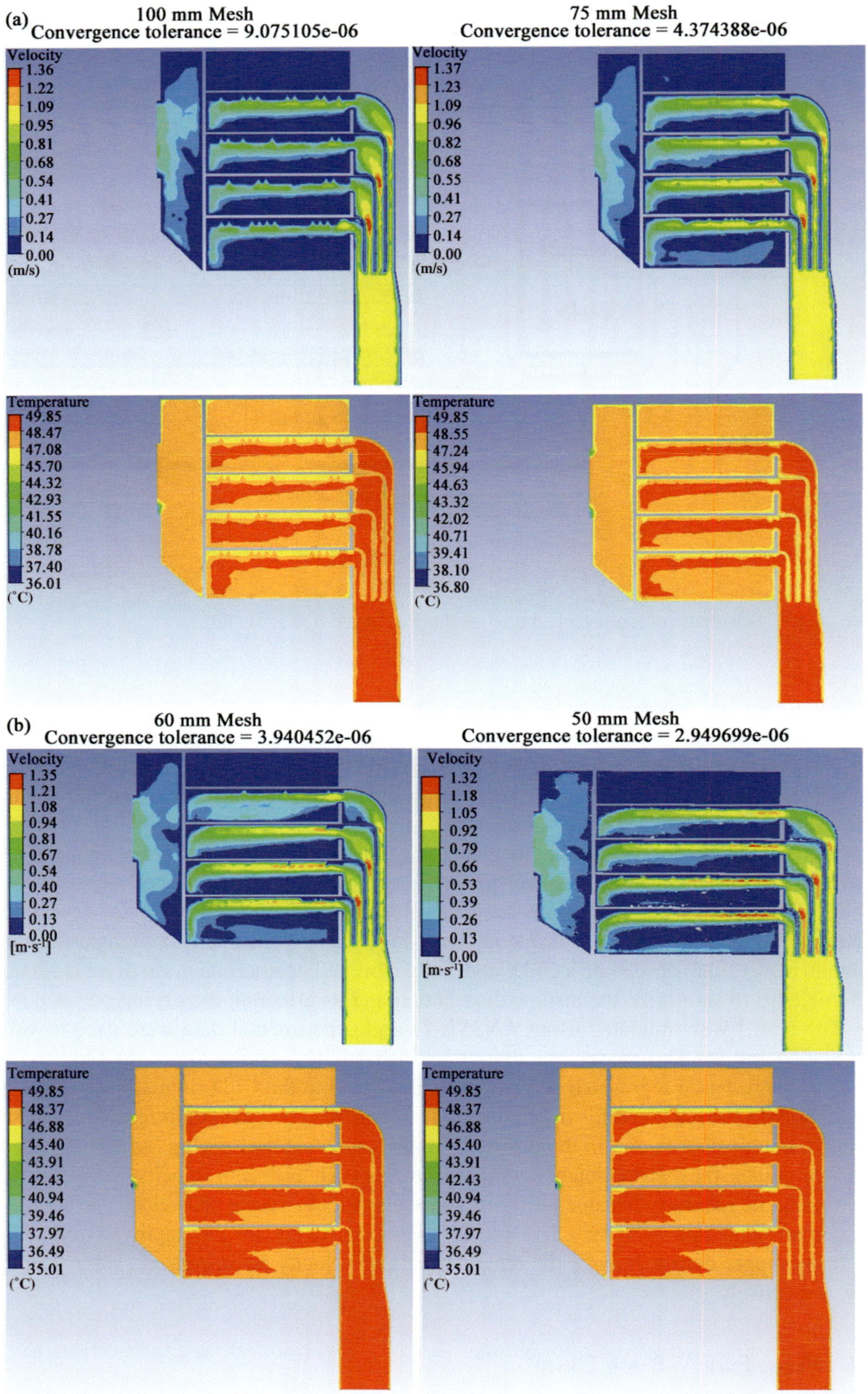

FIGURE 11.11 (a) Velocity and temperature contours for 100 and 75 mm mesh of the improved dryer. (b) Velocity and temperature contours for 60 and 50 mm Mesh of the improved dryer. (Kulundu et al., 2022.)

FIGURE 11.12 Developed prototype solar dryer with a wind generator. (Ndukwu et al., 2020.)

of electricity for heating and drying purposes, a close look at the potentials of solar and wind energy becomes essential.

Low-cost and small wind energy that can be harnessed locally by smallholder farmers, who cannot afford sophisticated and expensive drying technology, could serve as a means to rapidly dry their food produce at the farm level. Wind energy utilization can also reduce the dependence of farmers on fossil fuels for drying their food products and associated running costs of fuel and reduce carbon emissions to the environment. It is, therefore, both environmentally and economically a more sustainable means of reducing postharvest losses of agricultural products in rural societies.

Though forced convective wind energy applications have been seldom used in grain drying operations, it is extensively used for other high torque operations like grain milling and water pumping. Ndukwu et al. (2020) developed a low-cost wind-powered active solar dryer integrated with glycerol as thermal storage to evaluate the potato slice drying performance. A wind generator produces air power when the ambient wind speed is higher than the cut-in speed. The solar dryer (Figure 11.12) comprises a solar collector, drying chamber, drying trays, and the axial fan fixed to the wind air generator. The results indicate that drying with an active-mixed-mode wind-powered fan solar dryer integrated with glycerol showed shorter drying time than drying with a passive mixed-mode non-wind-powered solar dryer. The drying efficiencies of the system ranged from 25.031% to 31.5%. Advanced numerical modeling could be applied in the design, development, and prototype testing of a wind energy exploiting drying systems.

11.4 CONCLUSION

Appropriate preservations of food products are very important to minimize postharvest loss and increase food security and safety in societies. Renewable, sustainable, and environment-friendly energy sources are very crucial for the preservation of food products. Optimization of drying parameters and dryer design using numerical approaches is becoming a promising alternative to minimize the deterioration of the food product quality during sustainable drying processes. This chapter discussed sustainable food drying systems that use sustainable energy sources and the capacity of numerical modeling in the design, operation, and prototyping of optimum sustainable dryers. Numerical modeling approaches that range from simple theoretical and empirical models to advanced CFD and AI modeling techniques have been applied for optimizing the design and operating parameters of sustainable drying systems. Sustainable food dryers use renewable energy sources such as solar, biomass, geothermal, and wind energy. There is an encouraging progress in the application of advanced numerical models like CFD to get detailed information about the hydrodynamics and thermal performance, including the quality aspects of dried food products of the drying process. Similarly, there is a progress in the application AI in sustainable food drying systems. But still there is a need to put more focus on the application of numerical modeling techniques in the design, operation, and control of sustainable food drying systems, since it is a cheap alternative to experimental studies.

REFERENCES

Abebe, M., and Getahun, E. 2025. Drying kinetics and performance evaluation of solar bubble parboiled rice dryers through computational and experimental studies. https://doi.org/10.1080/23311916.2025.2572290

Abhay, L., Chandramohan, V.P., Raju, V.R.K., 2018. Numerical analysis on solar air collector provided with artificial square-shaped roughness for indirect type solar dryer. J. Clean. Prod., https://doi.org/10.1016/j.jclepro.2018.04.130

Afshari, F., Khanlari, A., Tuncer, A.D., Sözen, A., Şahinkesen, İ, Di Nicola, G., 2021. Dehumidification of sewage sludge using quonset solar tunnel dryer: An experimental and numerical approach. Renew. Energy., 171, 784–798. https://doi.org/10.1016/j.renene.2021.02.158

Alqadhi, A., Misha, S., Rosli, M.A.M., Akop, M.Z., 2017. Design and simulation of an optimized mixed mode solar dryer integrated with desiccant material. Int. J. Mech. Mechatronics Eng.

Amanlou, Y., Zomorodian, A., 2010. Applying CFD for designing a new fruit cabinet dryer. J. Food Eng., 101, 8–15. https://doi.org/10.1016/j.jfoodeng.2010.06.001

Anand, S., Mishra, D.P., Sarangi, S.K., 2020. CFD supported performance analysis of an innovative biomass dryer. Renew. Energy., 159, 860–872. https://doi.org/10.1016/j.renene.2020.06.039

Ananno, A.A., Masud, M.H., Dabnichki, P., Ahmed, A., 2020. Design and numerical analysis of a hybrid geothermal PCM flat plate solar collector dryer for developing countries. Sol. Energy., https://doi.org/10.1016/j.solener.2019.11.069

Bennett, A.S., Bern, C.J., Richard, T.L., Anex, R.P., 2007. Corn Grain Drying Using Corn Stover Combustion and CHP Systems. Trans. ASAE, 50, 2161–2170.

Calado, T., Venâncio, A., Abrunhosa, L., 2014. Irradiation for mold and mycotoxin control: A review. Compr. Rev. Food Sci. Food Saf., https://doi.org/10.1111/1541-4337.12095

Capossio, J.P., Fabani, M.P., Reyes-Urrutia, A., Torres-Sciancalepore, R., Deng, Y., Baeyens, J., Rodriguez, R., Mazza, G., 2022. Sustainable solar drying of Brewer's spent grains: A comparison with conventional electric convective drying. Processes., https://doi.org/10.3390/pr10020339

Catorze, C., Tavares, A.P., Cardão, P., Castro, A., Silva, M.E., Ferreira, D.W., Lopes, S., Brás, I., 2022. Study of a solar energy drying system: Energy savings and effect in dried food quality. Energy Rep., https://doi.org/10.1016/j.egyr.2022.01.070

Chandramohan, V.P., 2020. Convective drying of food materials: An overview with fundamental aspect, recent developments, and summary. Heat Transf., https://doi.org/10.1002/htj.21662

Chaudhari, V.D., Kulkarni, G.N., Sewatkar, C.M., 2021. Mathematical modeling and computational fluid dynamics simulation of cabinet type solar dryer: Optimal temperature control. J. Food Process Eng., https://doi.org/10.1111/jfpe.13643

Curcio, S., Aversa, M., Chakraborty, S., Calabrò, V., Iorio, G., 2016. Formulation of a 3D conjugated multiphase transport model to predict drying process behavior of irregular-shaped vegetables. J. Food Eng. 176, 36–55. https://doi.org/10.1016/j.jfoodeng.2015.11.020

Darabi, H., Zomorodian, A., Akbari, M.H., Lorestani, A.N., 2015. Design a cabinet dryer with two geometric configurations using CFD. J. Food Sci. Technol., 52, 359–366. https://doi.org/10.1007/s13197-013-0983-1

Defraeye, T., 2016. Towards more efficient intermittent drying of fruit: Insights from combined hygrothermal-quality modelling. Innov. Food Sci. Emerg. Technol., https://doi.org/10.1016/j.ifset.2016.10.003

Defraeye, T., Radu, A., 2017. Convective drying of fruit: A deeper look at the air–material interface by conjugate modeling. Int. J. Heat Mass Transf., 108, 1610–1622. https://doi.org/10.1016/j.ijheatmasstransfer.2017.01.002

Defraeye, T., Radu, A., 2018. Insights in convective drying of fruit by coupled modeling of fruit drying, deformation, quality evolution and convective exchange with the airflow. Appl. Therm. Eng., https://doi.org/10.1016/j.applthermaleng.2017.10.082

Demissie, P., Hayelom, M., Kassaye, A., Hailesilassie, A., Gebrehiwot, M., Vanierschot, M., 2019. Design, development and CFD modeling of indirect solar food dryer. Energy Procedia., https://doi.org/10.1016/j.egypro.2019.01.278

Dhanushkodi, S., Wilson, V.H., Sudhakar, K., 2017. Mathematical modeling of drying behavior of cashew in a solar biomass hybrid dryer. Resour. Technol., https://doi.org/10.1016/j.reffit.2016.12.002

Dincer, I., Hussain, M.M., 2002. Development of a new bi-di correlation for solids drying. Int. J. Heat Mass Transf., https://doi.org/10.1016/S0017-9310(02)00031-5

Dincer, I., Hussain, M.M., 2004. Development of a new biot number and lag factor correlation for drying applications. Int. J. Heat Mass Transf., 47, 653–658. https://doi.org/10.1016/j.ijheatmasstransfer.2003.08.006

Dincer, I., Hussain, M.M., Sahin, A.Z., Yilbas, B.S., 2002. Development of a new moisture transfer (bi-re) correlation for food drying applications. Int. J. Heat Mass Transf., 45, 1749–1755. https://doi.org/10.1016/S0017-9310(01)00272-1

Ekechukwu, O.V., Norton, B., 1999. Review of solar-energy drying systems II: An overview of solar drying technology. Energy Convers. Manag., 40, 615–655. https://doi.org/10.1016/S0196-8904(98)00093-4

El-Sebaii, A.A., Shalaby, S.M., 2012. Solar drying of food products: A review. Renew. Sustain. Energy Rev., 16, 37–43. https://doi.org/10.1016/j.rser.2011.07.134

Elsheikh, A.H., Sharshir, S.W., Abd Elaziz, M., Kabeel, A.E., Guilan, W., Haiou, Z., 2019. Modeling of solar energy systems using artificial neural network: A comprehensive review. Sol. Energy., 180, 622–639. https://doi.org/10.1016/j.solener.2019.01.037

Erbay, Z., Icier, F., 2010. A review of thin layer drying of foods: Theory, modeling, and experimental results. Crit. Rev. Food Sci. Nutr., https://doi.org/10.1080/10408390802437063

Ertekin, C., Firat, M.Z., 2017. A comprehensive review of thin-layer drying models used in food products. Crit. Rev. Food Sci. Nutr., 57, 701–717. https://doi.org/10.1080/10408398.2014.910493

Esfahani, J.A., Majdi, H., Barati, E., 2014. Analytical two-dimensional analysis of the transport phenomena occurring during convective drying: Apple slices. J. Food Eng., 123, 87–93. https://doi.org/10.1016/j.jfoodeng.2013.09.019

Farkas, I., 2013. Use of artificial intelligence for the modelling of drying processes. Drying Technol., 31(7), 848–855.

Fasina, O., Sokhansanj, S., Tyler, R., 1997. Thermodynamics of moisture sorption in alfalfa pellets. Dry. Technol., https://doi.org/10.1080/07373939708917307

Francik, S., Łapczyńska-Kordon, B., Francik, R., Wójcik, A., 2018. Modeling and Simulation of Biomass Drying Using Artificial Neural Networks., https://doi.org/10.1007/978-3-319-72371-6_56

Getahun, E., Delele, M.A., Gabbiye, N., Fanta, S.W., Demissie, P., Vanierschot, M., 2021a. Importance of integrated CFD and product quality modeling of solar dryers for fruits and vegetables: A review. Sol. Energy., 220, 88–110. https://doi.org/10.1016/j.solener.2021.03.049

Getahun, E., Delele, M.A., Gabbiye, N., Fanta, S.W., Vanierschot, M., 2021b. Studying the drying characteristics and quality attributes of chili pepper at different maturity stages: Experimental and mechanistic model. Case Stud. Therm. Eng., 26, 101052. https://doi.org/10.1016/j.csite.2021.101052

Getahun, E., Gabbiye, N., Delele, M.A., Fanta, S.W., Gebrehiwot, M.G., Vanierschot, M., 2020a. Effect of maturity on the moisture sorption isotherm of chili pepper (*Mareko fana* variety). Heliyon., https://doi.org/10.1016/j.heliyon.2020.e04608

Getahun, E., Gabbiye, N., Delele, M.A., Fanta, S.W., Vanierschot, M., 2021c. Two-stage solar tunnel chili drying: Drying characteristics, performance, product quality, and carbon footprint analysis. Sol. Energy., 230, 73–90. https://doi.org/10.1016/j.solener.2021.10.016

Getahun, E., Vanierschot, M., Gabbiye, N., Delele, M.A., Workneh, S., Gebreslasie, M., 2020b. Computational fluid dynamic modeling and simulation of red chili solar cabinet dryer, in: Lecture Notes of the Institute for Computer Sciences, Social-Informatics and Telecommunications Engineering (pp. 597–609), LNICST. https://doi.org/10.1007/978-3-030-43690-2_45

Ghaffari, A., Mehdipour, R., 2015. Modeling and improving the performance of cabinet solar dryer using computational fluid dynamics. Int. J. Food Eng., 11, 157–172. https://doi.org/10.1515/ijfe-2014-0266

Gill, J., Singh, J., Ohunakin, O.S., Adelekan, D.S., Atiba, O.E., Nkiko, M.O., Atayero, A.A., 2020. Adaptive neuro-fuzzy inference system (ANFIS) approach for the irreversibility analysis of a domestic refrigerator system using LPG/TiO_2 nanolubricant. Energy Rep., 6, 1405–1417. https://doi.org/10.1016/j.egyr.2020.05.016

Hashim, N., Daniel, O., Rahaman, E., 2014. A preliminary study: Kinetic model of drying process of pumpkins (Cucurbita moschata) in a convective hot air dryer. Agric. Agric. Sci. Procedia, 2, 345–352. https://doi.org/10.1016/j.aaspro.2014.11.048

Iranmanesh, M., Samimi Akhijahani, H., Barghi Jahromi, M.S., 2020. CFD modeling and evaluation the performance of a solar cabinet dryer equipped with evacuated tube solar collector and thermal storage system. Renew. Energy, 145, 1192–1213. https://doi.org/10.1016/j.renene.2019.06.038

Joardder, M.U.H., Masud, M.H., 2019. Food Preservation in Developing Countries: Challenges and Solutions. https://doi.org/10.1007/978-3-030-11530-2

Khin, M.M., Zhou, W., Perera, C., 2005. Development in the combined treatment of coating and osmotic dehydration of food : A review. Int. J. Food Eng., 1, 1–21. https://doi.org/10.2202/1556-3758.1005

Khoshhal, A., Dakhel, A.A., Etemadi, A., Zereshki, S., 2010. Artificial neural network modeling of apple drying process. J. Food Process Eng., 33, 298–313. https://doi.org/10.1111/j.1745-4530.2009.00435.x

Kostoglou, M., Chrysafis, N., Andritsos, N., 2013. Modelling tomato dehydration in a tunnel dryer using geothermal energy. Dry. Technol., https://doi.org/10.1080/07373937.2012.710694

Kulundu, L., Ndiritu, H., Kituu, G., Kimotho, J., 2022. Performance simulation of a modified geothermal grain dryer based at menengai well 3 in Kenya. World J. Eng. Technol., https://doi.org/10.4236/wjet.2022.101004

Liu, Z.L., Bai, J.W., Wang, S.X., Meng, J.S., Wang, H., Yu, X.L., Gao, Z.J., Xiao, H.W., 2020. Prediction of energy and exergy of mushroom slices drying in hot air impingement dryer by artificial neural network. Dry. Technol., https://doi.org/10.1080/07373937.2019.1607873

Martynenko, A., 2017. Artificial intelligence: Is it a good fit for drying? Drying Technol., DOI:10.1080/07373937.2017.1362153

Masud, M.H., Ananno, A.A., Ahmed, N., Dabnichki, P., Salehin, K.N., 2020. Experimental investigation of a novel waste heat based food drying system. J. Food Eng., https://doi.org/10.1016/j.jfoodeng.2020.110002

Mehta, S., Singh, A., 2006. Adsorption isotherms for red chilli (*Capsicum annum* L.). Eur. Food Res. Technol., https://doi.org/10.1007/s00217-006-0279-z

Menshutina, N.V., Goncharova-Alves, S.V., Matasov, A.V., 2019. Intelligent systems for design of drying processes. Drying Technol., 38, 1–11. DOI:10.1080/07373937.2019.1632344

Moghimi, P., Rahimzadeh, H., Ahmadpour, A., 2021. Experimental and numerical optimal design of a household solar fruit and vegetable dryer. Sol. Energy., https://doi.org/10.1016/j.solener.2020.12.023

Mutabilwa, P., Nwaigwe, K.N., 2020. Design, construction and CFD modeling of a banana-solar dryer with double pass solar air collector, in: ASME 2020 14th International Conference on Energy Sustainability, ES 2020. https://doi.org/10.1115/ES2020-1614

Ndukwu, M.C., Onyenwigwe, D., Abam, F.I., Eke, A.B., Dirioha, C., 2020. Development of a low-cost wind-powered active solar dryer integrated with glycerol as thermal storage. Renew. Energy., https://doi.org/10.1016/j.renene.2020.03.016

Novák, V., Perfilieva, I., Dvořák, A., 2016. Insight into Fuzzy Modeling. John Wiley & Sons, Inc. https://doi.org/10.1002/9781119193210

Prakash, O., Kumar, A., 2014. Solar greenhouse drying: A review. Renew. Sustain. Energy Rev., https://doi.org/10.1016/j.rser.2013.08.084

Qu, H., Masud, M.H., Islam, M., Khan, M.I.H., Ananno, A.A., Karim, A., 2022. Sustainable food drying technologies based on renewable energy sources. Crit. Rev. Food Sci. Nutr., https://doi.org/10.1080/10408398.2021.1907529

Raso, J., Barbosa-Cánovas, G.V., 2003. Nonthermal preservation of foods using combined processing techniques. Crit. Rev. Food Sci. Nutr., https://doi.org/10.1080/10408690390826527

Sagar, V.R., Suresh Kumar, P., 2010. Recent advances in drying and dehydration of fruits and vegetables: A review. J. Food Sci. Technol., https://doi.org/10.1007/s13197-010-0010-8

Sahdev, R.K. 2014. Open sun and greenhouse drying of agricultural. Int. J. Eng. Res. Technol., 3(3), 1053–1066.

Sandali, M., Boubekri, A., Mennouche, D., Gherraf, N., 2019. Improvement of a direct solar dryer performance using a geothermal water heat exchanger as supplementary energetic supply: An experimental investigation and simulation study. Renew. Energy., https://doi.org/10.1016/j.renene.2018.11.086

Sanghi, A., Ambrose, R.P.K., Maier, D., 2017. CFD simulation of corn drying in a natural convection solar dryer. Dry. Technol, 1–12. https://doi.org/10.1080/07373937.2017.1359622

Sanghi, A., Ambrose, R.P.K., Maier, D., 2018. CFD simulation of corn drying in a natural convection solar dryer. Dry Technol., 36, 859–870. https://doi.org/10.1080/07373937.2017.1359622

Sen, S., Ganguly, S., Das, A., Sen, J., Dey, S., 2016. Renewable energy scenario in India: Opportunities and challenges. J. African Earth Sci., https://doi.org/10.1016/j.jafrearsci.2015.06.002

Singh, R., Salhan, P., Kumar, A., 2021. CFD modelling and simulation of an indirect forced convection solar dryer, in: IOP Conference Series: Earth and Environmental Science. https://doi.org/10.1088/1755-1315/795/1/012008

Sinha, A., Bhargav, A., 2022. A simplified modelling approach for predicting shrinkage and sensitive material properties during low temperature air drying of porous food materials. J. Food Eng., https://doi.org/10.1016/j.jfoodeng.2021.110732
Sun, Q., Zhang, M., Mujumdar, A.S., 2018. Recent developments of artificial intelligence in drying of fresh food: A review. Crit. Rev. Food Sci. Nutr., https://doi.org/10.1080/10408398.2018.1446900
Tegenaw, P.D., Gebrehiwot, M.G., Vanierschot, M., 2019. On the comparison between computational fluid dynamics (CFD) and lumped capacitance modeling for the simulation of transient heat transfer in solar dryers. Sol. Energy., https://doi.org/10.1016/j.solener.2019.04.024
Thorpe, G.R., 2008. The application of computational fluid dynamics codes to simulate heat and moisture transfer in stored grains. J. Stored Prod. Res., 44, 21–31. https://doi.org/10.1016/j.jspr.2007.07.001
United Nations, 2019. World Population Prospects 2019, Department of Economic and Social Affairs.
Xiao, H.W., Bai, J.W., Xie, L., Sun, D.W., Gao, Z.J., 2015. Thin-layer air impingement drying enhances drying rate of American ginseng (*Panax quinquefolium* L.) slices with quality attributes considered. Food Bioprod. Process., https://doi.org/10.1016/j.fbp.2014.08.008
Zewdie, T.A., Delele, M.A., Fanta, S.W., Vanierschot, M., Alemayehu, M., Alemayehu, G., Adgo, E., Assaye, H., Verboven, P., Nicolai, B.M., 2022. Design and evaluation of a prototype solar energy driven onion curing system using CFD modeling. Front. Sustain. Food Syst., 6. https://doi.org/10.3389/fsufs.2022.1030915

12 Computational Modelling and Validations for Sustainable Drying Processes

M. Adnouni, X. J. Zhang, and L. Jiang
Key Laboratory of Refrigeration and Cryogenic Technology of Zhejiang Province, Institute of Refrigeration and Cryogenics, Zhejiang University, Hangzhou, China

12.1 INTRODUCTION

Theoretical models are mathematically complex and involve many parameters since they are established from fundamental drying physics [1], which couples transport equations with thermodynamically interactive fluxes and phase equilibrium expressions. This time-consuming modelling has led to the development of powerful numerical methods. In this regard, the food drying sector has increasingly adopted computational fluid dynamics (CFD) approaches in the quest to analyse processes, design, and optimise drying systems [2]. CFD is a discipline that has progressed from a mathematical curiosity to become an indispensable tool in nearly every branch of fluid dynamic [3]. The use of numerical solution methods is trendy, and the popularity of such an advanced modelling approach is due to its plentiful advantages. CFD modelling plays a pivotal role in understanding and solving complex heat and transport equations and other associated phenomena such as chemical reactions [4], providing an attractive visualisation of temperature fields, velocity vectors, shrinkage, and moisture distribution inside the food or within the drying chamber to improve final product quality [5, 6].

The user-friendly interface of the computer simulation tool allows assessing the performance of the drying system with a qualitative flow visualisation throughout the process. Data from CFD simulation can be analysed to improve efficiency and product quality and thus design a new optimised drying system. CFD results give more assured scale-up, better reliability, higher productivity, and the ability to investigate systems under hazardous conditions at and above their normal performance limits with a reduced workload, cost, and drying space would be required in laboratory experiments. However, numerical simulation should be combined with physical experiments for validation and analytical solutions to accurately depict the kinetics of the heat and mass transfer mechanisms. CFD modelling relies on involving a precise physical definition of variables and proper initial and boundary conditions, as well as a selection of mathematical equations such as Navier–Stokes equations in line with suitable turbulence models using numerical methods and empirical correlations [7].

This chapter predominantly highlights the recent advancements in CFD modelling of sustainable drying technologies over conventional drying methods, including physical field-based drying technologies, renewable drying technologies, and hybrid drying technologies. Such cost-effective alternative systems should be promoted to reap the benefits of sustainable drying systems with low costs and simple technologies.

Conjugated and nonconjugated approaches, along with appropriate numerical methods, namely, finite differences, finite elements, and finite volumes, are comprehensively outlined. Furthermore, new insights about embedding CFD into the Digital Twin paradigm or the concept of providing a virtual representation using simulation data are addressed. The modelling challenges, validation, future applications, and potential benefits of CFD modelling in terms of quality modelling, energy savings, and reduction of GHG emissions to enhance sustainability in the food drying industry are

DOI: 10.1201/9781003344087-12

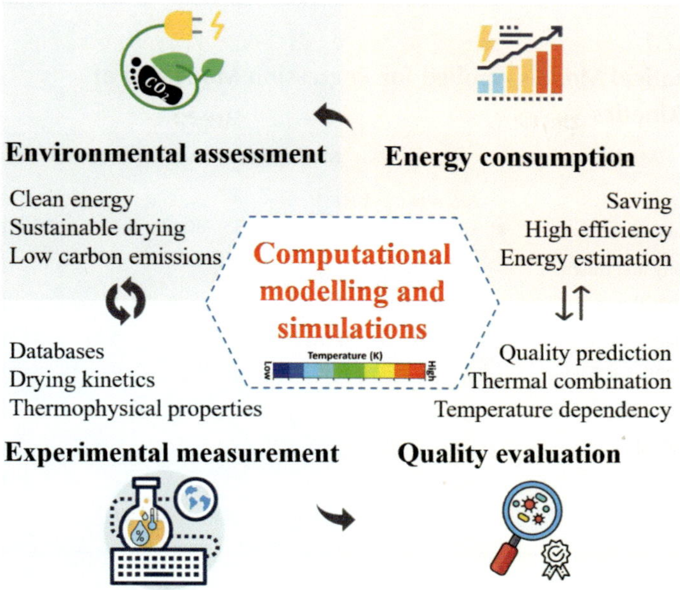

FIGURE 12.1 A conceptual schema of the interrelationships between the results of numerical modelling, drying parameters, and key goals of sustainable drying.

also discussed. The energy–quality–environment–modelling nexus related to the sustainable drying process of food products is illustrated in Figure 12.1.

12.2 A REVIEW OF THE MATHEMATICAL MODELLING EMPLOYED IN DRYING PROCESSES

Mathematical models for food drying are essential to understand better the underpinning heat and mass transfer processes governing the drying process. Further research are needed to optimise the drying system's design and operating conditions and maximise the product quality attributes preservation [8]. The mathematical modelling method leads to a straightforward construction of numerical simulation and, as a result, precise analysis and adjustment of the parameters during the process. There are theoretical, semi-theoretical, and empirical equations to predict the drying behaviour of different foodstuff products.

12.2.1 Semi-Theoretical and Empirical Models

Semi-theoretical and empirical models are practical and simple for describing drying kinetics. Most models, such as thin-layer models, are based on Fick's law of diffusion theory and Newtons' law of cooling. They simulate the drying curve and therefore predict the drying results, but they have little physical significance than theoretical models. These models neglect food shrinkage, geometry, and mass diffusivity/conductivity. To characterise dynamic drying behaviour of materials, it is crucial to choose the model that can grasp water diffusion mechanism throughout the drying process and properly fit material–moisture ratio curve. The moisture content data are converted into moisture content ratio expressions, and then mathematical models are subjected to a curve-fitting calculation with the drying time. Table 12.1 reports the most frequently cited models. Several statistical approaches can be used to assess the degree of fit of experimental data to the mathematical model with acceptable accuracy [9].

TABLE 12.1
Mathematical Models Applied for Regression Modelling of Drying Kinetics

No.	Model Name	Model Equation	Reference
1	Page	$MR = \exp\left(-k \times t^n\right)$	[10]
2	Lewis	$MR = \exp(-k \times t)$	[11]
3	Wang and Singh	$MR = 1 + a \times t + b \times t^2$	[12]
4	Logarithmic	$MR = a \times \exp(-k \times t) + c$	[13]
5	Diffusion approach	$MR = a \times \exp(-k \times t) + (1-a) \times \exp(-k \times b \times t)$	[14]
6	Midilli and Kucuk	$MR = a \times \exp\left(-k \times t^n\right) \times b \times t$	[15]
7	Two terms	$MR = a \times \exp(-k1 \times t) + b \times \exp(-k2 \times t)$	[16]
8	Weibull	$MR = \exp\left(-(t/\beta)^n\right)$	[17]
9	Henderson and Pabis	$MR = a \times \exp(-k \times t)$	[18]
10	Verma et al.	$MR = a \times \exp(-k \times t) + (1-a) \times \exp(-g \times t)$	[19]

12.2.2 Theoretical Models

Theoretical models are based on fundamental physics involved in the drying process, primarily expressed by coupling fundamental transport equations with thermodynamically interactive fluxes and phase equilibrium expressions. Mostly, these models require high mathematical complexity and an excessive number of parameters [20]. The approach involves several steps summarised in Figure 12.2, starting with conceptualisation and establishing the transport phenomena equations to depict actual systems. The partial differential equations (PDEs) describe heat, mass, and momentum transfer in two distinct subdomains (air–food). Then, thermophysical properties of materials and transfer coefficients are determined. Finally, depending on the complexity of geometry and equations, PDEs are solved either analytically or numerically through multiple methods,

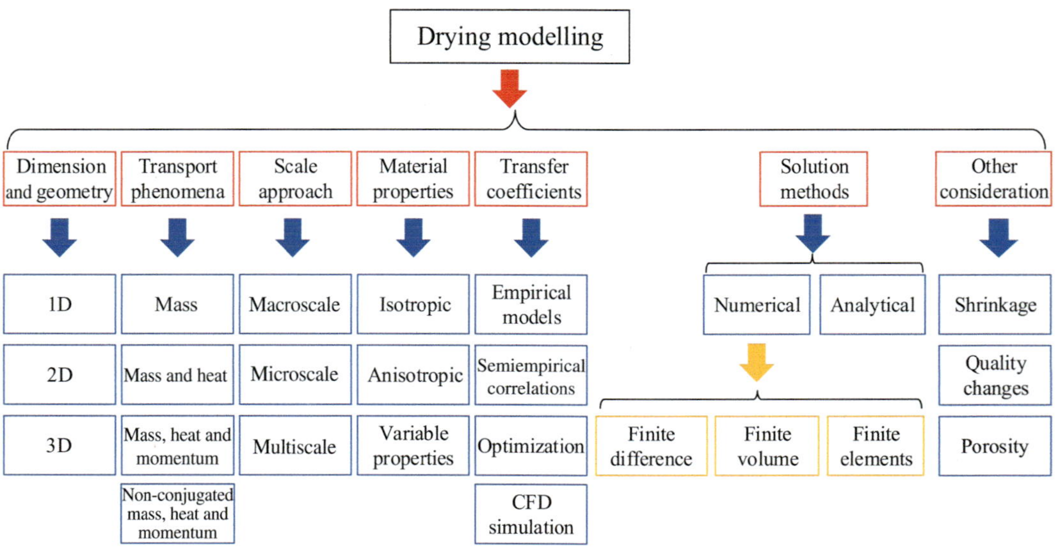

FIGURE 12.2 Drying modelling parameters of sustainable drying.

e.g., finite differences method (FDM), finite volume method (FVM), and finite elements method (FEM). Models are validated using experimental data to ensure that mathematical representation captures reality. Advanced numerical solutions have increased the use of theoretical models due to their relevance for food drying understanding by covering many coupled mechanisms and source terms.

In addition to experimental optimisation for industrial implementation, mathematical and numerical modelling are required to fully understand the drying mechanism and solve remaining problems such as drying uniformity and thermal runaway. Table 12.2 summarises some recent research on various sustainable drying technologies.

TABLE 12.2
Some Selected Works about Sustainable Drying Technologies

No.	Drying Method	Material	Highlights	Investigation Approach	Reference
1	Solar dryer	Figs	• High temperature (80°C) and velocity (300 m^3/h) decrease drying time (200 min) and improve slice quality • Specific energy (13.12 and 44.55 MWh/kg) inversely proportional to velocity and temperature • Energy efficiency was greater (3.98%) in drying conditions	• Experimental • Mathematical models	[21]
2	Solar dryer	Brewer's spent grains	• A natural convection solar dryer (NCSD) was compared to a conventional electric convective drying system (CECD) • The drying time for the CECD and NCSD were 30–85 min (for 363.15 K and 333.15 K) and 345–430 min (depending on the process starting) • NCSD is 40–45% cheaper than the CECD • NCSD offers a superior product quality and is eco-friendlier	• Experimental • Empirical models • ANN • CFD • Combined models	[22]
3	Ventilated solar-geothermal drying	Tomato paste	• Highest exergy and drying efficiency were achieved at 54% and 68.28% • Payback period of 0.169–0.325 years • Reduced drying time and better phenolic and flavonoids retention	• Experimental • Mathematical models	[23]
4	Microwave	Peppermint leaves	Specific energy consumption (SEC) is reduced compared to hot-air drying, but the essential oil yield was ten times lower	• Experimental • Mathematical models	[24]
5	Ultrasound (US)	Raspberry	Energy-saving and rehydration were improved. Porosity, total pore volume, and drying rate increased	• Experimental • Mathematical models	[25]
6	HPD	kiwi	Improved mass transfer gradient between low humidity air and the kiwifruit slices The addition of an auxiliary heat pump system improved drying kinetics and reduced energy usage from 21.7 kWh/kg to 7.5 kWh/kg	• Experimental • Mathematical models	[26]

(Continued)

TABLE 12.2 (Continued)
Some Selected Works about Sustainable Drying Technologies

No.	Drying Method	Material	Highlights	Investigation Approach	Reference
7	HPD	Banana slices	• The higher temperature reduced drying time and increased effective moisture diffusivity • The best drying performance was achieved at the highest measured temperature (430°C), consuming 3.586 kWh	• Experimental • Mathematical models	[27]
8	FD	Mushrooms	When water activity equalled 0.6, the ANN model properly predicted moisture content, primary and secondary drying temperatures, sample thickness, and time, resulting in a 34.9% reduction in energy usage	• Experimental • ANN • Semi-empirical models	[28]
9	Intermittent microwave and convective	Apple	Evaporation rate, capillary diffusion, effective thermal conductivity, and gas pressure can be characterised The influence of shrinkage on the accuracy of the predicted results was investigated. Energy efficiency and food quality were improved	• Mathematical models	[29]

12.2.3 Partial Differential Equations of CFD

CFD modelling is simply a numerical computation related to fluid mechanics, which consists of solving the fundamental equations using the law that governs fluid movement in a given geometry, which can optionally be coupled with heat and mass transfer equations. The solved equations are mainly the continuity or conservation of mass (Equation 12.1), Navier–Stokes or conservation of momentum (Equation 12.2), and conservation of energy equations (Equation 12.3) [30]. They are described as follows:

$$\frac{\partial \rho}{\partial t} + \nabla \cdot (\rho \vec{u}) = 0 \tag{12.1}$$

$$\frac{\partial(\rho u)}{\partial t} + \nabla \cdot (\rho u \vec{u}) = -\nabla P + +\mu \nabla^2 \vec{u} + \rho g \tag{12.2}$$

$$\rho C_p \frac{\partial T}{\partial t} + \rho C_p \vec{u} \cdot \nabla T = \nabla.(k \nabla T) + S \tag{12.3}$$

where ρ is the density (kg/m^3), $\vec{u}$ is the velocity components (m/s), P is the pressure (Pa), g is the acceleration of gravity (m/s), T is the temperature (°C) of at time t, C_p is the specific heat of the material (J/kg K), k is the thermal conductivity (W/m K), and S is the heat source (W/m^3).

12.3 METHODS OF SOLUTION OF THE GOVERNING DIFFERENTIAL EQUATIONS

Numerical methods provide an approximate solution for the investigated problem by discretising the computational domain into many finite elements or grids, transforming complex governing equations to simple linear or polynomial equations. The appropriate solution method may differ depending on the situation and the information that the scientist wants to obtain in a specific problem. The selection involves the characteristics of mathematical models, such as determining the domain geometry, initial and boundary conditions, and the classification of the differential equation. The most common numerical methods that have been successfully applied to solve those complex equations in convective drying are FDM (46% of previous works), FVM (39% of previous works), and FEM (15% of previous works) [31].

The finite differences approach is due to the work of several 18th-century mathematicians (Euler, Taylor, Leibniz), which consists of approximating PDEs by the domain discretisation in small elementary rectangles. The algorithm is classified as an explicit [32] or implicit system [33], linking the values of unknown functions at certain points sufficiently close to each other. Unlike the finite difference method, which involves approximations of the derivatives, the finite volume and finite element methods use integral approximations. However, the finite element method is based on a variable formulation of the equation (weak formulation), consequently working in non-polynomial spaces [34]. It is particularly well-suited to solve problems with complex shapes and solid mechanical structures involving a heterogeneous medium [35, 36]. The finite-volume method is also widely used in commercial CFD packages since they are simple to understand and program and have a high computation speed [37]. It also discretises the domain by generating various irregular grids such as polyhedra. The discrete unknowns are located inside the meshes and are not the ends of the meshes as in FDM.

Furthermore, modelling the convective heat and mass transfer of porous materials with their environment during the drying process is distinguished into two main categories: non-conjugated and conjugated models. The non-conjugated approach treats heat and mass transfer in the air and inside the material separately by assuming that airflow properties are unaffected by moisture and heat transfer within the product, so the effect of internal water vaporisation on the outside fluid flow is neglected [38]. The momentum transfer generally begins with determining temperature and velocity profiles to estimate convective transfer coefficients (CTCs) at air–product interface and then simulating heat and mass transport inside the product. CTC values are inputs to this model. Thus, an appropriate estimation of those coefficients is required to use such an approach. The study about the non-conjugated model conducted by Esfahani et al. [39] showed that the CTCs could be considered variable along the surface during drying simulation, allowing them to demonstrate the effect of front and rear faces of intended product.

While non-conjugated models are simple to formulate, flexible, and generally require less computational cost, conjugated models are complex, involve deep knowledge of mathematical equations, and are strong computational tools. In the conjugated approach, the heat and mass transport in the porous material and the surrounding airflow region are solved simultaneously in a transient way. Numerical modelling methods usually incorporate the heat diffusion equation inside the product to dry with external heat convection in the boundary conditions through CTCs, which are generally unavailable and increase the uncertainty in the computation. However, conjugated models maintain the continuity of heat and mass fluxes, temperature, and mass fraction at the air–food interface, excluding global and local CTCs in process simulation. They are instead evaluated as a part of the simulation process. The practical application of such an approach in the food drying industry can significantly improve the representation of morphological characteristics of the product, more insights, and description of the drying process, thus reducing energy consumption and laboratory testing cost.

Khan and Straatman [40] demonstrated the capability of conjugated methods to accurately predict the realistic situation of drying kinetics of porous materials, wherein mass, momentum, energy, and moisture transport equations are solved in parallel in two domains considering the water and vapour exchanges between the constituents. For each of the fluid and solid components of the porous material, separate energy and moisture transfer equations are used, which refers to non-equilibrium approach. Although simplifications are advantageous and only a few authors have employed the conjugated approach, the authors showed that using the complete formulation is more valuable to simulate non-equilibrium energy and mass transport in computational domains of all scales, and it is also necessary to take next step towards the dynamic phase-coupled approach.

The fluid flow can be considered as either laminar or turbulent. Several turbulence models and wall functions are often used when the flow is turbulent, which is often the case in modelling drying processes due to the complex geometry of dried products and high flow rates. The need for an accurate flow-field prediction makes selecting an appropriate model crucial and significantly impacts CFD performance. The most common turbulence models reported in the literature are k–ε models, k–w models, and Reynolds stress models. However, for further details about those models, the work of Malekjani and Jafari can be reviewed [41]. They have extensively investigated the application of different turbulence models in CFD simulations and multiphase numerical modelling methods such as Eulerian–Eulerian, Eulerian–Lagrangian, and volume of fluid [42]. A basic flowchart of the CFD simulation procedure is presented in Figure 12.3.

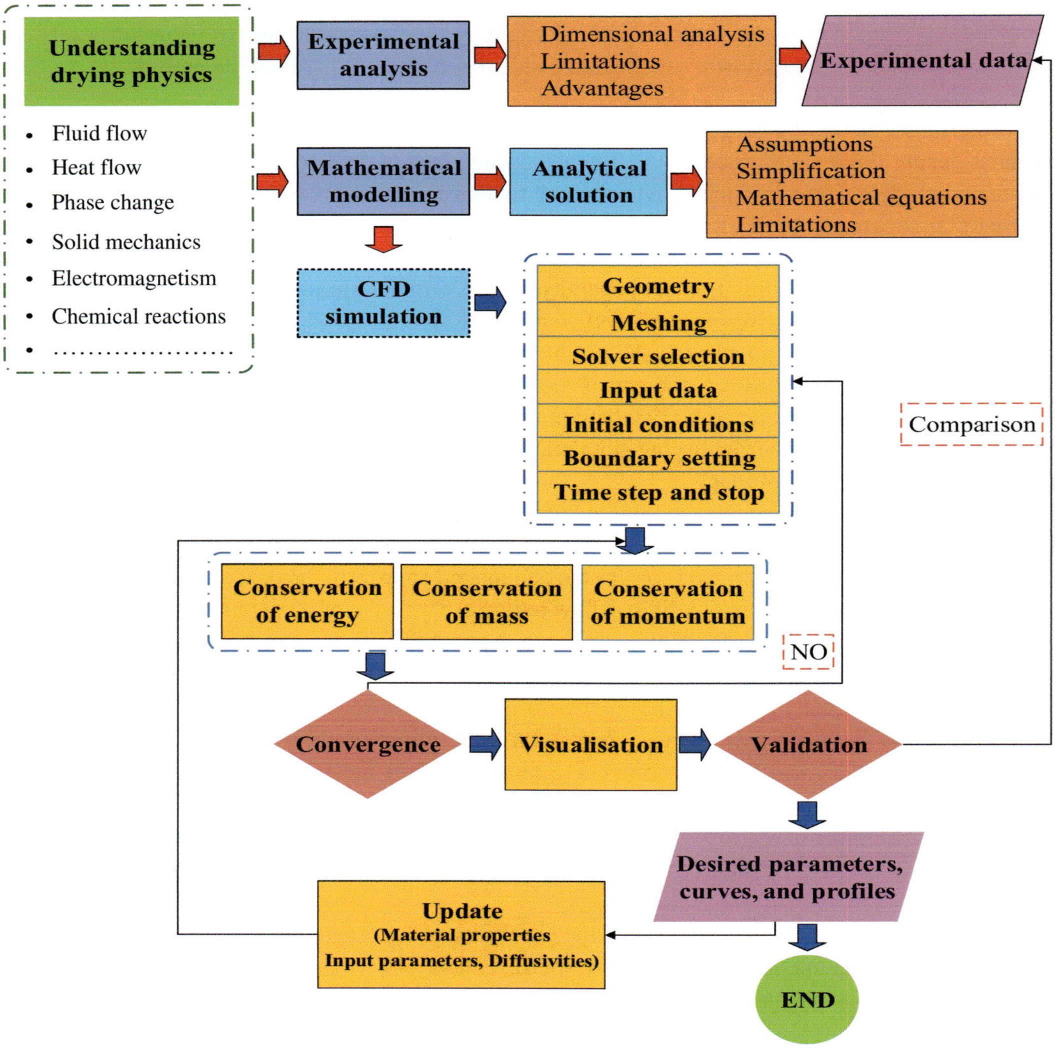

FIGURE 12.3 CFD simulation flowchart.

12.4 REVIEW OF CFD STUDIES IN DRYING PROCESSES

A broad selection of numerical models has been studied with their application to simulate heat and mass transfer processes inside the cabinet dryer and within the dried product. Therefore, based on the research scope, researchers decide which area to investigate.

12.4.1 CFD Investigations Inside the Drying Room

Modelling external flow is essential to increase the efficiencies of drying systems by reducing energy consumption and waste materials. CFD can help investigators to control drying conditions and find optimal parameters for better performance. In this respect, Chaudhari et al. [43] developed a mathematical model and performed a CFD simulation to determine the optimal air temperature in a solar drying room. The results unveiled that optimal performance was achieved at 45°C, where auxiliary power consumption was low, total energy consumption amount was 97.3 kWh per day, and solar energy utilisation was maximised. The results were in good agreement with the findings of the mathematical model. Besides, the numerical study indicated that the dried product should be placed farther away from the wall boundary because the air temperature was higher near the boundaries. Rahman et al. [44] performed a 3D Eulerian–Lagrangian multiphase model to optimise spray drying in a novel radial multizone dryer using ANSYS v14.5. They examined the influence of solids outlet location, droplet size, and relative hot/cold airflow ratio on residence time, separation efficiency of the product, G-acceleration, and heat and mass transfer characteristics to determine the most advantageous configuration and parameters based on the depicted correlations between the parameters mentioned above. A detailed insight into temperature and moisture fields inside the dryer was provided.

A sustainable solar tunnel dryer was designed by Afshari et al. [45] based on an experimental and numerical assessment. For three different structures and various process conditions, a 3D steady-state CFD simulation was performed to determine the adequate geometry for the tunnel dryer. CFD analysis was successfully carried out, and the results were in close agreement with the experimental investigation. The study demonstrated that a Quonset-type dryer with fins had better thermal performance than the rectangular configuration in terms of homogeneity in temperature and velocity distribution, energy efficiency, and drying time. Similarly, Babu et al. [37] presented a numerical simulation using ANSYS FLUENT to evaluate several geometrical configurations of drying chamber for thin-layer drying of leaves. The best configuration is characterised by the airflow uniformity and air temperature distribution, the highest mass transfer rate with short drying time, and minimal pressure drop. In a recent study, Tuncer et al. [46] had assessed experimentally and numerically the thermal performance of a quadruple-pass solar air collector combined with a pilot-scale greenhouse dryer. Evaluation results showed the high efficiency of the proposed design and its ability to improve the drying process of kiwi and red pepper. The highest percentage deviation between CFD and experimental was 10%.

Aktaş et al. [47] conducted a CFD analysis of heat and mass transfer over apricot slices to improve the performance of IRD. The authors demonstrated the effectiveness of integrating a heat recovery system into the drying unit to decrease total energy consumption. Approximately 58–62% of the consumed energy can be recovered. The use of low-velocity airflow resulted in successful moisture removal from the fruit. Selimefendigil et al. [48] studied the impact of inlet velocity, air temperature, size, and distance between moist objects on heat and mass transport. The drying time decreases at higher temperatures. When velocity increases, the size of the vortices expands, resulting in increased convection around the moist objects. Besides, because of the blockage effect, the larger the objects, the slower the mass transport. The interspacing effect is noticeable on temperature distribution in the back of the first block and the front of the second block. Figure 12.4 shows the effect of five different inlet velocity values on a generated 2D plane and 3D channel.

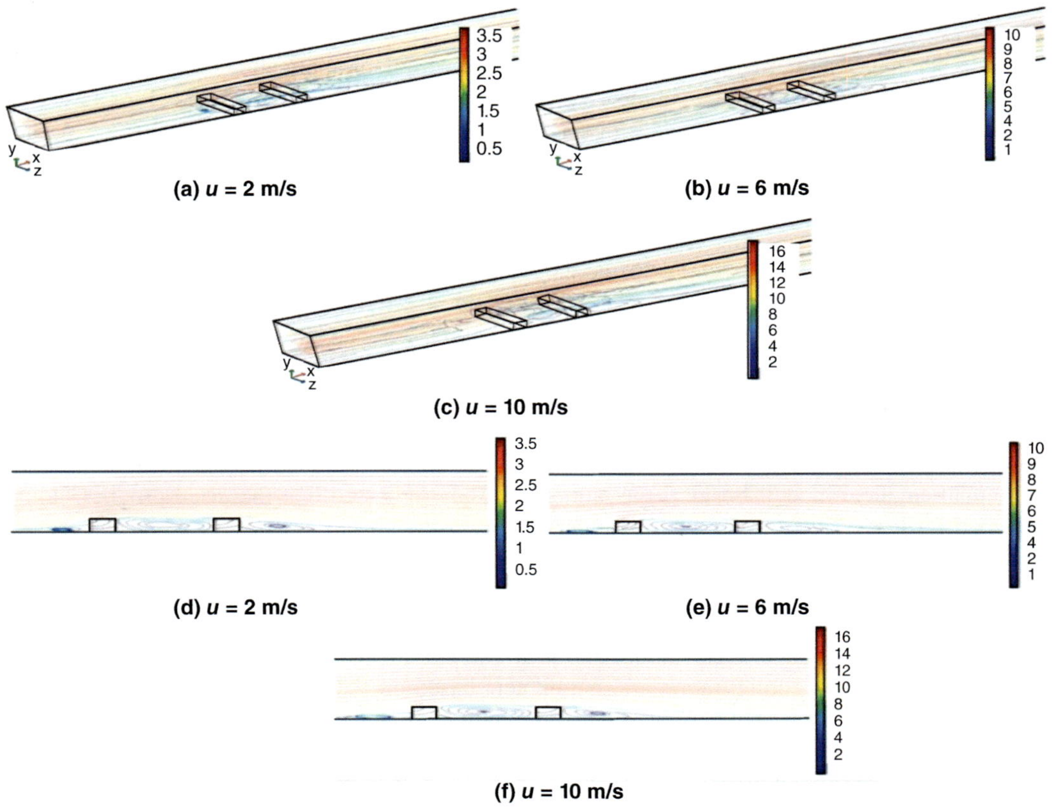

FIGURE 12.4 (a)–(f) Streamline distribution of airflow for different inlet velocities for 3D and 2D cut-planes ($h = 0.2{*}h$ (object height), $l = 5{*}h$ (object length), with h as the channel height, $t = 333.15$ K) [48].

12.4.2 CFD Investigations within the Dried Product

Several researchers carried out numerical simulations to investigate heat and mass transfer inside food materials during the drying process. In this respect, Zhu et al. [49] developed a multiphase media model with thermo-hydro and mechanical (THM) bidirectional coupling for food convective drying. A series of CFD simulations using COMSOL Multiphysics 5.4 was conducted to investigate the impact of shrinkage on the heat and mass transfer process during shiitake mushroom drying. Due to the changes in the flow area caused by shrinkage, the authors employed the Arbitrary Lagrangian–Eulerian dynamic grid approach for accurate predictions. The grid convergence index was used for grid-independent solution evaluation. The results revealed that the shrinkage greatly influenced the moisture content, the evaporation rate, and the temperature evolution (Figure 12.5). Thus, shrinkage consideration in numerical models can reliably simulate the drying characteristics of food materials to improve the drying process and enhance product quality. However, only a few studies on CFD modelling of fruit and vegetable shrinkage have been conducted. The challenges in establishing realistic models that take shrinkage into account have been addressed by Mahiuddin et al. [50].

Spatial distribution of air velocity around dried products is rarely incorporated in numerical studies, even though it influences CTCs that affect food drying kinetics. Khan et al. [51] recently tackled this research gap through a 3D model that integrated CFD with an IMCD model. For microwave volumetric heat generation modelling, Lambert's law was applied. Computational results were

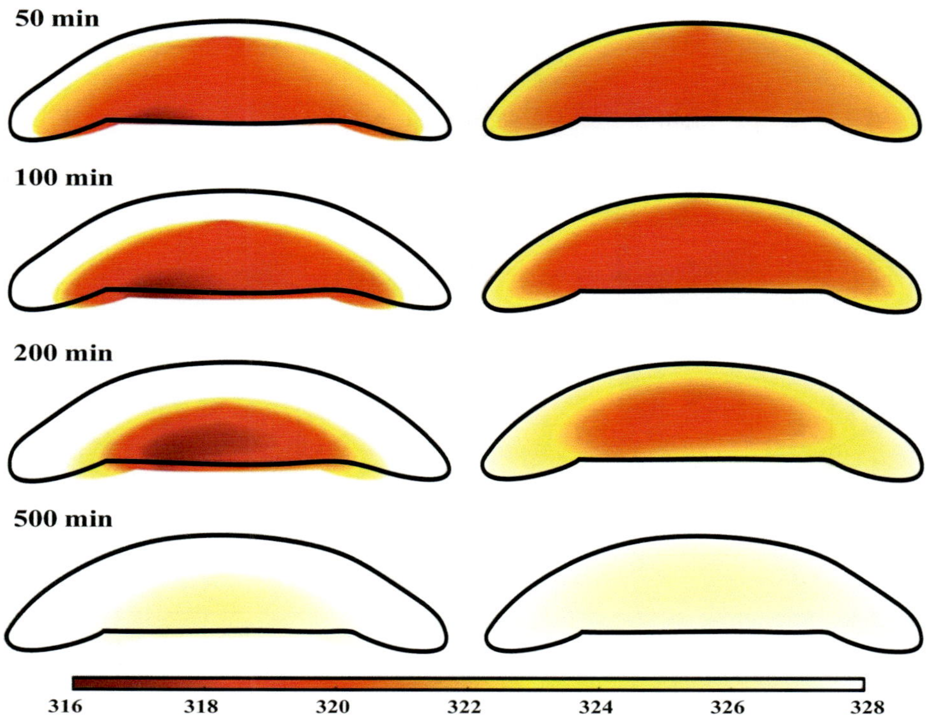

FIGURE 12.5 The comparison between the distribution of temperature within the shiitake mushroom from THM model (*left*) and TH model (*right*) at drying temperature of 328.15 K [49].

in good agreement with the experimental ones. It was found that spatial variation of CTCs due to air velocity changes significantly affected the moisture content and temperature distribution. Hence, without the presented model, over-prediction of drying kinetics may occur. Inhomogeneous flow distribution on the surfaces of the dried product is clearly observed in Figure 12.6. Besides, the non-uniform temperature distribution is found because of the uneven distribution of microwave power absorption.

Defraeye and Radu [52] carried out a numerical study to evaluate the ability of different modelling assumptions in analysing the effect of the peel on the drying behaviour of fruits. Within the fruit slice, considerable differences in internal moisture distribution were noticed, mainly due to the complex airflow field across the slice, which resulted in a spatial variation in CTCs over the different surfaces of the dried fruit. The discrepancies in numerical findings between the non-conjugated approach and those of the spatially resolved CTCs (conjugated) approach confirmed the necessity to impose a conjugated approach. The shrinkage consideration was also included, positively influencing the drying kinetics modelling despite the high cost of computation and solution convergence problems. The CTCs incorporation into heat and transfer models lead to a better prediction.

Many numerical works were devoted to non-conjugated modelling with assumptions of negligible shrinkage and constant thermophysical properties. Wei et al. [53] performed a 3D simulation using COMSOL 5.3a to determine corn kernels' temperature and moisture distribution. The geometric conceptualisation was based on three different configurations, which were one-component homogeneity (OH), multi-component heterogeneity of moisture diffusion coefficient (pericarp, soft endosperm, hard endosperm, and germ) (MD), and multi-component heterogeneity of initial moisture content and moisture diffusion coefficient (MMD). Numerical predictions were very close to the experimental observations of nuclear magnetic resonance imaging (MRI) and infrared thermal

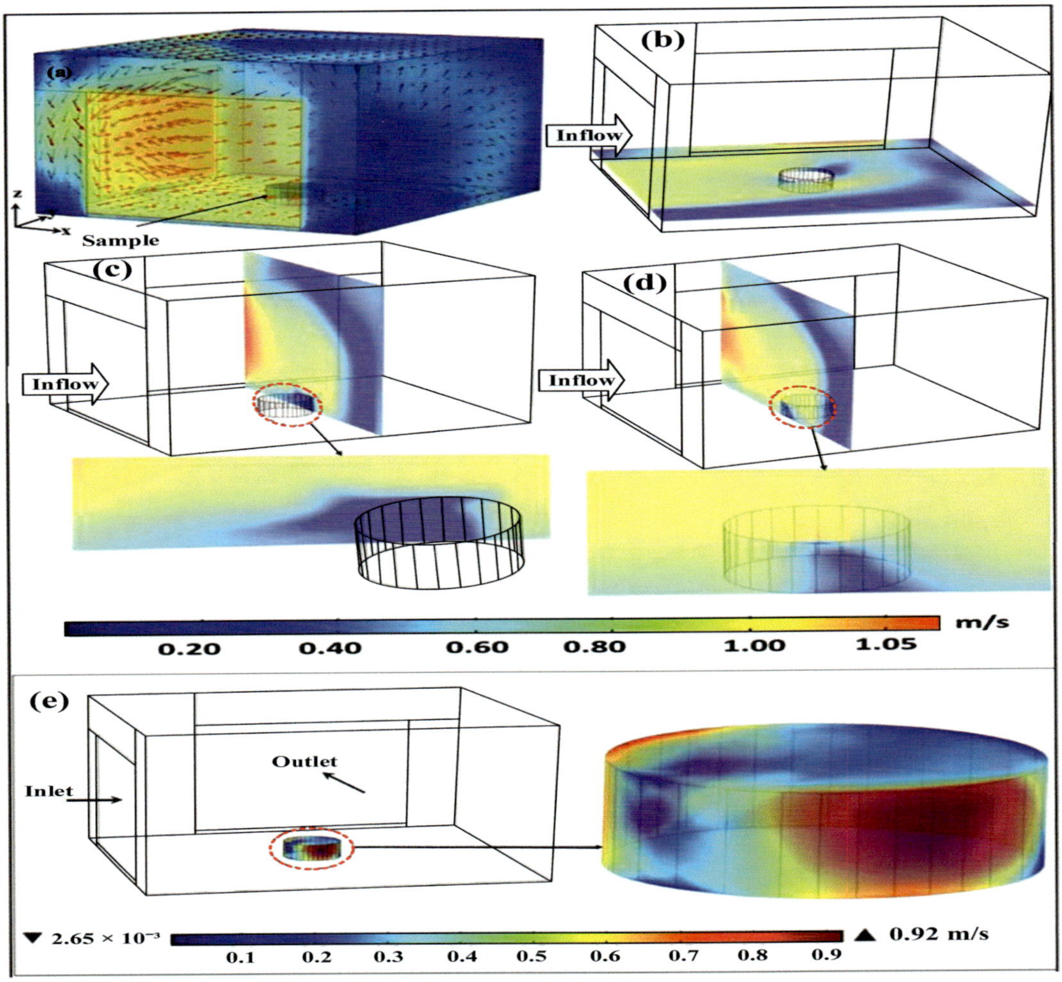

FIGURE 12.6 The fluid speed (m/s) distribution. (a) Isometric view of drying chamber with product. (b) Isometric view at *xy* plane. (c) Isometric view at *zx* plane of the sample with the drying chamber. (d) Isometric view at *zx* plane. (e) The spatial distribution [51].

imaging. A better description of the temperature and moisture distribution was attributed to the MMD configuration. Endosperm had lower moisture content, higher temperature than germ. Owing to the higher moisture and heat flux across the germ, the drying rate of germ was higher than that of the endosperm. The findings offered a more reliable alternative for predicting moisture and temperature distribution with multi-component food materials. Arunsandeep and Chandramohan [54] developed a 2D numerical model for spherical, cylindrical, and rectangular shapes to study the moisture and temperature behaviour during food drying. The authors determined the diffusion coefficient as a function of temperature, and then the estimated temperature-dependent diffusion coefficient was integrated into the transient mass transfer governing equation. The non-dimensionalised and discretised equations were solved simultaneously using MATLAB. The computed results were accurate and in good agreement with experimental values. Table 12.3 illustrates some selected research studies on numerical modelling drying systems by using various approaches. However, several unanswered concerns exist regarding the integration of quality and energy aspects in CFD-based Multiphysics models.

TABLE 12.3
Selected Numerical Modelling Research Works on Food Drying

Author(s)	Type of Model	Dimension	Fluid Flow	Software	Agro-Product	Key Highlights
Solomon et al. [35]	FEM	2D	Laminar	COMSOL v 5.5	Ethiopian fresh injera	Temperature and velocity had an impact on moisture reduction when drying was conducted ($p < 0.05$)
Selimefendigil [55]	FEM	3D	Laminar	COMSOL	Fruits and vegetables	The spacing between products influences the heat and mass transfer
Rani et al. [56]	FEM	3D	None	COMSOL v 5.3	Pineapple slice	Moisture reduced from 92% to 25% (wb) in 10.5 h. Spatial moisture distribution was visualised
Onwude et al. [57]	FEM	2D axisymmetric	None	COMSOL	Sweet potato	Infrared energy reduced the drying time, increased drying rate, and improved the product quality
Rufino et al. [58]	FVM	3D	None	Mathematica software	Rice grains	Comparison between continuous and intermittent drying for various operational conditions by simulation tests
Park et al. [59]	None	3D	Turbulent	COMSOL v5.2a	Soybeans	Mass transfer coefficient was estimated (0.0594 m/s)
Chandramohan [60]	FVM	2D	Laminar	FASTEST 3D	Rectangle-shaped moist object	Variable heat transfer and diffusion coefficients were integrated in numerical part
Curcio et al. [61]	FEM	3D	Laminar	COMSOL v4.4	Irregular-shaped vegetables	Conjugated model accurately predicted spatial moisture profiles at all times
Chilka et al. [62]	None	3D	None	ANSYS Fluent v14.5	Almonds	Effective diffusivity estimated from single-particle drying data predicted the moisture variation for a multiple particles system
Defraeye et al. [63]	FVM	3D	Turbulent	ANSYS Fluent 12	Apple	Microscale modelling of horticultural products for complex geometries
Cârlescu et al. [64]	FVM	2D	Laminar	ANSYS fluent 2006	Apricot	Blanched apricots dried faster than unblanched apricots. CFD simulation results can estimate the drying uniformity degree
Franco et al. [65]	FEM	3D	None	COMSOL v4.3	Yacon	Temperature and the thickness of the foam layers influenced process time, moisture content, and water activity

12.5 QUALITY EVALUATION AND CFD MODELLING OF DRYING

Although technological advancements in CFD have significantly contributed to modelling the food drying kinetics, few attempts have been made to predict the quality evolution of dried products. In this context, various drying parameters that influence food quality properties are presented in Figure 12.7. In a previous study, Devahastin and Niamnuy [66] briefly reviewed some advances in modelling quality changes of fruits and vegetables during drying, including approaches to model colour, nutritional, microstructural, and physical changes giving more details about the advantages and shortcomings of each modelling approach. They mentioned that the evolution of quality attributes could be assessed as a function of temperature and moisture content by reliable drying simulation techniques such as CFD and ANN. Quality modelling is a complex endeavour, involving physical deformations and microstructural changes in addition to heat and mass transfer.

According to Boekel and Safety [68], kinetics modelling can address various quality changes during the drying process, implying that changes can be captured in mathematical models comprising characteristic kinetic parameters, such as rate constants and activation energies. Therefore, for a single quality attribute at concentration C, the general rate law model for the quality loss is expressed as follows:

$$\text{Rate} = \frac{-\mathrm{d}C}{\mathrm{d}t} = k_r C^p \tag{12.4}$$

where p is the order of reaction and k_r is the rate constant. For a single component, the differential equation represents the dependence of rate r on concentration. The equation can be integrated with respect to time to acquire an equation that directly links concentration with time. The essence of kinetic modelling of quality changes during the drying process is to evaluate a quality attribute as

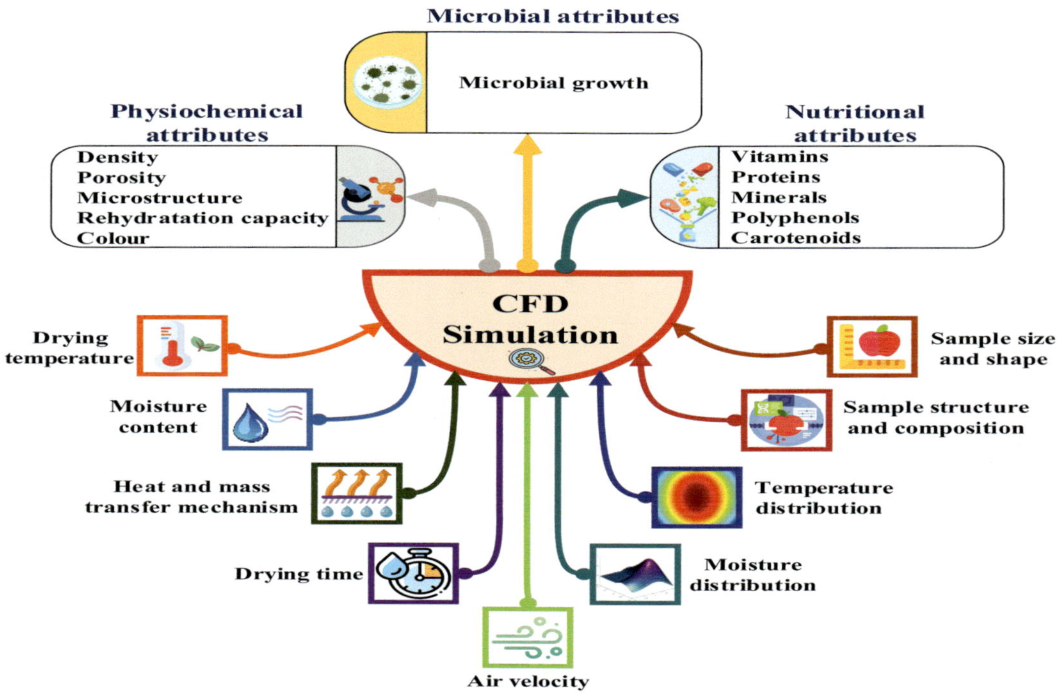

FIGURE 12.7 Drying process parameters and quality attributes relationship for CFD modelling [67].

a function of drying time at a specific temperature using temperature-dependent reaction rate constants after defining the order of reactions. The authors reported that quality changes are usually modelled using zero-, first-, or second-order reactions as follows:

$$C = C_0 - k_r t \text{ for zero-order reactions } (p = 0) \tag{12.5}$$

$$C = C_0 \mathrm{e}^{-k_r t} \text{ for first-order reactions } (p = 1) \tag{12.6}$$

$$\frac{1}{C} = k_r t + \frac{1}{C_0} \text{ for second-order reactions } (p = 1) \tag{12.7}$$

where C_0 is the initial value of concentration at $t = 0$. Zero-order reactions are the most frequently reported, especially for representing browning changes resulting from Maillard reactions. First-order reactions are also frequent, but for degradation reactions such as vitamin loss. Second-order reactions remain the least reported; they could be used to model the loss of amino acids due to Maillard reactions. The order of the reaction is often not based on a theoretical cause, and it is simply a fit parameter. Arrhenius law is often used to describe the temperature dependence on quality decay [69]:

$$k_r(T) = Ae^{\frac{-E_a}{RT}} \tag{12.8}$$

where A is a pre-exponential factor, E_a is the activation energy, R is the ideal gas constant (8.314 J/mol·K), and T is the absolute temperature. The activation energy is defined as the energy barrier required to initiate a chemical reaction. The number of molecules with energy higher than activation energy increases with temperature, which explains the influence of temperature on reaction rate qualitatively. Using a high order for the Arrhenius equation by including more terms that affect food quality during drying is recommended.

As of date, very few studies reported the application of kinetics modelling for quality evaluation. Recently, Prawiranto et al. [70] identified trade-offs between drying time, product, quality, and energy use for solar drying of apple fruit, performing a CFD simulation at different weather conditions to assess drying characteristics and quality deterioration (e.g., browning reaction, vitamin C decay). Defraeye [71] optimised the intermittent drying process conditions using numerical hygrothermal simulations and a temperature-dependent model for quality loss. For both zero- and first-order reaction kinetics, Defraeye and Radu [52] compared the evolution of the quality attribute over time for different numerical simulation assumptions. Moreover, Pham et al. [72] developed a CFD model to predict temperature and moisture distributions along with nutrient depletion, such as total phenolic content, ascorbic acid, and colour variability during drying. Similarly, the kinetics of colour changes in trellis dried sultanas was also simulated and validated by Sabarez [73]. The simulation models increased the ability to predict heat and mass transfer processes as well as product quality profiles in a spatiotemporal domain. Besides, solving heat and mass transfer and then Arrhenius equations separately allow the operator to monitor quality attributes during drying and adjust the drying inputs to save energy and improve quality.

While kinetic modelling seems to be simple, the reactions are more complicated. For example, changes including lipid oxidation, enzymatic degradation, and Maillard reaction cannot be given in a single basic equation. The previously mentioned models are deterministic because they generate the same output when given the same input from reliable drying simulations. The pitfall is that the real world is unsettled, since uncertainty and variability are inherent in biological materials. Therefore, deterministic models are unrealistic and can only be applied to quantify the quality attributes of an ideal food system. The full potential of the kinetic modelling approach has not been proven, and hence, further studies should focus on developing more realistic models for quality prediction during food drying.

Other types of models can predict changes in quality attributes during the drying of food materials. Wijerathne et al. [74] conducted a realistic simulation of morphological changes of apple tissues undergoing drying process using a novel coarse-grained multiscale numerical model, which integrated mesh-free features. Mesh-free-based numerical models are a recent and popular advancement in numerical modelling of microscale tissue deformations while drying, which have many advantages concerning performance and versatility over classical grid-based modelling approaches (FEM/FDM methods) [75]. These new numerical models have shown a high capability to provide more robust solutions for complex physical phenomena: excessive deformations of cell aggregates, intense moisture reductions, multiscale physics, and multiphase interactions. It should be noted that using a 3D numerical model addresses more details and describes better cellular-scale drying behaviour than previous 2D numerical models. However, the high computational cost is a substantial hurdle, especially in large tissue modelling. Overall, it can be concluded that identifying critical relationships and correlations between quality attributes and drying process parameters may provide new opportunities to improve drying quality, mainly through different modelling approaches. Additionally, the enhanced quality can be preserved more effectively without requiring additional energy.

12.6 ENERGY CONSUMPTION AND CFD MODELLING OF DRYING

Energy analysis is a primary step in improving drying system performance and energy efficiency. An effective assessment provides information on energy consumption and cost by analysing specific data at the system's input and output. Recently, more studies have focused on developing drying modelling methods, particularly CFD models. Extensive research has also been conducted on the energy evaluation of food drying. Previous researchers used various energy performance indicators to conduct an energy analysis [76–82]. In this context, Kudra [83] presented a detailed methodology of calculation of the most common indices that evaluate and benchmark different convective dryers from an energy standpoint. For instance, Aktaş et al. [84] experimentally investigated the effect of different drying air temperatures (45°C and 50°C) on the performance of a HPD of grape pomace system. Despite the high energy supply, SEC, drying efficiency, heat pump COP, and SMER calculations indicated that the drying system performed better at lower temperatures.

Furthermore, Tohidi et al. [85] assessed energy attributes for fixed deep-bed drying of paddy. For a moisture content of paddy from 20.4% to 12% ($kg_{water}/kg_{wet\ matter}$), the mean values of consumed thermal and mechanical energies were 94.48% and 5.52%, respectively. The study found that increasing air temperature and decreasing air velocity and relative humidity improved energy efficiency. On the other hand, Lingayat et al. [86] evaluated thermal efficiency and heat loss to portray energy variations during indirect natural convection solar drying of bananas. The authors pointed out that energy analysis is insufficient to encounter all the deficiencies influenced by thermodynamic phenomena during the drying process, thus an exergy analysis was needed. Exergy analysis is also a practical tool that can offer a more in-depth evaluation of the drying process efficiency by identifying the main points of energy degradation and irreversibility [87] in order to better understand exergy loss values and increase the energy efficiency [88, 89].

Experimental energy analysis can be a time-consuming process involving costly instruments with significant imprecisions. Hence, virtual representation is recommended to depict the drying process and supplement the available sensor data [90]. However, energy or exergy features are seldom integrated into numerical modelling. Most numerical studies in the literature are concerned with drying kinetics and food quality prediction to optimise and control the drying process conditions and equipment. Only a few researchers have made some steps towards including energy consumption in these advanced computational modelling methods. Majdi and Esfahani [91] solved coupled conjugated heat and mass transfer equations during convective drying of apple slices using the Lattice Boltzmann method (LBM). Optimal drying parameters were also statistically determined using response surface methodology (RSM) and Box–Behnken design. The comparison between computed LBM and forecasted RSM results revealed that the errors were 1.2% and 1.6%

for the minimal energy consumption and the drying time, respectively. The analysis of the effect of the drying parameters demonstrated that the thickness ratio of the moist product has the greatest impact on energy consumption compared to the effect of velocity and temperature.

A recent computer simulation was developed by Duc and Hyuk using mathematical modelling to predict the moisture, temperature, and energy consumption during concurrent-flow drying of rapeseeds [92]. The simulated values of fuel energy consumption for rapeseed drying were 4.62% and 8.57% less than the measured values. Zhang et al. [93] performed an engaging multiscale CFD simulation to study the mechanism of energy consumption in hot-air drying of corn kernels. Heat and mass transfer were simulated using an accurate 3D real-body model, which is first created by scanning corn grains with a high-precision medical computerised tomography machine, then edited by MIMCS and reconstructed in ANSYS, and finally exported for postprocessing in COMSOL. A dimensional analysis demonstrated that hot-air velocity and temperature are the main factors influencing energy consumption. Based on the dimensional homogeneity theorem, the authors used the following equation:

$$W = \phi \Delta T C_p G = \phi \rho_a \Delta T C_p u \tag{12.9}$$

where W is the energy consumption dimension, u is hot wind velocity, T is the hot wind temperature, ρ is the density, C_p is the specific heat, G is the air mass flow, and ϕ is a constant.

Air temperature has a major effect on the interior mass transfer, primarily hot-air drying process, while air velocity has virtually no impact. The computed values showed that air velocity and temperature are linearly related to energy consumption, which varies along with the diffused moisture. Energy consumption can be minimised at low moisture content and velocity. Modelling such processes leads to excellent energy consumption optimisation and convective drying time. It can alleviate the environmental and economic impact of the food drying process.

12.7 ENVIRONMENTAL ASSESSMENT AND CFD MODELLING OF DRYING

In recent decades, quality enhancement of dried goods led to intensive energy consumption. The growing awareness of global warming and the United Nations 2030 Sustainable Development Goals [94, 95] is driving interest in decarbonising food drying. It is a cutting-edge transformation in the food industry requiring substantial expenditure and coordination among companies, governments, and other stakeholders [96]. Efforts are currently directed towards curtailing GHG to mitigate climate change while improving industrial productivity and competitiveness [97]. Carbon footprint analysis is a valuable tool to evaluate and quantify the environmental impacts of products and processes at any life stage across the food chain from cultivation, processing, transport, and delivery. This methodology is based on the principle of life cycle assessment (LCA), which is used to assess the performance of the drying processes and highlight developments and improvements needed to meet sustainability criteria [98].

The most promising approaches that could bring industries GHG emissions to a minimal value are energy-efficiency improvements, automation and robotics (Industry 4.0 concept), the use of biomass or hydrogen fuel or feedstock, heat electrification, and carbon capture usage (CCU) or storage (CCS). For agricultural products drying processes, the decarbonisation options mentioned above are well summarised in Figure 12.8. The optimal combination of decarbonisation options and gains can differ depending on the facility. Options need to be evaluated on a site-specific basis to decide which ones seem more feasible and economical by closely examining local factors. A recent step towards implementing CCU and CCS into fertiliser production was done by Gonzalez-Diaz et al. [99]. The evaluation indicated that the percentage of GHG emissions was remarkably reduced by 6–7% in the rice supply chain and 14–16% in the wheat supply chain. Moreover, enhanced oil recovery seems profitable at an industrial scale and may provide an economic incentive. The existing literature is insufficient to draw steady conclusions about the impact of incorporating CCU and CCS in the crop drying process on GHG emission reduction. Hence, this method merits further research.

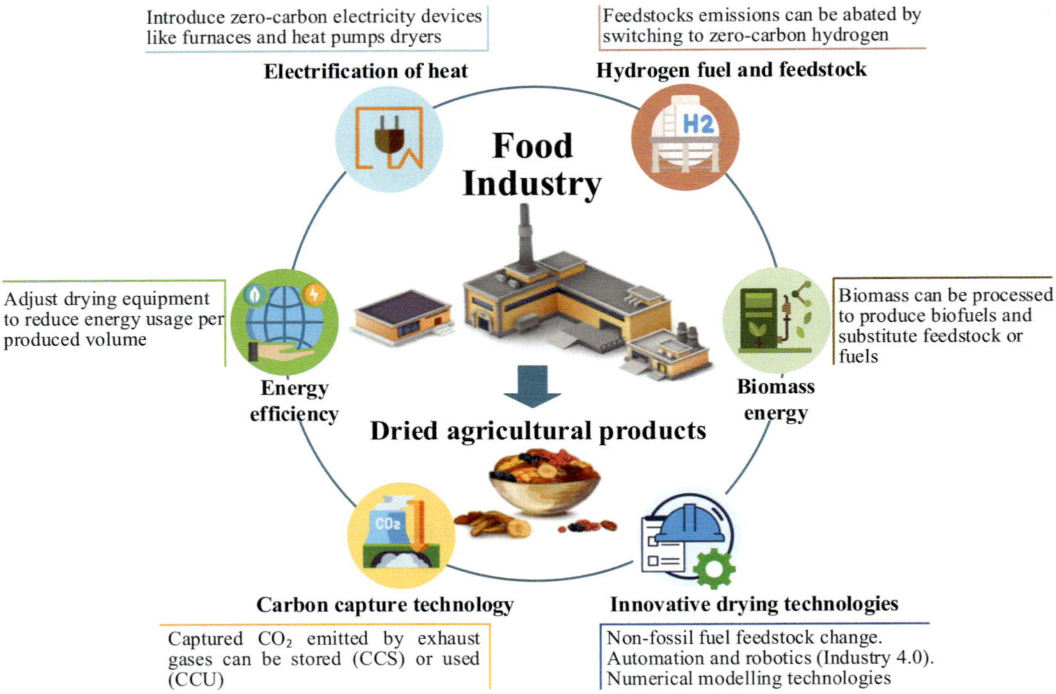

FIGURE 12.8 Options for decarbonising drying of agricultural products [67].

Grid decarbonisation would be critical in achieving long-term GHG reduction in food and beverage industry. Thus, the electrification of heat is considered one of the most important options. A recent review of technology options related to the decarbonisation of food processing by heat electrification was carried out by Atuonwu and Tassou [96]. They conducted for the first time a comparative energy analysis between different food processing technologies (e.g., drying, evaporation, cooking thawing, and tempering) at different unit operations levels and product types. The study proved that the electrified technologies significantly reduce food processing GHG emissions compared to conventional thermal treatment due to the combined effects of electricity grid decarbonisation, the end-use energy efficiency, and the processing time savings guaranteed by the emergent technologies. Excellent savings were found to be associated with phase change and structure formation processes on solid foods.

Furthermore, CFD modelling could be considered an efficient mitigation methodology elaborated from a technical perspective [100]. Numerical modelling approaches offer a powerful tool to simulate drying processes and alleviate the influence of intensives experimental methods in increasing carbon emissions. Efficient knowledge and control of the mechanism of the drying process are crucial to saving energy. Although a growing number of researches on reducing GHG emissions are conducted, no published paper shows the potential of CFD in decarbonising food manufacturing. Due to a lack of relevant details and comprehensive datasets, it is challenging to estimate the range of reduction achieved using such an approach.

12.8 CHALLENGES OF NUMERICAL MODELLING APPROACHES

Experiments suffer from certain shortcomings, such as the inability to acquire data at inaccessible points and at all time steps, the limitation in the number of tests and variation of process conditions, and the need for extensive laboratory areas with trained operators for maintenance. Previous studies have attempted to overcome the drawbacks mentioned above by modelling the drying process through CFD simulations. Despite their success, certain limitations associated with the numerical approach also exist. Hence, the simulation of such complex processes using CFD poses significant challenges, particularly given that the model might be required to estimate heat and mass transfer coefficients, temperature distribution, velocity, streamlines, and moisture content.

Food products are among the materials that stand out due to their complexity of structure, composition, and biological variability, making modelling the drying process difficult. As a result, geometry generation of such models is frequently challenging on smaller scales or for complex and irregular morphologies. Besides, drying involves the interaction of many operating parameters, which are strongly coupled and sometimes unusual. The more the incorporation of additional parameters into the model, the more complex the solution. As a result, writing a separate computer code or calling a user-defined function (UDF) becomes necessary and tedious, the computation time increases, and a powerful computer is needed. Such a drying modelling technique for low-value agricultural harvests might be economically questionable, given that considerable efforts from experts are required, and the computation facilities are expensive. Another major hurdle in using numerical modelling for drying analysis is the insufficient data on the material properties, particularly the physicochemical properties of food, which are highly temperature and moisture content dependent, change during drying, and show inter-and intra-sample variability. The challenge is to obtain the material properties optimally in a minimum of time and have the right skills to use the necessary specialised equipment. Moreover, the computational cost of modelling three-dimensional and complex realistic configurations requires user-friendly modelling software with a simple setup, simulation process, and straightforward implementation of the relevant multiphysics processes. Currently, there is a gap in the development of such tools due to different requirements of universities and industries, where industrial users need fast and user-friendly tools for realistic 3D engineering problems, unlike researchers who prefer to develop open-source codes and interfaces to study more fundamental aspects of drying processes that are currently beyond the reach of commercial tools.

The consideration of dehydration within porous media, airflow, and shrinkage can often be a bottleneck when modelling the drying process. Many researchers only considered mass transfer to simplify their models, and few worked on heat, mass, and momentum transfer simultaneously. In addition to multiphysics modelling, integrating energy sources and local weather conditions into current simulation software is neglected and scarce. The challenging task of better predicting and satisfactorily illustrating the realistic system with the actual occurring phenomenon during food dehydration using CFD is to determine the optimal balance between simplifying assumptions and the level of accuracy. A summary of the strengths and weaknesses of CFD modelling is provided in Table 12.4.

TABLE 12.4
Strengths and Weaknesses of CFD Modelling

Strengths	Weaknesses
• Product geometry and morphology are depicted in detail in a scaled model • Possibility of analysing various problems whose experiments are perilous and inaccessible • Provide the capability of studying the system under conditions that exceed its limits • Simulations provide measured data at any point (of the pattern), whereas experiments provide data only at a few selected points • Compared to experiments, the costs are usually much relatively low • Colourful visualisation and physical tracking of flow characteristics • Virtual storage of data and negligible drying space requirements	• Errors may occur as a result of simplified boundary conditions or flow models • Possible uncertainties caused by insufficient computing values per cell resulting in interpolation errors • Time-consuming computation for large and complex models • In comparison to experiments, the costs of incorrect consulting may be much higher • Untrained CFD users tend to believe that the computer's output is always correct

12.9 CONCLUSION

Understanding the physical and mathematical fundamentals of the drying process is vital for a sustainable production with a high-quality dried product. Therefore, computational modelling is highly advantageous to heat and mass transfer prediction and drying parameter optimisation. However, such modelling techniques cannot substitute for laboratory facilities, pilot plants, or full-scale experiments; they are applied to complement them in enhancing the performance of existing dryers and creating innovative designs suited to the needs of the future population.

In this chapter, various approaches, including conjugated models, multiscale models, shrinkage consideration, 3D conceptualisation, and models for emerging drying systems, were discussed. Comparative analyses between different approaches will be beneficial and recommended. Determining whether the model is applied to gain scientific insight into the occurring phenomena during drying or to achieve engineering goals is elementary and precedes the definition of a suitable approach. Indeed, complex and rigorous simulations are more reliable and accurate but remain challenging to solve, costly, and time-consuming. Thus, depending on the research aim, the simplest models could successfully depict reality in some instances.

Such interrelation modelling can be done using a single CFD software or multiple simulation tools [101]. According to the Food And Agriculture Organization of the United Nations (FAO), water, energy, and food security are intrinsically linked, which means that actions in one area can often have an impact on one or both of the others [102]. Therefore, computational modelling should move towards integrating the food–energy–water concept [103]. Under the nexus approach, the inextricable linkages, synergies, and trade-offs require further analysis and investigation to ensure long-term development and allowing experts to take further steps forward into next-generation dying technology.

Despite challenges and limitations of numerical modelling techniques, these models are subject to improvement to develop energy-efficient dryers that are environment-friendly with low GHG emissions, cost-effective, and can handle the retention of quality. This chapter can assist the reader in comprehending in-depth the topics and issues covered in other chapters.

REFERENCES

1. Kudra, T., and A. Martynenko, *Electrohydrodynamic drying: Theory and experimental validation.* Drying Technology, 2019. **38**(1–2): p. 168–175.
2. Mhd Safri, N.A. et al., *Current status of solar-assisted greenhouse drying systems for drying industry (food materials and agricultural crops).* Trends in Food Science & Technology, 2021. **114**: p. 633–657.
3. Kuriakose, R., and C. Anandharamakrishnan, *Computational fluid dynamics (CFD) applications in spray drying of food products.* Trends in Food Science & Technology, 2010. **21**(8): p. 383–398.
4. Sharma, S.K., and V.R. Kalamkar, *Computational fluid dynamics approach in thermo-hydraulic analysis of flow in ducts with rib roughened walls: A review.* Renewable and Sustainable Energy Reviews, 2016. **55**: p. 756–788.
5. Caccavale, P., M.V. De Bonis, and G. Ruocco, *Conjugate heat and mass transfer in drying: A modeling review.* Journal of Food Engineering, 2016. **176**: p. 28–35.
6. Sinha, A., and A. Bhargav, *A simplified modelling approach for predicting shrinkage and sensitive material properties during low temperature air drying of porous food materials.* Journal of Food Engineering, 2022. **317**: p. 110732.
7. Getahun, *Importance of integrated CFD and product quality modeling of solar dryers for fruits and vegetables: A review*, Solar Energy, 2021. **220**: p. 88–110.
8. Mathew, A.A., and V. Thangavel, *A novel thermal energy storage integrated evacuated tube heat pipe solar dryer for agricultural products: Performance and economic evaluation.* Renewable Energy, 2021. **179**: p. 1674–1693.
9. Ertekin, C., and M.Z. Firat, *A comprehensive review of thin-layer drying models used in agricultural products.* Critical Reviews in Food Science and Nutrition, 2017. **57**(4): p. 701–717.
10. Page, G.E., Factors influencing the maximum rates of air drying shelled corn in thin layers. 1949. Unpublished M.S. thesis, Purdue University, Lafayette, IN.

11. Lewis, W.K., *The rate of drying of solid materials.* The Journal of Industrial and Engineering Chemistry, 1921. **13**(5): p. 427–432.
12. Babalis, S.J. et al., *Evaluation of thin-layer drying models for describing drying kinetics of figs (Ficus carica).* Journal of Food Engineering, 2006. **75**(2): p. 205–214.
13. Chandra, P.K., and R.P. Singh, *Applied Numerical Methods for Food and Agricultural Engineers.* 2017. CRC Press.
14. Kassem, A. *Comparative studies on thin layer drying models for wheat.* In 13th International Congress on Agricultural Engineering. 1998.
15. Midilli, A., H. Kucuk, and D. Yapar, *A new model for single-layer drying.* Drying Technology, 2002. **20**(7): p. 1503–1513.
16. Henderson, S., and S.J.J.A.E.R. Pabis, *Grain drying theory: IV. The effect of air flow rate on the drying index.* Journal of Agricultural Engineering Research, 1962. **7**(2): p. 85–89.
17. Uribe, E. et al., *Characteristics of convective drying of pepino fruit (Solanum muricatum* Ait.*): Application of Weibull distribution.* Food and Bioprocess Technology, 2011. **4**(8): p. 1349–1356.
18. Akpinar, E.K., Y. Bicer, and F.J.J. Cetinkaya, *Modelling of thin layer drying of parsley leaves in a convective dryer and under open sun.* Journal of Food Engineering, 2006. **75**(3): p. 308–315.
19. Verma, L.R., et al., Effects of drying air parameters on rice drying models. Transactions of the ASAE, 1985. **28**(1): p. 296–301.
20. Sabarez, H.T., *Modelling of drying processes for food materials.* Modeling Food Processing Operations, 2015. p. 95–127.
21. Hssaini, L. et al., *Kinetics, energy efficiency and mathematical modeling of thin layer solar drying of figs (Ficus carica* L.*).* Scientific Reports, 2021. **11**(1): p. 21266.
22. Capossio, J.P. et al., *Sustainable solar drying of brewer's spent grains: A comparison with conventional electric convective drying.* Processes, 2022. **10**(2): p. 339.
23. Hadibi, T. et al., *Effect of ventilated solar-geothermal drying on 3E (exergy, energy, and economic analysis), and quality attributes of tomato paste.* Energy, 2021. **243**: p. 122764.
24. Torki-Harchegani, M. et al., *Dehydration behaviour, mathematical modelling, energy efficiency and essential oil yield of peppermint leaves undergoing microwave and hot air treatments.* Renewable and Sustainable Energy Reviews, 2016. **58**: p. 407–418.
25. Szadzińska, J. et al., *Microwave- and ultrasound-assisted convective drying of raspberries: Drying kinetics and microstructural changes.* Drying Technology, 2019. **37**(1): p. 1–12.
26. Mohammadi, I., R. Tabatabaekoloor, and A. Motevali, *Effect of air recirculation and heat pump on mass transfer and energy parameters in drying of kiwifruit slices.* Energy, 2019. **170**: p. 149–158.
27. Tunckal, C., and İ Doymaz, *Performance analysis and mathematical modelling of banana slices in a heat pump drying system.* Renewable Energy, 2020. **150**: p. 918–923.
28. Tarafdar, A. et al., *Artificial neural network modeling of water activity: A low energy approach to freeze drying.* Food and Bioprocess Technology, 2018. **11**(1): p. 164–171.
29. Joardder, M.U.H., C. Kumar, and M.A. Karim, *Multiphase transfer model for intermittent microwave-convective drying of food: Considering shrinkage and pore evolution.* International Journal of Multiphase Flow, 2017. **95**: p. 101–119.
30. Yadav, A.S., and J.L. Bhagoria, *Heat transfer and fluid flow analysis of solar air heater: A review of CFD approach.* Renewable and Sustainable Energy Reviews, 2013. **23**: p. 60–79.
31. Chandramohan, *Convective drying of food materials: An overview with fundamental aspect, recent developments, and summary.* Heat Transfer, 2020. **49**(3): p. 1281–1313.
32. Singh, P., V. Shrivastava, and A. Kumar, *Recent developments in greenhouse solar drying: A review.* Renewable and Sustainable Energy Reviews, 2018. **82**: p. 3250–3262.
33. Arunsandeep, G. et al., *A numerical model for drying of spherical object in an indirect type solar dryer and estimating the drying time at different moisture level and air temperature.* International Journal of Green Energy, 2018. **15**(3): p. 189–200.
34. Montanuci, F.D. et al., *Experimental analysis and finite element simulation of the hydration process of barley grains.* Journal of Food Engineering, 2014. **131**: p. 44–49.
35. Solomon, A.B. et al., *Modeling and simulation of heat and mass transfer in an Ethiopian fresh injera drying process.* Heliyon, 2021. **7**(2): p. e06201.
36. Zhao, C.-J. et al., *A review of computational fluid dynamics for forced-air cooling process.* Applied Energy, 2016. **168**: p. 314–331.
37. Babu, A.K. et al., *CFD studies on different configurations of drying chamber for thin-layer drying of leaves.* Energy Sources, Part A: Recovery, Utilization, and Environmental Effects, 2019. **42**(18): p. 2227–2239.

38. Sun, D.-W., Computational Fluid Dynamics in Food Processing. 2018. CRC Press.
39. Esfahani, J.A., H. Majdi, and E. Barati, *Analytical two-dimensional analysis of the transport phenomena occurring during convective drying: Apple slices*. Journal of Food Engineering, 2014. **123**: p. 87–93.
40. Khan, F.A., and A.G. Straatman, *A conjugate fluid-porous approach to convective heat and mass transfer with application to produce drying*. Journal of Food Engineering, 2016. **179**: p. 55–67.
41. Malekjani, N., and S.M. Jafari, *Simulation of food drying processes by computational fluid dynamics (CFD); Recent advances and approaches*. Trends in Food Science & Technology, 2018. **78**: p. 206–223.
42. Jamaleddine, T.J., and M.B. Ray, *Application of computational fluid dynamics for simulation of drying processes: A review*. Drying Technology, 2010. **28**(2): p. 120–154.
43. Chaudhari, V.D., G.N. Kulkarni, and C.M. Sewatkar, *Mathematical modeling and computational fluid dynamics simulation of cabinet type solar dryer: Optimal temperature control*. Journal of Food Process Engineering, 2021. **44**(2). DOI: 10.1111/jfpe.13643
44. Jamil Ur Rahman, U. et al., *Numerical study toward optimization of spray drying in a novel radial multizone dryer*. Energies, 2021. **14**(5): p. 1233.
45. Afshari, F. et al., *Dehumidification of sewage sludge using quonset solar tunnel dryer: An experimental and numerical approach*. Renewable Energy, 2021. **171**: p. 784–798.
46. Tuncer, A.D. et al., *Thermal performance analysis of a quadruple-pass solar air collector assisted pilot-scale greenhouse dryer*. Solar Energy, 2020. **203**: p. 304–316.
47. Aktaş, M. et al., *Experimental analysis and CFD simulation of infrared apricot dryer with heat recovery*. Drying Technology, 2016. **35**(6): p. 766–783.
48. Selimefendigil, F., S. Çoban, and H.F. Öztop, *Investigation of time dependent heat and mass transportation for drying of 3D porous moist objects in convective conditions*. International Journal of Thermal Sciences, 2021. **162**: p. 106788.
49. Zhu, Y. et al., *Multiphase porous media model with thermo-hydro and mechanical bidirectional coupling for food convective drying*. International Journal of Heat and Mass Transfer, 2021. **175**: p. 121356.
50. Mahiuddin, M. et al., *Shrinkage of food materials during drying: Current status and challenges*. Comprehensive Reviews in Food Science and Food Safety, 2018. **17**(5): p. 1113–1126.
51. Khan, Md. M.H. et al., *Modelling of simultaneous heat and mass transfer considering the spatial distribution of air velocity during intermittent microwave convective drying*. International Journal of Heat and Mass Transfer, 2020. **153**: p. 119668.
52. Defraeye, T., and A. Radu, *Insights in convective drying of fruit by coupled modeling of fruit drying, deformation, quality evolution and convective exchange with the airflow*. Applied Thermal Engineering, 2018. **129**: p. 1026–1038.
53. Wei, S. et al., *Simulation and experimental studies of heat and mass transfer in corn kernel during hot air drying*. Food and Bioproducts Processing, 2019. **117**: p. 360–372.
54. Arunsandeep, G., and V.P. Chandramohan, *Numerical solution for determining the temperature and moisture distributions of rectangular, cylindrical, and spherical objects during drying*. Journal of Engineering Physics and Thermophysics, 2018. **91**(4): p. 895–906.
55. Selimefendigil, F., and H.F. Öztop, *Three dimensional unsteady heat and mass transport from six porous moist objects in a channel under laminar forced convection*. Applied Thermal Engineering, 2021. **183**: p. 116100.
56. Rani, P., and P.P. Tripathy, *Modelling of moisture migration during convective drying of pineapple slice considering non-isotropic shrinkage and variable transport properties*. Journal of Food Science and Technology, 2020. **57**(10): p. 3748–3761.
57. Onwude, D.I. et al., *Experimental studies and mathematical simulation of intermittent infrared and convective drying of sweet potato (Ipomoea batatas* L.*)*. Food and Bioproducts Processing, 2019. **114**: p. 163–174.
58. Rufino Franco, C.M. et al., *Intermittent drying of rice grains with husk: Modelling and experimentation*. Diffusion Foundations, 2020. **25**: p. 9–36.
59. Park, H.W., and W.B. Yoon, *Prediction of the intermittent drying behavior of soybeans [Glycine max* (L.)*] using novel multilayered mass transfer simulation with an image analysis*. Drying Technology, 2018. **37**(10): p. 1228–1238.
60. Chandramohan, V.P., *Experimental analysis and simultaneous heat and moisture transfer with coupled CFD model for convective drying of moist object*. International Journal for Computational Methods in Engineering Science and Mechanics, 2016. **17**(1): p. 59–71.
61. Curcio, S. et al., *Formulation of a 3D conjugated multiphase transport model to predict drying process behavior of irregular-shaped vegetables*. Journal of Food Engineering, 2016. **176**: p. 36–55.

62. Chilka, A.G., and V.V.J.I.C.E. Ranade, *Drying of almonds II: Multiple particles.* Indian Chemical Engineer, 2018. **60**(4): p. 338–355.
63. Defraeye, T. et al., *Convective heat and mass exchange at surfaces of horticultural products: A microscale CFD modelling approach.* Agricultural and Forest Meteorology, 2012. **162–163**: p. 71–84.
64. Cârlescu, P.-M. et al., *CFD simulation of heat and mass transfer during apricots drying.* LWT – Food Science and Technology, 2017. **85**: p. 479–486.
65. Franco, T.S. et al., *Foam mat drying of yacon juice: Experimental analysis and computer simulation.* Journal of Food Engineering, 2015. **158**: p. 48–57.
66. Devahastin, S., and C. Niamnuy, *Invited review: Modelling quality changes of fruits and vegetables during drying: A review.* International Journal of Food Science & Technology, 2010. **45**(9): p. 1755–1767.
67. Adnouni, M. et al., *Computational modelling for decarbonised drying of agricultural products: Sustainable processes, energy efficiency, and quality improvement.* Journal of Food Engineering, 2023. **338**, p. 111247.
68. Van Boekel, M.A.J.S., *Kinetic modeling of food quality: A critical review.* 2008. Comprehensive Reviews in Food Science and Food Safety, **7**(1): p. 144–158.
69. Ling, B. et al., *Kinetics of food quality changes during thermal processing: A review.* Food and Bioprocess Technology, 2014. **8**(2): p. 343–358.
70. Prawiranto, K., J. Carmeliet, and T. Defraeye, *Physics-based digital twin identifies trade-offs between drying time, fruit quality, and energy use for solar drying.* Frontiers in Sustainable Food Systems, 2021. **4**: p. 606845.
71. Defraeye, T., *Towards more efficient intermittent drying of fruit: Insights from combined hygrothermal-quality modelling.* Innovative Food Science & Emerging Technologies, 2016. **38**: p. 262–271.
72. Pham, N.D., M.I.H. Khan, and M.A. Karim, *A mathematical model for predicting the transport process and quality changes during intermittent microwave convective drying.* Food Chemistry, 2020. **325**: p. 126932.
73. Sabarez, H.T., *Mathematical modeling of the coupled transport phenomena and color development: Finish drying of trellis-dried sultanas.* Drying Technology, 2014. **32**(5): p. 578–589.
74. Wijerathne, W.D.C.C. et al., *A coarse-grained multiscale model to simulate morphological changes of food-plant tissues undergoing drying.* Soft Matter, 2019. **15**(5): p. 901–916.
75. Rathnayaka Mudiyanselage, C.M. et al., *Novel trends in numerical modelling of plant food tissues and their morphological changes during drying: A review.* Journal of Food Engineering, 2017. **194**: p. 24–39.
76. Prakash, O. et al., *Review on various modelling techniques for the solar dryers.* Renewable and Sustainable Energy Reviews, 2016. **62**: p. 396–417.
77. Motevali, A., S. Minaei, and M.H. Khoshtagaza, *Evaluation of energy consumption in different drying methods.* Energy Conversion and Management, 2011. **52**(2): p. 1192–1199.
78. Motevali, A. et al., *Comparison of energy consumption and specific energy requirements of different methods for drying mushroom slices.* Energy, 2011. **36**(11): p. 6433–6441.
79. Motevali, A. et al., *Comparison of energy parameters in various dryers.* Energy Conversion and Management, 2014. **87**: p. 711–725.
80. Baker, C.G.J., and K.A. McKenzie, *Energy consumption of industrial spray dryers.* Drying Technology, 2005. **23**(1–2): p. 365–386.
81. Kuan, M. et al., *Numerical simulation of a heat pump assisted solar dryer for continental climates.* Renewable Energy, 2019. **143**: p. 214–225.
82. Luo, W.-J. et al., *Performance enhancement of a sludge continuous feed heat pump drying system by air deflectors and auxiliary cooling subsystems.* Energies, 2020. **13**(24): p. 6651.
83. Kudra, T., *Energy performance of convective dryers.* Drying Technology, 2012. **30**(11–12): p. 1190–1198.
84. Aktaş, M. et al., *Heat pump drying of grape pomace: Performance and product quality analysis.* Drying Technology, 2019. **37**(14): p. 1766–1779.
85. Tohidi, M., M. Sadeghi, and M. Torki-Harchegani, *Energy and quality aspects for fixed deep bed drying of paddy.* Renewable and Sustainable Energy Reviews, 2017. **70**: p. 519–528.
86. Lingayat, A., V.P. Chandramohan, and V.R.K. Raju, *Energy and exergy analysis on drying of banana using indirect type natural convection solar dryer.* Heat Transfer Engineering, 2020. **41**(6–7): p. 551–561.
87. Aghbashlo, M. et al., *A review on exergy analysis of drying processes and systems.* Renewable and Sustainable Energy Reviews, 2013. **22**: p. 1–22.
88. Aktaş, M. et al., *Performance analysis of heat pump and infrared-heat pump drying of grated carrot using energy–exergy methodology.* Energy Conversion and Management, 2017. **132**: p. 327–338.

89. Gungor, A., Z. Erbay, and A. Hepbasli, *Exergetic analysis and evaluation of a new application of gas engine heat pumps (GEHPs) for food drying processes.* Applied Energy, 2011. **88**(3): p. 882–891.
90. Thijs, D., and D.I. Onwude, *The future of digital twins for drying.* Drying Technology, 2020. **39**(1): p. 1–2.
91. Majdi, H., and J.A. Esfahani, *Energy consumption and drying time optimization of convective drying for performance improvement: Response surface methodology and lattice Boltzmann method.* Journal of Heat Transfer, 2018. **140**(10): p. 102009.
92. Duc, L.A., and Hyuk, K.D., *Mathematical modeling and simulation of rapeseed drying on concurrent-flow dryer*, in *Current Drying Processes.* 2020, IntechOpen.
93. Zhang, S. et al., *3D model-based simulation analysis of energy consumption in hot air drying of corn kernels.* Mathematical Problems in Engineering, 2013. **2013**: p. 1–12.
94. Zakari, A. et al., *Energy efficiency and sustainable development goals (SDGs).* Energy, 2022. **239**: p. 122365.
95. Smith, P. et al., *Land-management options for greenhouse gas removal and their impacts on ecosystem services and the sustainable development goals.* Annual Review of Environment and Resources, 2019. **44**(1): p. 255–286.
96. Atuonwu, J., and S. Tassou, *Decarbonisation of food manufacturing by the electrification of heat: A review of developments, technology options and future directions.* Trends in Food Science & Technology, 2021. **107**: p. 168–182.
97. Rissman, J. et al., *Technologies and policies to decarbonize global industry: Review and assessment of mitigation drivers through 2070.* Applied Energy, 2020. **266**: p. 114848.
98. Engler, N., and M. Krarti, *Review of energy efficiency in controlled environment agriculture.* Renewable and Sustainable Energy Reviews, 2021. **141**: p. 110786.
99. Gonzalez-Diaz, A. et al., *The potential of decarbonising rice and wheat by incorporating carbon capture, utilisation and storage into fertiliser production.* Green Chemistry, 2020. **22**(3): p. 882–894.
100. Xu, Z. et al., *Research developments in methods to reduce the carbon footprint of the food system: A review.* Critical Reviews in Food Science and Nutrition, 2015. **55**(9): p. 1270–1286.
101. Martinez-Hernandez, E., M. Leach, and A. Yang, *Understanding water–energy–food and ecosystem interactions using the nexus simulation tool NexSym.* Applied Energy, 2017. **206**: p. 1009–1021.
102. Owen, A., K. Scott, and J. Barrett, *Identifying critical supply chains and final products: An input–output approach to exploring the energy–water–food nexus.* Applied Energy, 2018. **210**: p. 632–642.
103. Wang, K. et al., *Understanding the impacts of climate change and socio-economic development through food–energy–water nexus: A case study of Mekong River delta.* Resources, Conservation and Recycling, 2021. **167**: p. 105390.

Index